# Waste Sites as Biological Reactors

●●●

## Characterization and Modeling

Percival A. Miller
with
Nicholas L. Clesceri

LEWIS PUBLISHERS

A CRC Press Company
Boca Raton London New York Washington, D.C.

Catalog record is available from the Library of Congress


Lewis Publishers is an imprint of CRC Press LLC

No claim to original U.S. Government works
International Standard Book Number 1-56670-550-9
Printed in the United States of America 1 2 3 4 5 6 7 8 9 0
Printed on acid-free paper

# Dedication

*To*
*My parents,*
*Ruthie V. and Leonard A. Miller*

*"Life is like a mountain railroad . . . ."*

***M.E. Abbey, Charles D. Tillman, 1890***

# Preface

The simplicity, ugliness and beauty of a waste disposal site confront us with a microcosm of nature at its most basic, yet it is gratifyingly functional. The functionality of a disposal site arises from the capacity to accept large numbers of items no longer useful and contain them until they eventually disappear. This expectation has been part of human activity since earliest human settlements. Where and how wastes disappear and how the environment is affected have, in recent times, become issues of greater concern to cities and towns around the world. Investigations of recent decades have convincingly shown that waste sites pose water, air and public health dangers that necessitate highly efficient engineered controls. Engineers and scientists have responded to these challenges with successful solutions for sites where disposal is permitted, such as bottom liners allowing collection and treatment of seepage, methods of manipulating decomposition rates and efficient gas management.

What goes on at or inside these sites is exceedingly complex. The scientific studies have, however, made it clear that impacts of waste sites originate from the interplay between wastes and the environment. The pursuit of an intimate understanding of the dynamics of waste sites has perhaps lagged behind the development of engineered controls, but engineers, managers and scientists with responsibility for these sites must be willing to delve further, with the goal of developing better tools for assessing and reducing risks. The need for more-rapid site stabilization through enhanced decomposition rates, site amelioration and remediation of existing water and soil contamination impacts makes the development of affordable, appropriate management tools extremely useful.

One of the less expensive (yet no less effective) methods for assessing impacts and risks of a system and devising management plans has been to develop mathematical and quantitative models that are sufficiently representative to allow examination of physical systems as units subject to environmental factors. The focus of this work is the examination of the interaction of biological, chemical and physical factors at the border between wastes and the soil or atmosphere and how it can be represented in modeling.

The waste site, as a physical environmental system, offers a particular challenge to attempts at modeling because it is inherently unorthodox. To reduce unorthodoxy, it is necessary to develop choices of how to represent the site as a waste management system that retains its function. The similarity of conversion of wastes in this system to formally designed biological treatment reactors thus presents an opportunity to model the waste site as a biological reactor.

Modeling this type of system requires a willingness to resort to multidisciplinary thinking and to use the tools of various applied science disciplines. Training has provided the environmental engineer, the applied scientist and environmental modeler with many of the tools needed. For instance, exposure to mathematics, statistical sciences, soil physics, biology and biochemistry, chemistry, hydrology, thermodynamics, heat and mass transfer and familiarity with computer languages and model development is excellent preparation for understanding or development of these types of models.

This book is also a first attempt to deconstruct the mystery of a waste site in such a way that it can be modeled using the familiar tools of waste treatment and chemical or environmental engineering, as well as to represent the waste site as a coherent and functional system. It provides a detailed discussion of the elements needed to develop a comprehensive waste site model, though it neither recommends a particular approach nor presents a specific model. Development and refinement of efficient models can be the subject of future efforts. A preliminary approach, developed in Chapters 1 and 2, is to represent the site as a matrix of soils and wastes of known physical properties and behavior. Properties and constituents of wastes and site effluents are further explored in Chapter 3. Chapter 4 examines the roles of surface and subsurface animals and microorganisms in decomposition at waste sites. Chapter 5 addresses the dominating influence of moisture on decomposition processes, while Chapter 6 takes a look at the influence of heat on system dynamics. Chapter 7 develops and discusses the various methods of representing kinetic processes likely to be appropriate to waste site modeling. Chapter 8 combines topics in Chapters 2, 5 and 6 to address mass transfer. Chapter 9 reviews model sensitivity analysis and provides some conclusions and recommendations.

On a more personal note, the contents of this book originate mainly from background research material developed for my dissertation as a doctoral student of environmental engineering at Rensselaer Polytechnic Institute in Troy, New York. It also follows from a timely suggestion by a distinguished member of my dissertation team, Dr. Henry Bungay, now Professor Emeritus of Biochemical Engineering, and by Dr. Nicholas Clesceri, head of the Environmental Engineering Department and my dissertation advisor, that the information might be useful in book form.

Many thanks to all of the persons who have encouraged or in some way supported or assisted this somewhat unusual undertaking, including Rick Morse for his patience and encouragement; Dr. N. Clesceri for his editing, advice and encouragement; the office staff at the Environmental Engineering Division at Rensselaer Polytechnic; Mari Ann Shake for unwavering love and support; friends Jack Wandell, Deborah Nason and Dr. C.C. Jon-Nwakalo; and editors Brian Kenet and Pat Roberson at CRC Press for their timely suggestions.

**Percival A. Miller**

# About the Authors

**Percival A. Miller** earned his B.Sc. in Civil Engineering at Southern University; his M.Sc. in Civil Engineering at the University of Colorado with a specialty in hydraulic engineering and hydrology; and his Ph.D. in Environmental Engineering at Rensselaer Polytechnic Institute. He is currently a senior research analyst at the New York State Legislative Commission on Solid Waste Management, where he has authored or contributed to commission papers, reports, scientific reviews and policy opinions, and has provided technical background for state legislation. He has also authored papers on landfills, solid waste and health in developing countries, biological reactors and the role of moisture in decomposition processes; and has chaired environmental conference sessions on biological and hazardous waste treatment in Colombia. He has also co-consulted on oily sludge and biological sludge land farms liner system design, construction and operation in Venezuela. Dr. Miller is a member of the American Society of Civil Engineers, where he is vice chair of the International Border Water Quality Committee (IBWC) and a member of the Environmental Documents Committee of the Environmental and Water Resources Institute.

**Dr. Nicholas Clesceri**, currently on leave as a program director for environmental sciences at the National Science Foundation (NSF), has been an environmental engineering professor and researcher at the Rensselaer Polytechnic Institute (RPI) in Troy, New York, for more than 35 years. Under an early NSF award of an Engineering Initiation Grant that allowed exploration in new directions in research, Dr. Clesceri and colleagues founded a research activity on Lake George, New York, leading to an NSF-sponsored Analysis of Ecosystems Program of the (U.S.) International Biological Program. As the designated coordinator for Lake George research, Dr. Clesceri interacted with five universities in research planning and prioritization and development of lakeside research facilities. He has served on the Environmental Advisory Board (EAB) of the U.S. Army Corps of Engineers, where he was a co-advisor to the chief of the corps on environmental matters relating to civil works projects. One such visit of the EAB to the U.S. Military Academy (USMA) at West Point involved discussion of the advisability of formal education in environmental engineering, as future corps officers would play prominent roles in environmental decision-making. This discussion led to the establishment of the ABET-accredited B.S. environmental engineering degree at USMA.

As Environmental Engineering Program Director at RPI, Dr. Clesceri has assisted a number of sister institutions in initiation of undergraduate degree programs in environmental engineering. He has been the principal investigator on an industry-sponsored multiyear project on industrial waste treatment with heavy metal recovery at a metal-treating facility, for which the major issue was chromium conversion coating of aluminum. As an outgrowth of this study, Dr. Clesceri and RPI researchers participated in an industry-focused study on nonchromium conversion coating of aluminum that was sponsored by the National Center for Manufacturing Sciences (NCMS).

Over the past 5 years, Dr. Clesceri has served as deputy project director on the Water Resources Development Act (WRDA) Project on New York/New Jersey Harbor Sediment Decontamination, where he has provided technical support on testing and evaluation of bench- and pilot-scale investigations of dredged material cleanup. Since 1999, after appointment by New York Governor George Pataki, Dr. Clesceri has served as member and chairman of the Technical Advisory Committee (TAC) of the New York City Watershed Protection and Partnership Council, where he works closely with technical staff from New York State's Department of State.

Dr. Clesceri earned his B.S. in civil engineering from Marquette University, and his M.S. and Ph.D. degrees in civil/sanitary engineering from the University of Wisconsin/Madison. He was a PHS/WQO postdoctoral research fellow at the EAWAG/ETH in Zurich, Switzerland. He has served on the boards of several organizations, including the Association of Environmental Engineering (now AEESP). He is a registered professional engineer in Wisconsin, and is a fellow of the American Society of Civil Engineers.

# Table of Contents

# 1 Introduction

## THE NATURE AND CONTROL OF WASTE DISPOSAL SITES

Waste disposal sites located near settlements, towns and cities have been a feature of public sanitation for human societies for hundreds, even thousands, of years. In a sanitary role these disposal sites have, until the latter half of the twentieth century, served as the final destination for a wide range of organic and inorganic materials considered of little further use. With the permanency and growth of populations in towns and cities, the noxious effects of disposal sites became familiar; but scientific control of these effects has attracted limited attention from trained professionals until the past several decades. Effects of these disposal sites, however, include the presence of a variety of animals and insects that feed on organic materials, intense odors generated from biological decomposition, the volatilization and release of stress-causing or hazardous chemicals to air and seepage into the soil — and frequently into groundwater or surface water (ponds, rivers and estuaries) — of a mixture of chemicals that can impair water resource quality.

The extent to which a disposal site impairs the surrounding environment depends upon several factors. These include the types of wastes disposed, biological organisms present, soil chemistry, the looseness or density of the soil and wastes, how well a soil cover keeps out water, how well soil under the site drains leachate, location and use of ground and surface water affected, concentrations of chemicals released into the air and whether capture and treatment controls are in place. A key goal in improving engineering practices for waste site control is thus the development of in-depth knowledge of how waste types, soils, site biology and climatic factors combine to affect the breakdown of wastes buried — and thus the rates of releases of chemicals from a site into water and air. Understanding of waste disposal sites has increased considerably through many studies conducted for government and city agencies. However, knowledge of the interior dynamics of waste sites and implications for site control still suffers a lack of in-depth analysis, due to the complexity of interactions and costs of exploratory investigations.

Environmental management goals applicable to disposal sites usually include the control of air and water impacts. Control, from an engineering point of view, means the development and use of site designs and equipment that reduce or eliminate air and water contaminant outputs. There are a number of ways of accomplishing site control. For instance, leachate quantity can be reduced by limiting water input; leachate quality can be improved by limiting the types of wastes disposed or recirculating leachate; and the amount of leachate seeping to ground or surface water can be reduced by installation of low-permeability bottom liners and leachate collection. Volatile compounds and gas releases can also be directed through special cover designs to vents and to combustion units to reduce their adverse impact on air quality or the health of site personnel or nearby residents. Modern landfill design and operation using these controls have been found to be quite effective.

There has, however, been no effective method of controlling the mixture of wastes entering the typical disposal site. In the dumps of old, in invisible or hard-to-find locations in every country of the world, it might be possible to find items from old industrial containers to building demolition materials, food and agricultural wastes, animal remains, hazardous wastes and tree branches. Today's disposal sites or landfills, though not as likely to contain drums of disposed chemicals, can still be the depositories of an increasing variety of modern waste materials, ranging from metals to manufactured plastics and glass. In countries where food and agricultural processing is not as advanced, landfills can contain a high percentage of organic, decomposable wastes. In advanced or more industrialized countries, landfills are likely to contain higher percentages of plastics and packaging wastes. In many countries there are also variations between the contents of urban and rural landfills, with the latter tending to contain more organic waste. Thus there has been little intrinsic change in the variety of materials disposed in waste sites. The problem this poses for the engineer, scientist or modeler is that the complex mix of wastes makes accurate prediction of decomposition processes extremely difficult. Many of the mathematical models developed for site control thus use overall rates, which are often derived indirectly and the accuracy of which cannot be ascertained. However, this does not reduce the effectiveness of the site control methods listed; it merely indicates major unknowns in how these sites are characterized.

Other types of sites involving wastes are also of interest to engineering for environmental control. These include sites where soil and groundwater have become polluted through the incidental or intentional disposal of persistent and hazardous industrial organic chemicals or heavy metals. Other sites of interest also include those where organic wastes are mixed for the sake of processing to useful or environmentally safe end products — for example, aerobic and anaerobic composting operations. For these sites, proven methods and technologies are used to generate finished composts, for gardening, horticulture and for forestry or agricultural soil amendments. Engineering control in this instance can take the form of manipulation of the biological system, through appropriate combinations of waste, water content and aeration rate if aerobic composting is involved. For composting operations, how the mix of wastes influences decomposition and release of gases and liquids is also a complicated but relatively unexplored area of study.

From a general point of view, a key control of chemicals released from waste sites is interior biology and chemistry. Given any mix of organic or inorganic wastes, solid or liquid, major agents affecting outputs to air and water are the aerobic and anaerobic organisms present, chemical conditions in the soil or wastes and climate-influenced factors such as soil temperature and moisture input.

Especially in the case of solid or liquid organic wastes mixed with soil, interior biology can be considered the major influence over waste decomposition. Though the role of site biology has been recognized for some time, it has been difficult to develop thorough engineering approaches based upon the interactions of wastes and organisms, as these interactions are often considered too complex for practical applications.

Sources of complexity include (1) no way to represent or analyze the range of interactions between wastes and biological organisms, and (2) heterogeneity of wastes plus influences of soils and climate. Extensive and expensive sampling, testing and statistical studies would be needed in most cases. In most cases these expenditures are considered unnecessary because control air and water inputs from landfills, for instance, are so well developed. For this reason, even the most advanced approaches to sound environmental management of disposal sites rely upon empirical methods and regard disposal site wastes and soil heterogeneity, and the subsequent roles of organisms in waste breakdown and conversion, as a somewhat impenetrable but tolerable unknown.

## THE BIOREACTOR CONCEPT

Efficient environmental treatment has been shown to develop from simple and straightforward representation of the complexities of environmental systems. Wastewater and organic wastes treatment systems are examples of how sophisticated technology can develop from relatively simple methods of representation of the interaction between waste streams and organisms.

The similarity of waste sites to the biological reactor systems developed for industrial wastes and wastewater treatment suggests that waste disposal could possibly benefit from straightforward representation of the remaining complexities of these sites. An analogy from the wastewater treatment shows how useful it is to consider the interaction between waste type, media and organisms present, including sewage and industrial wastewater treatment systems. Wastewater treatment for industrial and publicly owned treatment works (POTWs) has become a very successful and sophisticated undertaking through the use of experiential information on interaction between a single waste stream and organisms, within a particular environment, to design treatment systems that can rapidly reduce pollutants present in incoming wastewater. Wastewater treatment and enclosed solid waste treatment units are considered biological reactor systems, with performance linked to organisms present and manipulation of the metabolic rates of these organisms. Compared with other types of biological reactors used for liquid and sold waste treatment, however, a waste disposal site can be regarded as a more "open" reactor, with fewer controls at its boundaries.

Can waste disposal sites be assessed as biological reactors? As noted, immediate complexities include heterogeneity of biodegradable wastes, a disordered mix of soils, moisture content variation in soils and wastes, leachate with a variety of unusual chemicals and a decomposition gas mixture with trace volatile substances present. This is particularly the case for a landfill or dump. These complexities have been considered enough to make bioreactor approaches somewhat impractical. The exception is the use of controlled landfill cells with collection and gas control systems as biological reactors. In these landfill cells, well-designed water controls allow manipulation of rates of biological degradation inside the landfills such that waste decomposition, methane gas generation and cell settlement are more rapid than in ordinary landfills. A facet of these systems is that the desired environmental control result can be attained without needing to assess the mix of wastes or biological interactions involved. Thus, and despite the success of recent landfill cell bioreactors, the landfill or dump interior remains for all practical purposes a 'black box.'

A key element of the success of wastewater systems science and design is that the waste stream, and its interaction with the microorganisms present, has been well characterized. Characterization of waste streams, site media and interactions can thus be regarded as a valuable and perhaps essential step toward representation of these systems as biological reactors.

It can then be hypothesized that redefining wastes disposal sites as bioreactors rather than merely as dumps or landfills has the essential but daunting task of removing as far as possible the black-box nature of the site, thereby simplifying its representation as a knowable system that can possibly be controlled as well as a wastewater treatment system. Synthesis of information from various areas of applied sciences and analysis methods is needed to accomplish this task, as data available are not necessarily specific to waste sites and must be adapted.

Experiential information useful to establishing waste sites as biological reactors includes air, soil and leachate quality data collected from studies at a variety of landfills, composting sites and sites undergoing remediation. For instance, soil studies done at landfills provide a practical range for parameters such as porosity, density, average water content and its variation with depth. Landfill and waste composting studies have established the types and concentrations of organisms likely in landfills. Gas generation and flow studies have also established the types of volatile compounds and gases likely to be present in soil pores and at the upper surface of landfills.

One of the most efficient approaches to using this information is the development of mathematical models to represent the environmental systems of interest, within a practical range, as closely as possible. Good representation requires (1) appropriate description of the reactor, (2) availability of methods and data to characterize the system flows and biological kinetics and (3) the use of mathematical models that can be solved with current methods. Bailey and Ollis (1986) also note the importance of time and length scales in descriptions of bioreactors and that bioreactors can have quite large length scales. The importance of this is that — whatever the scale the modeler uses as a first representation of wastes and soils and organisms in a disposal site — these must be incorporated into a large-scale representation.

In environmental systems evaluation and engineering design for treatment, mathematical models have been found to be invaluable in reducing time and expense, as models reduce the cost of analysis by reducing the need for experimental studies. The concept of a waste site as a bioreactor is thus explored in this work as a means to simulation methods for decomposition processes and effects at sites with and without controls.

## REACTOR CONFIGURATIONS OF RELEVANCE TO PRACTICAL DESCRIPTION OF A WASTE SITE

A first attempt at definition of the waste site problem must include the difficult question: what kind of bioreactor is a waste disposal site? Known characteristics of the waste site must be compared with what is known from classical definitions of reactors.

The definition must be reasonable yet sufficiently general to allow simulation. Waste treatment system design science has provided a significant variety of reactor types, each applicable to particular treatment objectives. It is useful to review these reactor definitions as a preliminary to choosing a suitable reactor representation for waste sites.

Biological reactors can be described as ideal or nonideal. Within the class of ideal biological reactors is the well-mixed reactor, where mixing of substrates and catalyst flow is sufficiently uniform and energetic that reactions and catalyst levels do not differ considerably throughout the length of the reactor. Thus average, uniform conditions can be assumed for mathematical models of the ideal reactor. For large reactors, complete mixing is difficult; thus mixing configurations must be designed or assumed — for example, several ideal reactors in series or parallel to reduce uncertainty related to spatial nonuniformity. Ideal reactors defined in engineering literature include the batch flow, continuously stirred tank reactor (CSTR) and plug flow reactor. There are also nonideal forms and variations of these types of reactor principles. Other reactor types are often variations of the ideal reactor and include the packed tower, fixed film and trickling filter reactor.

For the *ideal flow batch reactor*, organism populations are not added or removed; but the concentrations of nutrients, organisms and reaction products change with time. The volume and system pressure of the ideal batch reactor can be constant or variable. The disappearance of nutrients also equals the accumulation of products, and the ideal batch reactor is well mixed and uniform in composition at all times. Thus, time-based mass flow relations can be developed for reaction and product generation rates from differential or integral relationships.

For the *ideal CSTR*, mixing is supplied by energy from outside the system, making the concentration uniform for gas, liquid and solid phases. The liquid volume is maintained constant. Reactor performance is controlled by the bacterial cell residence time. The CSTR system is considered isothermal, and the dissolved oxygen level is uniform for the liquid phase due to mixing. Configuration and operation of the CSTR thus allows a steady environment for its microorganism population and balanced organism growth for optimal reactor performance. A general need for good CSTR reactor performance is a suitable concentration of bacterial enzyme, supplied by maintaining the mass of organisms. However, the energy, equipment and control requirements for large-scale CSTR reactor operations can make batch reactor operation more practical for small-scale treatment.

The *ideal plug flow reactor* is a tube or channel reactor in which the fluid flow has a relatively constant velocity across any cross section along the tube. Because the fluid velocity is constant, an element of fluid at any distance along the plug flow reactor is unaffected by the fluid element before or after it; but the concentration of substances in each fluid element changes, in the same manner, as the flow progresses through the reactor. Change in concentrations of substrates, organisms and products thus relates to the length of the reactor. If a mass balance is taken at any cross section, steady-state conservation principles can be applied. Relations between inflow and output are free of the residence time factor. The residence time, in the case of a plug flow reactor, is usually included as the ratio of column length to fluid velocity. It is the time required for fluid to move a distance $z$ through the reactor, though the time element is included in the ratio of column length over velocity. The ideal plug flow reactor has as its analogies the long tube, the packed bed, the multistage and the counterflow reactors. It can be considered a mobile batch culture, as each channel cross-section of fluid considered can be analyzed according to batch culture kinetics and principles.

For the *packed tower (or packed bed) reactor*, the support medium, which is usually a fixed granular media such as plastic, rock or soil, is not submerged in liquid as is usually the case for a pure CSTR or batch reactor. Instead, the liquid phase moves as a thin sheet or film over the support media through the packed tower. Reactions occur at the interface of the medium and, within the thin film of liquid, remove substrates from the liquid. Removal of substrates or organic compounds from the liquid depends upon rates of diffusions of these compounds out of the bulk liquid into the organisms in the biofilms attached to the surfaces of the solid media. If the flow rate changes, so do the reaction rate and concentrations of product along the whole length of the packed tower. Recirculation of flow can result in mixing, suspension of some organisms and solids, and can provide an environment with both suspended and attached organisms, which affect the removal rates for compounds. As recirculation increases, the reaction rates become more uniform (Grady, Daigger, and Lim, 1999). If there is no recirculation of flow, the reaction environment changes according to tower length (for example, because of accumulation of products).

A *packed bed reactor* can be described using essentially the same approach as for a plug flow reactor. The exception is that the catalyst or enzyme source is fixed within the bed or medium, and the flow velocity through a cross section of the porous medium is the flow volume divided by the total cross-section area times the void fraction. The packed bed reactor obviously has features similar to unsaturated waste sites, wherein the solid media is a support for microorganisms growing on surfaces.

The *fixed film reactor* and the packed bed reactor operate essentially according to the same principles. Characteristic features of the fixed film reactor make it of interest to comparison to a disposal site subject to precipitation inputs. Though the fixed film reactor is similar in concept to the plug flow reactor, flow in the system is relatively complex. Definition of a fixed film reactor thus requires various modifications from theoretical definitions for plug flow or batch reactors. In both the fixed film and packed bed reactor, the microorganisms that remove or convert mixed or heterogeneous substrates grow in a thin film (biofilm) on the surface of a fixed, solid medium. Microorganisms are fed by and remove substances from the liquid flowing past the biofilm.

Transport of the dissolved substrates from their sources to microorganisms on the solid medium involves generation and release of substrates into the bulk liquid flow and movement of the bulk liquid and diffusion of these substrates to cell surfaces of the microorganisms. Thus, mass transfer principles become important in theoretical descriptions of fixed film reactors. However, for thin and periodically stagnant films of water, such as in unsaturated soil in landfills, underground pollution sites and compost piles, there may be little resistance to mass transfer of liquid compounds between the bulk water film and microorganisms on the solid surfaces of wastes and soils. In this case the observed reaction rate at a location in the reactor may be nearly or actually equal to the reaction rate in the biofilm formed by the fixed microorganisms. Nevertheless, kinetic reaction rates for fixed film reactors benefit from inclusion of a mass transfer coefficient to represent flux of substances across the water film to microorganisms. The biofilm is likely to be thin and gelatinous, with a network of tortuous pores through which substrates can diffuse to microorganisms fixed to its external and internal surfaces. The simplest method of representing this substrate transport (flux) to microorganisms is diffusion according to Fick's law:

$$Flux = -D\frac{ds}{dx} \tag{1.1}$$

where $D$ = diffusion rate (area units/time unit), and $S$ = substrate concentration. Grady, Daigger, and Lim (1999) have noted the similarity of the structure of the biofilm to a porous catalyst with interconnected channels through substrate diffuses for reaction with its active surface — and that for this case of a porous, 'catalytic' support (natural biofilm), it is typical to represent diffusion rate $D$ in terms of the effective diffusion rate $D_{diff}$, where the two are related by the void volume $V_v$ and pore tortuosity $\tau$ of the catalyst:

$$D_{diff} = D\frac{\text{void volume}}{\text{tortuosity}} = D\frac{V_{void}}{T} \tag{1.2}$$

Approximate range values for these terms may be obtained from biofilm studies.

The mass transfer coefficient $k$ represents fluid properties as well as diffusion and convection of the compound present and the air–water film interface. An important consideration, especially for solid media likely to be present in waste sites, is that definition of the mass transfer coefficient $k$ allows consideration of the shape and porosity of the media on trickling or stationary flows. For the fixed film reactor system, rates of substrate removal are typically expressed in terms of amount consumed per unit of film surface area involved. This is because the mass concentration of organisms is difficult to determine except in terms of reactor surface area covered. Grady, Daigger, and Lim (1999) state that the thickness of biofilms in these types of reactors is typically small, and the active mass of organisms per unit area is also relatively constant. This means that the maximum substrate removal rate includes the mass of organisms per unit area. For less structured reactor media such as in landfills and soil-waste reactors, biofilm and flowing water films are likely to assume the shape of soil grain (stones and other particles). The presence of an active biofilm is also likely to depend upon moisture availability, suggesting that the biofilm is likely to be patchy in distribution, rather than well distributed as in the case of wastewater treatment packed beds or trickling filters. There would thus be no direct means of characterizing the surface area of the media covered by biofilm except in terms of relations between mass of active microorganisms involved, biofilm thickness and substrate removal rate, and surface area available to seepage flow through the solid media of the fixed film reactor.

The *trickling filter reactor* is also a form of the packed tower reactor. In it the microorganisms grow in a thin film on the surface of fixed, solid media. Liquid flowing over this film contains substrates that, upon diffusion into the biological film, are used by the organisms for metabolic processes. Removal of substrates from the flow as it passes through the trickling filter results in a substrate reduction rate proportional to the distance from the flow inlet. The depth of the filter is such that a certain substrate removal percentage can be attained by the time the fluid reaches the underdrain. Other characteristic features of the trickling filter are that the support media are mechanically, chemically and biologically stable (do not degrade), the surface area available to microorganism growth is large, fluid flow over the microorganism is in a thin, even sheet, voids in the trickling filter are not filled with liquid so free flow of air is possible and organic solids washed off the biological film can be carried forward rather than accumulating in the voids. Varieties of the trickling filter have been used in sewage and industrial wastewater treatment. In these varieties, stone and plastic media of two- to four-inch size have been used to provide both the surface area and the flowthrough needed; the inflow is designed to provide even fluid distribution over the media, and the underdrain is designed such that air (oxygen) can reach the microorganisms in the biofilm covering the media. A goal of inflow design is the wetting of the total surface of the trickling filter for best performance. Additionally, the concentration of the substrates in the flow, or organic loading, must be matched to the liquid inflow rate — otherwise the trickling filter can be clogged by the biofilm buildup. To avoid this, the hydraulic loading is adjusted by recirculating a portion of the flow such that excess film buildup is washed out and collected through sedimentation. High concentrations of organics in wastewaters may reduce trickling filter performance and result in odor due to anaerobic conditions caused by exceeding the oxygen transfer rate to the biofilm. Concentrations of organics at toxic levels can often be handled if the toxic level in the flow is only of short duration. A trickling filter with smaller sized media has greater surface area but can be subject to flow interference resulting from buildup of biofilm in the pores.

## THE WASTE SITE AS A BIOLOGICAL REACTOR

The waste site as a biological reactor has several features similar to the reactors described, with the exception of the CSTR. From the reactor descriptions above, a waste site is likely to have features of packed beds or towers, nonideal plug flow reactors, fixed film reactors and trickling filter reactors. The substrates likely involved are initially solids, thus imposing a need for representation of conversion of solid substances to liquid, or hydrolysis, before packed bed and fixed film reactor analysis can be applied. The conversion of solids to liquid and gas reduces the sizes of particles, imposing a shrinking particle effect on the reactor, as evidenced in settlement of ground surface as waste disposal sites age.

Preliminary comparisons suggest that, ideally, the waste site can be compared to a packed bed system, as the solid media is granular and unlikely to be submerged in liquid at any time. Waste sites are rarely saturated, and saturation conditions severely retard substrate removal rates. Granular media in the waste site (as a packed bed) is likely to be disordered and irregular with regard to shape, size, density and compressive strength. Thus, characterization of the granular media requires modification of the typical representation of solid media for packed beds.

Packed bed reactors are physical systems in which bulk fluid flow moves over the solid medium as a thin sheet or film. This description is likely to be applicable to significant portions of the interior of a waste site, for which water from precipitation and decomposition of organic wastes may slowly trickle over the media (stones, soil or waste particles) under gravity forces. With unsaturated conditions and irregular granular particles, the bulk flow of liquids would be through the most permeable paths, including large pores and channels, or the most permeable soil and waste fractions. The waste site would also contain stagnant water in films and in fine pores. With periodic moisture inflows from precipitation or top surface flood flow, the waste site as a packed bed will only occasionally experience the (saturated) bulk liquid flow characteristic of classical packed bed reactor operation. Thus, the modeler would need to define how bulk fluid flow and water distribution in a waste site varies from typical packed bed fluid flows.

The packed bed reactor was previously a system for which rate of removal of compounds (substrates) from bulk liquid flow depends upon their diffusion to organisms attached to the solid medium biofilms. In a waste site, the generation of liquid substrates depends upon the hydrolysis of solid organic materials as a primary step in decomposition. Removal of liquid substrates such as hydrolysis products from decomposer microorganisms (bacteria, fungi, actinomycetes) is accomplished by hydrolytic and a second line of organisms, such as acidotrophs and acetotrophs. Thus, packed bed, fixed biofilm kinetics can properly be applied to activities of hydrolytic and mutually dependent organisms if the waste site has been defined as a biological reactor. If a fraction of the bulk flow is recirculated for strategic reasons (improvement of decomposition rates, for instance), flow in the upper part of the waste site might be more representative of classical packed bed operation, with comparable kinetics.

The packed bed reactor was described as having an average flow velocity, through any horizontal section, equal to the volume of the bulk flow ($V_{bf}$) divided by the cross-section area ($A_{cs}$), times the void fraction of the cross section area ($\varepsilon$),

$$\text{Average flow velocity} = V_{bf} A_{cs} \varepsilon \tag{1.3}$$

where $v$ = average velocity or *conductivity*, $V_{bf}$ = bulk flow, $A_{cs}$ = cross-sectional area of the packed bed and $\varepsilon$ = void fraction of the packed bed. The velocity of the bulk flow through porous media can be stated in terms of Darcy flow, wherein the flow rate is proportional to permeability and pressure head. This will be discussed in Chapter 5.

The void fraction $\varepsilon$, as stated above, can be interpreted as the total void fraction of a cross section of the packed bed. However, liquid flow that completely occupies the void fraction of the packed bed happens only in the case of saturated flow. For a waste site, this would be an extreme condition because a site is likely to be mostly unsaturated. The bulk fluid occupancy of larger pores of waste is likely to be annular — i.e., distributed in a thin film on the interior surface of the pores — and movement of the bulk liquid in the system will involve the fraction of the water film involved in free drainage. Only a fraction of the total cross-section area of the voids would thus be involved. The flow, in thin films or sheets, is also likely to be subturbulent in velocity, or laminar. When outflow is minimal for the waste site, the low bulk fluid flow rate is likely to be reflected by a combination of stagnant liquid films, or thinner films, and minimal flow velocities of the same. Stagnant liquid films represent nondraining moisture, retained in fine pores and between contact points of the waste site or packed bed. The thinness of these films and the high energy needed to remove them from the waste or soil surface ensures moisture sorbed in the system under any circumstances. The volume of these films decreases with evaporation and other drying forces. The *velocity relationship for the waste sites as a system undergoing packed bed flow must thus be modified to reflect bed drainage as annular rather than as saturated liquid film flow*:

$$\text{Average flow velocity} = V_{bf} A_{af} \varepsilon \tag{1.4}$$

where = $V_{bf}$ = bulk flow rate, as before, but $A_{af}$ = cross-section area of the trickling film flow = (average wetted perimeter of cross-sectional area in flow × film thickness). The cross-sectional area of the moisture film on the surface of a theoretical pore can be represented by $t_{tot}$, where $t_{tot}$ = thickness of mobile trickling flow film ($t_{mf}$), plus the thickness of (stagnant) moisture film ($t_{sf}$) retained on the pore surface by capillary forces:

$$t_{total} = t_{mobile} + t_{stagnant} \tag{1.5}$$

As a packed bed or packed tower reactor — with laminar bulk film flow and a fixed biofilm growing on the supports — the waste site can be described as a fixed film reactor, with plug flow approaches to system modeling applicable. A general description of the waste site as a fixed (biological) film reactor or biofilter, however, is that the *bulk liquid film flow may be periodic and unevenly distributed; that the fixed biofilm might or might not cover most of the media, i.e., might be patchy; and that concentrations of substrates in the liquid film moving over the surface of the solid media will change with inflow rate, with age of or reduction in volume of substrate sources and with distance from the reactor inlet.*

Inactive or active landfills and dumps can thus be regarded as reactors with an essentially unsaturated soil system, subject to less than saturated flow of moisture. The overall particle size distribution of the waste site would also include a high percentage of fine soil particles — in contrast to commercial-scale trickling filters and packed bed reactors — and thus be subject to slow or trickling flow. Fluid flow is likely to be laminar, except for temporary short-circuiting of flows through large subsurface channels or pores. The removal of liquid substrates is likely to be due mainly to microorganisms in the biofilm.

These descriptions can be combined to describe the inactive landfill or dump, lined or not, as a trickling biological filter, or trickling film, packed bed biofilter. Because the biological activity in a landfill is controlled as much by environmental factors as by applied controls such as soil covers, landfills or dumps without these controls can be described as *open* biological trickling filter reactors.

Other investigators have also suggested descriptions of waste sites as reactors. For example, Knox and Gronow (1990) state that, biologically, landfills can be considered *fixed film reactors*. In considering MSW landfills as *anaerobic* bioreactors, however, Knox and Gronow (1990) also stated that these partially saturated containment sites are amenable to analysis; but that for useful analyses, the reactive capacity of the (wastes) should be defined. These authors also called for investigation of effects of landfill design and hydraulics on reactor performance, as in sewage treatment.

Baccini et al. (1987) have also described landfills as chemical and biological fixed bed reactors, with municipal solid waste (MSW) and water for inputs, and with gas, leachate and residual solids as outputs.

The *reactive* capacity of waste disposal sites has in the past been defined in terms of average or overall decomposition or conversion values for organic wastes and substances, based upon grouping of representative waste factions likely to be present in the site and average kinetic rates of microorganisms likely to be present in these sites. These rates are developed from field or laboratory information from waste decomposition and landfill studies. While these values are practical and have been successfully used in predicting landfill leachate and gas generation, many of the simplifications incorporated in these models do not fully address physical heterogeneity of solid wastes and site media. Wastes and media heterogeneity are, however, important factors influencing bioreactor performance. Thus, the inclusion of a representative variety of solid wastes and their estimated specific reactive or physical properties, and a representative variety of solid media into waste site bioreactor models, could at least partially address bioreactor performance uncertainties arising from physical heterogeneity of wastes and media inside a waste site.

The importance of moisture dynamics to assessing the impacts of waste sites has been recognized throughout landfill research. Recent studies have confirmed the importance of moisture dynamics as a major control of biological activity in waste sites. In recent years moisture input modeling, developed mainly from the vast research and mathematical modeling of flow through soil columns, has been applied to landfill modeling. Of recent moisture flow models, the EPA model of hydrologic evaluation of leaching procedure (HELP) is perhaps the most developed and is widely used for landfill leachate estimations. The HELP model permits real-time moisture flow values to be applied based upon site geographic location, local precipitation, temperature, soil type and cover or liner design.

In a biological reactor study, the focus is upon phenomena of the biologically reactive segment of the reactor. If other reactor system controls are well defined, it is convenient to use a relatively reliable moisture flow model, such as the EPA HELP model, to allow estimation of the time-varied moisture inputs to the biologically active segment of the landfill when an organized landfill is being modeled. Incorporation of a range of typical moisture controls — for example, varying cover (or liner) types — also permits consideration of normal management options without compromising the reliability of predictions. Modeling of the biologically active segment of the landfill as a bioreactor, with reactor model simplifications, could provide insights of practical value to management efforts.

While models have been successfully developed and calibrated to predict liquid flows through waste sites, their focus upon gravity drainage through waste sites limits direct consideration of the effects of moisture content and distribution upon decomposition and conversion of substances. The bioreactor model, by contrast, must consider moisture supply and distribution inside the reactor (or waste site), because, as mentioned, moisture availability to the reactive segment of the reactor is an important determinant of biological decomposition rates and thus reactor performance. Further consideration of internal moisture dynamics could thus improve definition of a waste site as a biological reactor.

In summary, the above definitions allow waste sites to be described as biofilm reactors of configurations similar to plug flow and fixed bed, trickling flow reactors, in some cases (e.g., landfills) subject to leachate and gas flows, gas–liquid interface heat and water fluxes and microorganism-mediated reactions at the surface of the solid media.

Issues to be addressed by waste site description as a biological reactor include:

- Characterization of the granular media present requires modification of typical representation of solid media for packed beds.
- Definition of how bulk liquid flows through a waste site varies from flow in typical packed beds is needed.
- Packed bed, fixed biofilm kinetics may only properly apply to the activities of the nonhydrolytic organisms in the packed bed, if defined as a waste site bioreactor.
- The velocity relationship for bulk liquid flow through a waste site undergoing packed bed flow must be modified to reflect annular rather than saturated liquid film flow.
- Bulk liquid film flow could be periodic and unevenly distributed.
- The fixed biofilm might or might not cover most of the media, i.e., might be patchy; concentrations of substrates in the liquid film moving over the surface of the solid media change with inflow rate, age of or reduction in volume of substrate sources and with distance from the reactor inlet.
- Fluid flow is likely to be laminar, except for temporary short-circuiting of flows through large subsurface channels or pores.
- The removal of liquid substrates is likely to be effected solely by fixed organisms in the biofilm.
- For useful analyses, the reactive capacity of the waste site should be defined.
- Investigation of effects of landfill design and hydraulics on reactor performance is needed.
- The inclusion of a representative variety of solid wastes, their estimated specific reactive or physical properties and a representative variety of solid media into waste site bioreactor models could at least partially address bioreactor performance uncertainties arising from physical heterogeneity of wastes and media inside a waste site.
- Modeling of the biologically active segment of the landfill as a bioreactor, with the foregoing considerations, and with other reactor modeling simplifications, can thus permit practical insight into reactor effects.
- Incorporation of a range of typical moisture controls, for example varying cover (or liner) types, permits consideration of normal management options without compromising the reliability of predictions.
- Further consideration of internal moisture dynamics of a waste site could thus improve definition of a waste site as a biological reactor.

# 2 Physical Characteristics of Waste Sites

## WASTE SITE BIOLOGICAL REACTOR CONCEPTS

Biological reactor geometry, biofilm distribution, mode of operation and substrate distribution parameters are important criteria in defining what type of biological reactor a system is (Moser, 1988), yet description of the physical system is useful. Initial concepts for bioreactor definition of a waste site are influenced by how liquid flow is influenced by the site's solid media, characterization of the soil and wastes and development of mass transfer parameters and their relationships for this unique system. Though information is lacking for validation of these concepts in any great detail, a careful comparison with other reactors such as those previously discussed, and the use of data from studies of waste sites, can allow development of simplified but efficient characterization.

In Chapter 1, liquid flow through a waste site as a packed bed was indicated to be laminar and dominated by film flow. The disordered media (stones, waste fragments, gravel, various soils) are likely to be randomly packed. The soils and solid wastes making up the solid medium are probably to vary between types, in volume, characteristic shape, density, heat capacity, surface area, water sorption properties and susceptibility to biodegradation by soil microorganisms. Flow through and drainage of the media will occur when the liquid content is greater than what can be held in the soil under a pressure gradient, and it will occur at the smallest percolation threshold anywhere in the system. These variables affect system flows and must be examined for their relationships to bioreactor functions. Definition of a waste site as a packed bed simplifies the approach to model development, but does not limit the complexity of this system or the large number of important variables that require consideration. The type of modeling proposed by this work assumes waste site system function as a bioreactor, thus it limits, but does not ignore, consideration of the system as primarily a generator of effluents. Models of landfills, for instance, predominantly focus on modeling of leachate generation and flow because of the importance of control of water pollution impacts.

### BASIC PHYSICAL CHARACTERISTICS OF SOLID MEDIA

As noted in the discussion of applicable types of reactors, a waste disposal site may contain solid media dominated by inert particles such as stones, gravel, sand, silt or clay and large to small particles or fragments of solid waste. In landfills (and composting operations) the solid wastes can make up much of the solid media. A major fraction of these solid waste particles or fragments can also be considered reactive rather than inert, as the organic fraction of the solid waste materials can be decomposed in the presence of microorganism enzymes.

However, three intrinsic (bulk) properties of the porous media for physical characterization of a waste site as a packed bed are *mean particle size, porosity* and *surface area*. Other basic properties of a site include *specific and bulk density of solids* and *specific and average water saturation*.

## DETERMINATION OF MEAN PARTICLE SIZE

In general, the size distribution of solid particles in a waste disposal site such as a landfill would be gap-graded — meaning irregular weighting of size fractions — or at least quite different from a site where only native soils are present. Also, in comparison with commercial or industrial packed bed or packed tower reactors, the solid particles would not have a mean grain size chosen for optimum flow and substrate removal.

The mean particle size is likely to be an artifact of the particular mix of wastes and soils. For instance, the study by Powrie et al. (2000) showed the mix of particle sizes in landfills varying from over 165 mm to under 10 mm, with particle size ratios varying with age. This suggests that mean particle size for a waste site is also site-specific, as would be expected. Nevertheless, it can be assumed that bulk flow through the site can be characterized through an effective mean particle size, $D_m$, of a value that can be incorporated in expressions for the average saturated flow permeability through a waste site. This is an idealization of the particle size distribution of the granular media and thus can be derived by various means — for example, from sieving soil samples from a waste site. These tests lead to site-specific size distributions, but the approach is useful for characterizing this basic physical property of a granular media.

## PARTICLE SIZE DISTRIBUTION APPROACHES TO FINDING MEAN SIZE OF POROUS MEDIA

An important determination for theoretical analysis of a packed bed representation of an environmental system is the mean size of all solid particles in the bed. Design for monitoring and control of hydraulic flow in many packed bed and filter bed treatment systems is based upon empirical or mathematical relationships that include the mean particle size of the granular bed as a key parameter. While there are many approaches to determining mean particle size, in the case of a system of mixed and irregular particles, practical considerations constrain this determination to indirect methods.

A relatively simple approach to particle size determination of a waste site hypothesized as a packed bed of solid particles is obtaining data from careful screening of soil samples, from various site locations and at specified site depths. While the accuracy of this method is not likely to be the highest, with careful choice of other parameters, bed characterization may be sufficiently accurate to provide a physical basis for liquid and gas flow considerations. Mean grain size can also be determined by more theoretical methods, some of which will be indicated.

**TABLE 2.1**
**Particle Diameter and Wentworth Sieve Scale**

| Lower Phi Value Limit | Soil Grade | Soil Class | Minimum Size, Millimeters | Microns ($10^{-6}$ meters) |
|---|---|---|---|---|
| −8 | Boulder | Gravel | 256 | 256,000 |
| −6 | Cobble | | 64 | 64,000 |
| −2 | Pebble | | 4 | 4,000 |
| −1 | Granule | | 2 | 2,000 |
| 0 | Very Coarse | Sand | 1 | 1,000 |
| 1 | Coarse | | 0.5 | 500 |
| 2 | Medium | | 0.25 | 250 |
| 3 | Fine | | 0.125 | 125 |
| 4 | Very Fine | Sand | 0.0625 | 62.5 |
| 5 | Coarse | Silt | 0.0313 | 31.3 |
| 6 | Medium | | 0.0156 | 15.6 |
| 7 | Fine | | 0.0078 | 7.8 |
| 8 | Very Fine | | 0.0039 | 3.9 |
| > 8 | | Clay | < 0.0039 | < 3.9 |

Simplified methods of particle size distribution analysis are indicated to provide mean particle size values of medium to low accuracy. Accuracy of the analysis depends upon both soil screening method and appropriateness of the statistical methods applied to data analysis.

Sampling of a waste or soil site can involve excavating or auguring soil portions at spaced locations across the site and at depths varying from surface level to above the groundwater level. Each sample is collected, weighed and analyzed separately. Soil screening typically involves measuring the total mass, weight or volume fraction of a soil sample retained between two screens of specific size — for example, weight of soil between quarter-inch and half-inch screens.

A range of sieves can be chosen, for example, according to the Wentworth scale, to cover particle sizes across the total range of statistical interest. In the Wentworth sieve scale, sequential screen sizes are designed to result in a constant geometric ratio between particle size scales. The weights of soil left between screens are measured and grain sizes (d) taken from the screens can be converted to phi ($\phi$) units, where $\phi = \log_2$ (d, in millimeters). The Uden–Wentworth scale is illustrated by the values in Table 2.1.

Histograms can then be prepared showing frequency of particle size occurrence vs. mass or volume, and plots can be prepared of cumulative weights vs. particle size. The cumulative weight vs. size plot shows what percentage by weight of a soil sample is coarser or finer. For example, the following table (Table 2.2) and plots (with landfill particle size ranges taken from the study by Powrie et al. (2000)) in Figures 2.1 and 2.2 show how mean particle size $D_m$ may be defined, in terms of discrete size distributions of screened material, from a waste site.

**TABLE 2.2**
**Soil Screening Values**

| Size Range, Screen, mm | % Weight Left on Lower Screen, kg | Mass, Kg (140 kg total) | Cumulative % Retained | Cumulative % Passing |
|---|---|---|---|---|
| +160 | 4 | 5.6 | 0 | 100 |
| −160 to +80 | 38 | 53.2 | 42 | 58 |
| −80 to +40 | 24 | 33.6 | 24 | 66 |
| −40 to +20 | 17 | 23.8 | 83 | 17 |
| −20 to +10 | 9 | 12.6 | 92 | 8 |
| −10 | 8 | 11.2 | 100 | 0 |

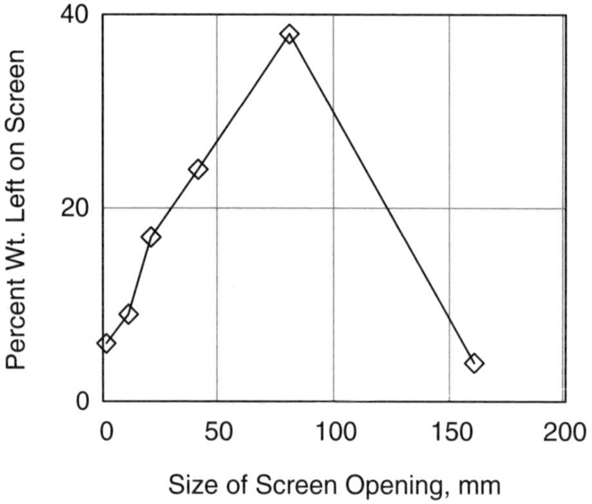


**FIGURE 2.1** Soil-size characteristics from screening

Figure 2.3 plots the data developed from soil sampling and screening as a graph. Grain sizes of mixed soil and wastes such as at disposal sites can cover several orders of magnitude, thus graphic representation is a convenient means of establishing basic parameters of a site soil with a wide range of particle sizes. Figure 2.4 shows the same data as Figure 2.3 — but with grain diameter, typically millimeters (mm), represented in logarithmic phi units (Pfannkuch and Paulson, 2000), where:

$$Phi = \phi = -\log_2(d) = -\frac{\log_{10}(d)}{\log_{10}(2)} = -3.222\log_{10}(d) \tag{2.1}$$

The log-phi graph approach, shown in Figure 2.4, allows calculation of the geometric mean, median value or average particle size ($D_m$) and standard deviation of

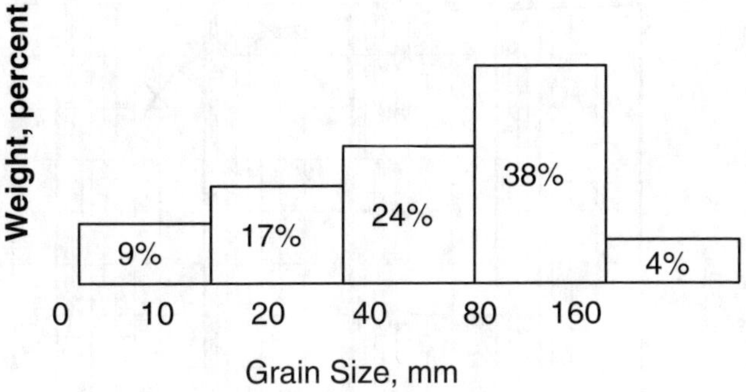


**FIGURE 2.2** Weight percent per grain size class

**TABLE 2.3**
**Expanded Soil Screen Values — Interpolated**

| Grain Size, mm | Weight of Size Fraction, kg | Percent Weight | Cumulative % Retained | Cumulative % Passing |
|---|---|---|---|---|
| 0.563 | 2.1 | 1.5 | 98.5 | 1.5 |
| 3.54 | 4.2 | 3.0 | 95.5 | 4.5 |
| 7.03 | 4.9 | 3.5 | 92 | 8 |
| 14.1 | 5.6 | 4 | 88 | 12 |
| 19.9 | 7.0 | 5 | 83 | 17 |
| 28.1 | 9.8 | 7 | 76 | 24 |
| 39.8 | 14.0 | 10 | 66 | 34 |
| 56.3 | 15.4 | 11 | 55 | 45 |
| 79.3 | 18.2 | 13 | 42 | 58 |
| 112.5 | 30.8 | 22 | 20 | 80 |
| 159 | 22.4 | 16 | 4 | 96 |
| 225 | 5.6 | 4 | 0 | 100 |

the range of particle sizes encountered in the site sample (Pfannkuch and Paulson, 2000):

$$\text{Geometric mean} = (\bar{\phi})_{GM} = \frac{(\phi_{16} + \phi_{50} + \phi_{84})}{3} \tag{2.2}$$

Median size $= \phi_{50} = 50\%$ value for cumulative frequency distribution curve.

$$\text{Standard deviation} = \sigma_\phi = \frac{(\phi_{84} - \phi_{16})}{4} + \frac{(\phi_{95} - \phi_5)}{6.6} \tag{2.3}$$

If waste site sampling and screening data are available, other statistical parameters related to particle size distribution data, such as skewness, can be determined from

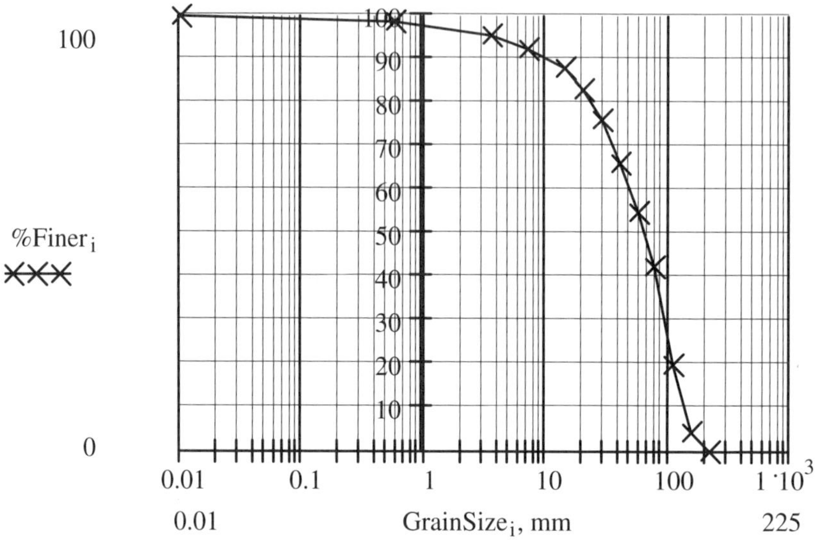


**FIGURE 2.3** Soil grain size distribution

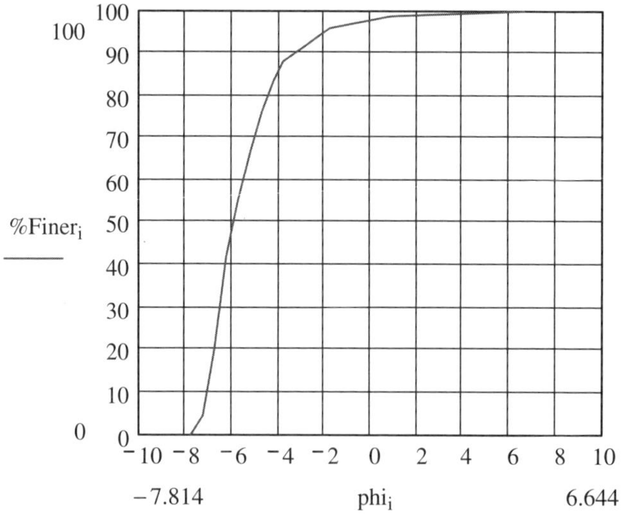


**FIGURE 2.4** Log-phi size distribution of soil

the phi distribution curves. The skewness of the data represents how symmetrical the data are about the mean.

For instance, if the soil grain size distribution curve for samples from the waste site is symmetrical, it would have a skewness of 0.0; if the soil has an excess of fine grains, it would have a skewness of positive value, e.g., 0.8; and if the soil has an excess of coarse grains, it could have a negative skewness value, e.g., –0.6. A general skewness formula (McBride, 1971) is given by:

$$\text{Skewness} = Sk_{\phi} = \left[ \frac{\sum_i f_i (m_i - D_m)^3}{100\sigma_{\phi}^3} \right] \tag{2.4}$$

where $f_i$ = weight percent of grains in the $i$th class of solid material, $m_i$ = midpoint value of each grain size class, in phi (or $\phi$) values, $D_m$ = mean grain size and $\sigma_{\phi}$ = standard deviation from phi values (see Equation 2.3). The Folk (1968) approach to skewness of a soil grain size distribution, often used in soil engineering, is also considered an efficient method of determining 'skewness,' as it not only determines the skewness of the mean value but also the skewness of the left and right tails of the distribution curve. It is independent of soil sample sorting:

$$\text{Skewness} = Sk_1 = \frac{\phi_{16} + \phi_{84} - 2\phi_{50}}{2(\phi_{84} - \phi_{16})} + \frac{\phi_5 + \phi_{95} - 2\phi_{50}}{2(\phi_{95} - \phi_5)} \tag{2.5}$$

The skewness of particle sizes (of wastes and soils) is likely to vary according to particle age. It is well known, for instance, that as waste particles rot in a landfill they decrease in size — thus, *older* wastes should have a higher percentage of fine grains. This would suggest, for instance, that samples from deep into a waste site should have a positive skewness value reflecting the higher mass percentage of fine grains, while fresh waste samples, usually in the upper part of a waste site or dump, should have a higher percentage of coarse grains and thus should have a negative skewness value. This also highlights the importance of composite sampling for simpler representations of a waste site.

The kurtosis of a soil sample can also be determined from its particle size distribution. This measure describes the peakedness of the particle distribution curve or how well behaved the sorting of particles is over the range of sizes. For instance, if the soil were well sorted throughout (no gaps in the weight versus size distribution), the plotted curve of phi ($\phi$) versus grain size would have a normal Gaussian distribution and a kurtosis of 1.00. If the soil is better sorted in the middle range than at the fine or coarse range (left and right tails), the curve would be considered peaked or leptokurtic, in which case the kurtosis exceeds 1.00 (e.g., 1.0–3.0). If the tails (fine and coarse size distributions) of the phi ($\phi$) curve are better sorted than its central portion, the curve is considered flat-peaked or platykurtic; and kurtosis is less than 1.00 (e.g., 0.5–1.0). Highly leptokurtic curves are highly peaked, while the peaks of highly platykurtic curves can flatten to two peaks (bimodal sorting distribution), with the sag in the middle representing poor sorting in the middle size range of the soil sample and good sorting in the tails or coarse and fine ranges (Pfannkuch and Paulson, 2000). The kurtosis $K_s$ of soil sampling data can be represented as proposed by Folk (1968):

$$\text{Kurtosis} = (\text{Kur})_{\phi} = \frac{\phi_{95} - \phi_5}{2.44(\phi_{15} - \phi_{25})} \tag{2.6}$$

Waste site material is likely to be strongly leptokurtic, considering that grain sizes in the middle range may make up the bulk of the waste materials, which is typically poorly sorted in terms of size.

Use of these statistical measures, mean particle size $D_m$, standard deviation $\sigma_\phi$, skewness $Sk_\phi$ and kurtosis $\mathrm{Kur}_\phi$ allow insight into how mixed solid particle sizes may be distributed by weight or volume in a waste site represented as a packed bed.

## Grain Size Statistics vs. Age of Wastes

Representation of soil size distribution at a waste site encounters the issue of particle age. As solid waste particles age and are converted to liquid or gas or are mineralized through biological decomposition or chemical reaction, they will decrease in size. The age of waste in a site is often represented by a size range that decreases with depth (or time of deposition). The data from the study of particle sizes in various landfills by Powrie et al. (2000) suggest that landfill particle size distributions become weighted toward a higher fine particle content as the landfill ages and the wastes rot.

If age vs. depth is to be considered as a modeling factor, one approach may be to model the waste site bioreactor as made up of a series of connected segments (layers or arbitrary volumes) of increasing age, where simple sampling methods can inform the modeler of grain size statistical parameters such as mean size, standard deviation, skewness and kurtosis. Series or parallel connectedness of the segments enables water and heat flow modeling consideration. These hypothetical segments can be initialized with data for water content, waste composition and heat capacity.

The sampling data for a waste should also reflect particle size variation with (horizontal) location of a sample, as well as size variation with depth at that location. To accomplish this efficiently, horizontally spaced sampling and depth-varied sampling are useful. Since this mainly involves sample excavation, screening and sorting, as done for the study of landfills by Powrie et al. (2000), cost of providing this type of information should not be prohibitive; also, waste soil samples taken at a few sites in a region should represent climatic and perhaps organism impacts on the age wastes at landfills or dumps in a region, provided these sites have been operated under the same conditions. As noted, composite sampling can reduce the need for age-related waste site grain size characterization.

## Packed Bed Porosity, Hydraulic Conductivity and Permeability

Waste disposal sites as packed bed may have porosities that fall into a relatively limited range and a measurable hydraulic conductivity $K_h$ that falls into an expected range. Various investigators have reported *porosity* and *hydraulic conductivity* ranges, usually for landfills.

For instance Oweis et al. (1990) reported a hydraulic conductivity range for solid waste landfills of $2 \times 10^{-2}$ cm/sec, to $5.1 \times 10^{-3}$ cm/sec, with effective porosity of about 0.41, porosity of landfill clays at over 0.45 of volume, sand porosity at 0.25 to 0.4 of volume, sand and gravel mix porosity at 0.15 to 0.25 of volume and municipal solid waste (MSW) porosity at 0.45 to 0.50 of volume.

Landva and Clark (1990) indicated that interparticle porosity in landfills ranges from 30% to 60% of volume ($\varepsilon = 0.3$ to 0.6), depending upon degree of compaction. McBean, Rovers and Farquhar (1995) reported hydraulic conductivities from $10^{-1}$ to $10^{-9}$ cm/second for gravel, sand, silt, clay and municipal solid waste (MSW) in various degrees of compaction; and *permeability* of landfills ranging from $10^{-11}$ to $10^{-12}$ m$^2$ in the vertical direction and $10^{-10}$ m$^2$ in the horizontal direction. Powrie et al. (2000) have also reported the hydraulic conductivity of landfilled material as varying from $10^{-1}$ to $10^{-6}$ cm/sec. These ranges of site soil porosity, conductivity and permeability can be useful for waste site characterization. They are summarized in Table 2.4.

## POROSITY OF A WASTE SITE

The porosity of a site system or packed bed, represented earlier as $\varepsilon$, is a measure of the void space in the porous bed that may be occupied by fluid. In the case of a saturated bed, the void space may be totally occupied by fluid — for example, water. For the unsaturated condition, void space may be partially occupied by water (as a mobile or stagnant liquid film on the surface of the pores) and partially by gas (air, landfill or decomposition or soil gases). For dry conditions, the void space is filled by gas. The total void space or *porosity* of the packed bed system can be represented by volume of voids $V_{void}$, divided by the total system volume $V_{tot}$:

$$\text{Porosity} = \varepsilon = \frac{V_{void}}{V_{tot}} \tag{2.7}$$

If the total volume of the particles is represented by $V_p$ and $V_w$ represents the total mobile or stagnant water volume in the packed bed or waste, then

$$\text{Porosity} = \varepsilon = \frac{V_{tot} - V_p - V_w}{V_{tot}} = 1 - \frac{V_p + V_w}{V_{tot}}$$

$$\text{Volume of particles} = V_p = (1 - \varepsilon)V_{tot} - V_w \tag{2.8}$$

The total volume $V_{tot}$ can be assumed or estimated. For instance, the volume of a waste site might be 2000 or 200,000 cubic meters (m$^3$). Assumptions for the other values in porosity relationship (equation 1.13) must be more carefully chosen. The porosity $\varepsilon$ values listed in Table 2.4 represent *bulk porosity*, which is determined from bulk density, found by oven drying of a weighed packed sample as indicated below:

Soil water content = (moist soil weight – oven dry soil weight)/oven dry soil weight
Soil bulk density = oven dry weight/soil volume
Volume of water in soil (g/cm$^3$) = soil water content (g/g) × bulk density (g/cm$^3$)
Soil water filled pore space (%) = (100 × volume of water in soil)/(soil porosity)
Soil porosity = 1 – (bulk density/dry density of soil type)

The bulk porosity $\varepsilon$ is thus a measure of the porosity of the (oven) dried soil and the water it contains. When the water content is added, the density of the *unsaturated* soil and the pore fraction occupied by water can be found.

**TABLE 2.4**
**Hydraulic Conductivity, Porosity and Permeability of Waste Site Soils**

| Soil Status | USCS Soil Class | Hydraulic Conductivity (K, cm/sec) | Porosity ($\varepsilon$) | Permeability ($m^2$) | Reference |
|---|---|---|---|---|---|
| MSW Landfills | NA | $2 \times 10^{-2}$–$5.1 \times 10^{-3}$ | 0.41 | | Oweis et al., 1990 |
| MSW Landfills | NA | | 0.3–0.6 | | Landva and Clark, 1990 |
| MSW Landfills | NA | $10^{-1}$–$10^{-6}$ | | | Powrie et al., 2000 |
| MSW, as Placed | NA | $10^{-3}$ | | $10^{-11}$–$10^{-12}$ vertical | McBean et al., 1995 |
| MSW, Shredded | NA | $10^{-2}$–$10^{-4}$ | | $10^{-10}$ horizontal | |
| MSW, Baled, Dense | NA | $7 \times 10^{-4}$ | 0.3–0.4 | | |
| MSW, Baled, Loose | NA | $1.5 \times 10^{-4}$ | | | |
| Compost Pile | NA | NA | 0.3–0.8 | | Woods End Lab., 2000 |
| Biofilter | NA | NA | 0.3–0.4 | | US Army Corps of Engineers, June 1999 |
| Gravel | GP | $10^{-1}$ | | | McBean et al., 1995 |
| | GW | $10^{-2}$ | 0.3–0.4 | | |
| | GM | $5 \times 10^{-4}$ | | | |
| | GC | $10^{-4}$ | | | |
| Sand | SP | $5 \times 10^{-2}$ | | | McBean et al., 1995 |
| | SW | $10^{-3}$ | | | |
| | SM | $10^{-3}$ | 0.35–0.4 | | |
| | SC | $2 \times 10^{-4}$ | | | |
| Silt | ML | $10^{-5}$ | | | McBean et al., 1995 |
| | MH | $10^{-7}$ | | | |
| Clay | CL | $3 \times 10^{-8}$ | 0.45–0.55 | | McBean et al., 1995 |
| | CH | $10^{-9}$ | | | |

## An Approach to Determining Porosity of a Packed Bed of Mixed Particle Types

The issue posed by bulk porosity values for a waste site is that, as a system of irregular particles of varying density, shape and water content, overall porosity is only a rough measure of the volume of water held in the pores of the system. Oven drying of a soil sample leaves percentage of water, considered irreducible saturation or residual water. In most cases, irreducible water content determination requires different methods than oven drying, thus the absolute water content of wastes and field capacity is unlikely to be known. In a practical sense, this does not matter because the water volume likely to be of modeling interest is the mobile water content, which at maximum is the amount that would drain from the pores under compression conditions similar to soil plus the amount of water held in pores up to field capacity.

Under field conditions, water is retained by the soil up to a critical volume referred to as *field capacity*. In terms of water or gas flow, which is often the main interest of soil porosity measurements, movement under gravity forces does not begin until the field capacity or critical volume has been reached. Water loss between field capacity and oven-dry conditions is considered the amount the system can retain inside pores or on surfaces. Saturation flow measurements, as are usually done for hydraulic conductivity determination, are often quite accurate for the amount of water required to fill the pores, per unit volume of system beyond field capacity. Total or bulk system porosity is thus the average field capacity plus the remaining volume filled by water at saturation. The irreducible water content is thus useful only as the lower bound of soil volume available for water uptake and flow.

If properties of each type of material in the site are generally known — for instance, waste or soil particle average size, dry density $\rho$ and saturated or average water content $w$ — for waste site conditions, bulk properties of the bed useful to bioreactor considerations can be derived.

For modeling purposes, a bed of mixed materials can be considered a combination of soil types $i = 1 \ldots n$. For the mixture or soils, each type has a total weight fraction $w_i$, an average water content $wc_i$ and a size distribution of $j = 1 \ldots m$, where $m$ represents the number of particle size intervals from fine to coarse, developed from soil screening data. If for example the total weight $W_i$ of a soil type — for instance, particles of waste paper — buried in a waste site represents 20% by weight or volume of all wastes buried in the site and soil screening has developed nine different particle size intervals from waste site sampling, the total weight of paper particles would fall into nine size intervals $j$, ranging from $j = 1$ to $j = n$ and paper as a type of solid material in the site would have a class designation of the number of types of solid materials in the waste site. Summing these particle interval weights would result in the total waste paper weight $W_i$:

$$\text{Total } \textit{moist} \text{ weight of soil type } i = W_{i(moist)} = \sum_{j=1}^{j=m} (W_{s_{i,j}}) \tag{2.9}$$

Assuming that the $i$th material has an initial (average) water content dependent upon its degree of saturation inside the waste site and would be oven-dried to determine its water content:

$$\text{Water content of } i\text{th soil material (g/g)} = \frac{W_{i(moist)} - W_{i(dry)}}{W_{i(dry)}} = wc_i \tag{2.10}$$

$$\text{Total } dry \text{ weight of soil type } i = W_{i(dry)} = \sum_{j=1}^{m} (W_{s_i} - wc_i) \tag{2.11}$$

$$\text{Total soil wet weight } = \sum_{i=1}^{n} (W_{i(moist)}) \tag{2.12}$$

$$\text{Total soil dry weight } = \sum_{1}^{n} (W_{i(dry)}) \tag{2.13}$$

$$\text{Total water content of soil bed (g/g)} = \sum_{i=1}^{n} (wc_i) = \sum_{i}^{n} \left[ \frac{w_{i(moist)} - w_{i(dry)}}{w_{i(dry)}} \right] \tag{2.14}$$

$$\text{Total water volume in the soil bed } = \frac{\sum_{i=1}^{n} (wc_i)}{\rho_{water}} \tag{2.15}$$

At or near bed saturation or when the bed has reached free drainage water-holding capacity, the total water volume in the soil also represents the water held in various fractions of the soil on internal and external surfaces of the soil materials. The actual volume of the $i$th soil material is also its oven dry weight divided by its dry density $\rho_i$:

$$\text{Volume of } i\text{th soil material} = vs_i = \left( \frac{W_{i(dry)}}{\rho_{i(dry)}} \right) \tag{2.16}$$

$$\text{Total soil volume in the packed bed} = \sum_{i=1}^{n} (vs_i) \tag{2.17}$$

$$\text{Soil bulk dry density} = \text{soil dry weight/soil volume} = \rho_{bulk} = \frac{\sum_{i=1}^{n} (W_{i(dry)})}{\sum_{i=1}^{n} (vs_i)} \tag{2.18}$$

The waste site soil porosity $\varepsilon$ is given by:

Porosity = (site volume – volume of soils – volume of water contained)/site volume,

or:

$$\text{Site soil porosity } = \varepsilon = \frac{V_{site} - \sum_{i=1}^{n} (vs_i) - \left( \frac{\sum_{i=1}^{n} (wc_i)}{\rho_{water}} \right)}{V_{site}} \tag{2.19}$$

Data for estimation of the above properties can be provided from sieve screening of soil samples combined with water content determinations; but other methods can be used for greater accuracy. For example, sieve screening of a sample from a waste site might determine that there are 13 categories of waste and three categories of soil, or $I = 1 \ldots 16$. Wastes and soils densities $\rho_I$ as-received or dry and unit dry weights $w_i$ can be determined or might be available from previous studies. Size classes of particles are developed from the coarsest and finest fractions plus the average size between two screens of standardized opening size; for instance, if six screens are used, $j = 1 \ldots 7$.

While this approach is not the most accurate method of determining specific particle properties such as shape, mean diameter and arrangement, it provides a simple and efficient determination of properties useful for applying reactor principles.

## DENSITY AND OTHER PROPERTIES OF MIXED SOIL AND WASTE MATERIALS

The site property estimation methods listed above also require information on specific weights and densities of various materials in a waste site. Separation and identification methods applied to each size fraction can allow specific weights and densities of various materials to be found. However, identification of waste fractions in each size category, particularly above fine soil sizes, is useful and may increase sorting effort. This effort can be reduced to some degree by careful sorting and identification of the middle and coarse fraction and providing enough information from these examinations to interpolate to the finer fraction, as suggested by Figure 2.5. Information available from studies of solid wastes or compostable materials and soils may also be used — for instance, the typical data listed in Table 2.5 compiled from various sources (Miller, 1998).

These properties can be utilized in calculations of other basic properties. For instance, the average density of all materials in the waste site can be found from the weighted mass–volume ratio:

$$\rho_{\text{ave}} = \frac{\sum_{i=1}^{n} (\frac{M_i}{V_i})(f_m)_i}{\sum_{i=1}^{n} (\frac{M_i}{V_i})} \tag{2.20}$$

where $M_i$ = mass of $i$th material, $V_i$ = volume and $f_{m,i}$ = mass fraction of $i$th material in the total volume of solid materials. If data are available for the saturated water content or field capacity (moisture content of the material at which it begins to drain water under gravity) of the various materials is available, the average saturated water content and field capacity can be similarly estimated.

## APPLICABILITY OF CONDUCTIVITY AND PERMEABILITY RELATIONS FOR PACKED BEDS

Granular porous media is made up of many particles. The spaces or pores between these particles are, in many cases, interconnected, allowing the free passage of gas or liquid flow. The average liquid flow velocity under saturated conditions is called

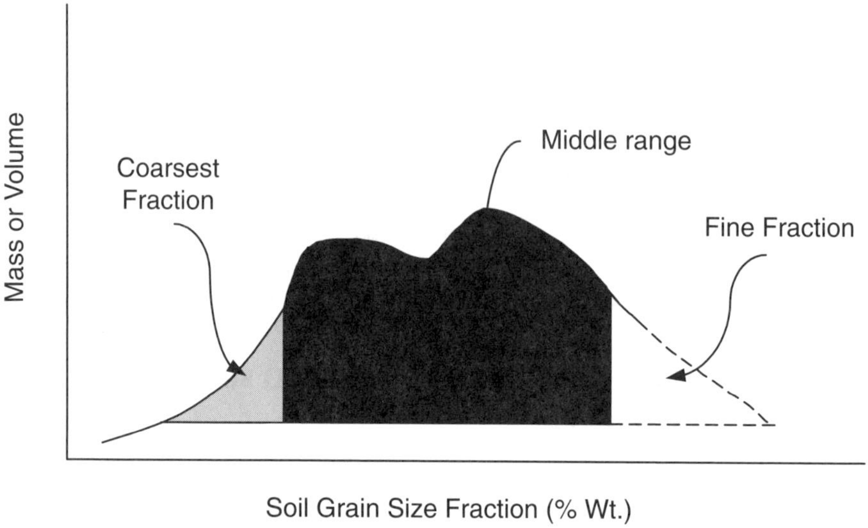


**FIGURE 2.5** Waste site soils mass distribution

the *hydraulic conductivity*, $K$. As saturated flow encounters the pore space quantity, pore space arrangement and packing arrangement of soil particles, the hydraulic conductivity $K$ reflects flow resistances of the granular bed, including surface friction, available flow volume and moisture uptake capacity of the soil. Hydraulic conductivity is thus a constant property of the soil material. For porous media such as soil, Darcy's Law (H.P.G. Darcy, 1830) establishes an empirical relationship between soil physical properties and liquid flow rate for water flow through a sand bed:

$$Q = \left(\frac{KA}{L}\right)\frac{\Delta(p+h)}{\rho g} \tag{2.21}$$

where $Q$ = fluid flow rate ($m^3$/sec), $K$ = hydraulic conductivity (m/sec), $p$ = pressure, $h$ = drop in piezometric pressure head across a (soil) column of length or depth $L$, $\rho$ = fluid density and $g$ = gravitational acceleration. The Darcy's Law relationship implies that the flow $Q$ is proportional to pressure change with soil bed depth and is inversely proportional to soil bed depth (a thinner bed releases more flow). Because it was developed from experiments with saturated sand, Darcy's Law is considered appropriate for homogenous media, such as sands and isotropic media where water retention and flow properties do not change with direction. It applies for low Reynolds' numbers or for slow or laminar flow. For instance, the conditions of Darcy's saturated sand flow experiments are considered laminar. However, *mixed solid materials in a waste site are likely to be both anisotropic and heterogeneous.* Darcy's Law thus has to be modified for use in characterizing fluid flow through a waste site system.

Darcy's Law also has upper and lower limits of application. The analogies established by soil water flow investigators to define the limits of Darcy's Law include

**TABLE 2.5**
**Density and Saturation Properties of Waste Site Materials**

| Site Material | Dry Density | Water Content Max. Wt. Fraction (2) | Water Content, Fraction, As Received (3) |
|---|---|---|---|
| Mixed paper | 0.92 (0.7–1.15) | 0.124 | 0.24 |
| Newspaper | 0.92 | 0.113 | 0.22 |
| Corrugated paper | 0.92 | 0.117 | 0.13 |
| Cardboard | 0.92 | 0.23 | 0.06 |
| Plastic film | 0.93 | 0.02 | 0.07 |
| Other plastic | 0.95 | 0.01 | 0.07 |
| Leather | | 0.34 | 0.07 |
| Rubber | 1.04 (1.0–2.0) | 0.01 | 0.01 |
| Textiles | 1.55 | 0.33 | |
| Leaves (dry) | 0.3 | 0.8 | 0.4 |
| Grass | 0.3 | 0.77 | 0.41 |
| Brush | 0.3 | 0.66 | 0.41 |
| Lumber | 0.45 | 0.26 | 0.17 |
| Food waste | 1.1 | 0.11–1.8 | 0.51 |
| Sludge | 1.02 | 0.4 | |
| Ferrous metal | 7.8 | 0.02 | |
| Aluminum | 2.7 | 0.02 | |
| Other metals | 6.0 | 0.02 | |
| Glass | 2.5 | 0.017 | 0.02 |
| Ash | 1.8 | 0.2 | 0.10 |
| Dirt | 1.8 | 0.2 | 0.08 |
| Concrete | 2.3 | 0.13 | 0.08 |
| Masonry | 0.67 | 0.11 | 0.08 |
| Rock | 1.8 | 0.2 | |
| Gravel | | 0.07 | |
| Sand | 2.65 | 0.4 (0.06) | |
| Silt | 2.6 | 0.2 | |
| Clay | 2.7 | 0.6 (0.4) | |

Source: Miller, P.A., 1998.

use of a dimensionless Reynolds' number Re, which is the ratio between viscous and inertial forces:

$$\mathrm{Re} = \frac{\rho v D}{\mu} \tag{2.22}$$

where $\rho$ = fluid density, $v$ = average flow velocity, $D$ = effective pipe or pore diameter, and $\mu$ = dynamic viscosity of the fluid. Experimental work has shown that Darcy's Law is valid for $\mathrm{Re} < 1.0$ and is well behaved up to $\mathrm{Re} = 10.00$. In the range $\mathrm{Re} < 10$ the flow thus can be considered Darcian or laminar. A Re value of 10 thus represents the upper limit for Darcy's Law validity. According to Bear (1972), Darcy's Law holds as long as the Reynolds' number (Re), *based upon average soil grain diameter* ($D_m$), does not exceed the range $1 < \text{Re} < 10$. Reynolds' number in terms of average grain diameter is:

$$\text{Re} = \frac{\rho q D_m}{\mu} \tag{2.23}$$

where $q$ = specific flow per unit cross-section area = Q /A ((m$^3$/sec)/m$^3$) = m/sec, Q = total flow through soil and A = cross-section area of the soil system (column or packed bed). Notably, the units of $q$ are of velocity; so there is no inconsistency between the relations shown above for Reynolds' number Re.

The stated Reynolds' number range for Darcy's Law validity, Re < 10, is considered the creeping flow regime, where flow is both laminar and linear. Above Re = 10 the flow regime changes from laminar to nonlinear laminar flow and then to turbulent flow.

The issue of where Darcian or creeping (laminar flow) ends has been investigated since Darcy and is reflected in the relations developed by Ergun (1952), McDonald et al. (1979), Comiti and Renaud (1989), Seguin et al. (1990) and Comiti et al. (2000). These studies of flow through granular media and packed beds of particles of various shapes and sizes showed that the creeping flow regime extends past Re = 10, even though Darcy's Law might not hold for this range. Also, for randomly packed beds of variously shaped and arranged particles such as (idealized) waste sites, it is likely that structural inhomogenieties would result in both linear and nonlinear laminar flow regimes. It should also be noted that Re represents the ratio of inertial to viscous forces on flow (Re = inertial/viscous forces) and that for (linear) laminar flow, the inertial forces are relatively small compared to viscous forces (on the order of 5%). Recent work by Comiti et al. (2000), examining the role of inertial vs. viscous forces on flow, has shown that the nonlaminar flow regime ends at about Re = 180. Ranges of laminar flow of interest to the packed bed type previously described should thus be:

$0 < \text{Re } 10$ : Creeping flow regime ~ laminar flow: Darcy's Law for porous medium holds

$10 < \text{Re} < 180$ : Nonlinear creeping flow regime ~ laminar flow ~ effects of inertial forces

Darcy's Law also does not hold when flow velocity is very small, because the liquid does not move until a particular (small) water potential is reached. In the case of surfaces of particles in a packed bed of soil, the potential to be exceeded varies with the amount of water absorbed by the particle. McBean et al. (1995) have noted that the lower limit of validity for Darcy's Law is reached for (water) flow through fine-grained materials of low fluid conductivity, such as clays, where pores of the soil may be so small that liquid molecules are held to the pore surface by electrical charges of the soil particles. This liquid binding force would add to the viscosity of the flow, thus further reducing the Reynolds' number Re. As this behavior is not likely to be accounted for by Darcy's Law, the Darcy relationship is unlikely to be well-behaved for flow through packing of very fine soil grains. Clay soil type flow velocities and grain sizes could thus be considered the lower limits for applying Darcy's Law for mobile liquid flow through soil beds.

In fine materials such as clays, the flow velocity and liquid discharge rates are very small, thus liquid discharge considerations may be insignificant for hydraulic flow modeling. However, slowly flowing or stagnant liquid in fine material in a packed bed could be of interest to soil system modeling as a bioreactor because microorganism activity is limited only by access to pores and availability of water and nutrients and continues, at levels decreasing with water content of the soil system, well below the condition where there is significant fluid discharge from a soil e.g., practically dry conditions.

Liquid flow through granular porous media systems such as waste disposal sites should thus be regarded laminar for the most part, although liquid (water) flowing solely through soil macropores may or may not be laminar. For most natural subsurface flow, Re $< 1.0$ (Todd, 1980), except for macropores, in rock fissures and through limestone or other rock openings suggesting laminar flow. For waste sites of sufficient depth and compaction, macropore flow may occur through connected large cavities or pores, but is less than likely to persist throughout the site. This suggests that macropore flow would be masked by the flow discharge and flow velocity from a waste site. The soil bulk velocity or hydraulic conductivity $K$ would thus still apply; and Darcy's Law can still be used for modeling of average hydraulic fluid flow through a waste site.

Hagen (1839) and Poiseuille (1841) studied laminar flow in pipes and concluded that potential energy loss due to friction along the flow path was proportional to flow velocity. Darcy's Law recognizes the similarity between laminar flow in saturated soils and laminar flow in pipes. The Hagen-Poiseuille relationship for laminar flow in tubes can thus be applied to soils, with the assumption that the soil is a porous medium with pores or channels conducting fluid flow, similar to a bundle of tubes conducting fluid at laminar velocity. The Hagen-Poiseiulle relationship relating flow of a viscous fluid through a tube to flow rate $Q$, pressure drop $\Delta p$ and tube length $L$ and the Darcy relationship are:

$$\text{Hagen-Poiseuille flow: } Q = \frac{\pi D^4}{128\mu}\frac{\Delta p}{L} = \frac{AD^2}{32\mu}\frac{\rho g}{L}\left(\frac{\Delta p}{\rho g}\right) \tag{2.24}$$

$$\text{Darcy flow: } Q = -KA\frac{dh}{dl} \tag{2.25}$$

where $dh/dl$ = hydraulic gradient along the flow path. Similarity and comparison between the Hagen-Poiseuille and the Darcy relationships for flow indicates that hydraulic conductivity ($K$) should depend upon pore size $D$ and kinematic viscosity $\mu$. Separating $K$ between the two relationships,

$$K = \frac{k\rho g}{\mu} \tag{2.26}$$

where $k$ = *intrinsic permeability* of the porous media. The intrinsic permeability $k$ is considered a proportionality constant specific to the media involved, i.e., its value depends only upon characteristics of the porous media. For water flow through a waste site, it is thus necessary to define permeability.

## PERMEABILITY k OF A MIXED POROUS MEDIA

The Hagen-Poiseuille relationship has been the basis of many models relating the internal structure of porous media to its permeability or flow resistance. In these models, the *intrinsic permeability* $k$ is provided as a function of media porosity and pore radius $r$, squared.

$$k = f(\varepsilon, r^2) \tag{2.27}$$

Dimensional analysis shows that the definition of *intrinsic permeability* has units of area:

$$k = \frac{(M/LT)(L/T)}{(M/L^3)(L/T^2)(L/L)} = (L^2) \tag{2.28}$$

Permeability relations thus take the form:

$$k = cD^2 \tag{2.29}$$

where $D$ is a characteristic length and $c$ is a dimensionless coefficient combining physical structure parameters of the media such as pore and solid material characteristics, e.g.:

$$k = f_{solid}\, f_{pore}\, d_m^2 \tag{2.30}$$

where $f_s$ = particle or pore shape factor, $f_p$ = porosity factor and $d_m$ is the equivalent mean particle size (Todd, 1980).

The relation between intrinsic permeability $k$ and *superficial velocity* $U_0$ — which has the same meaning as *hydraulic conductivity* — for single-phase flow through a porous medium ($U_0 = K$) is given by:

$$U_0 = -\frac{k}{\mu}\frac{\Delta P}{H} \tag{2.31}$$

In terms of the pressure gradient and allowing for both linear and nonlinear laminar flow, the pressure gradient $\Delta P/H$ had been presented as of the form (Comiti et al., 2000):

$$-\frac{\Delta P}{H} = k\mu U_0 + b\rho U_0^2 \tag{2.32}$$

where $k$ = intrinsic permeability.

However, the interstitial velocity or wetting front velocity is the actual laminar flow velocity and is the ratio of the superficial velocity to medium porosity:

$$\nu_{\text{int}} = \frac{U_0}{\varepsilon} \tag{2.33}$$

The Hagen-Poiseuille relation for *interstitial velocity* $v_{\text{int}}$ for laminar flow through porous media, with pores of length $L$ represented as a bundle of tubes, is:

$$\nu_{\text{int}} = \frac{d_{pore}^2}{32\mu}\frac{\Delta P}{L} \tag{2.34}$$

where the representative pore diameter $d_{pore}$ is:

$$\begin{aligned} d_{pore} &= 4\frac{\text{cross-section area of flow}}{\text{wetted perimeter}} = 4\frac{\text{volume of free space}}{\text{interfacial area}} \\ &= 4\frac{\text{porosity}(=\varepsilon)}{a_v = \text{interface area}} \end{aligned} \tag{2.35}$$

For the representative pore of diameter $d_{\text{pore}}$ of a porous medium, the *interface area* $a_v$ is represented as:

$$a_v = \frac{\textit{unit surface area of solid}}{\textit{volume of solid}} = \frac{a}{(1-\varepsilon)} \tag{2.36}$$

where $a$ = specific surface area of solid/volume of solid. The term $a_v$ has the definition:

$a_v$ = surface area presented to laminar flow by packed bed particles/volume of solid material. As a collective term for surface area, $a_v$ represents interface area per unit volume of total bed pore undergoing flow or $a_v$ = *interfacial surface* area of void fraction. Thus, in terms of the mean pore diameter $d_{pore}$, the mean particle diameter $D_m$, the Reynolds number Re and the interstitial velocity $v_{\text{int}}$ are:

$$v_{\text{int}} = \frac{U_0}{\varepsilon} \tag{2.37}$$

$$d_{pore} = 4\frac{\varepsilon}{a_v(1-\varepsilon)} \tag{2.38}$$

$$\text{Re} = \frac{\rho d_{pore} v_{\text{int}}}{\mu} = \frac{\rho}{\mu}\frac{4\varepsilon}{a_v(1-\varepsilon)}\frac{U_0}{\varepsilon} = \frac{4\rho U_0}{\mu a_v(1-\varepsilon)} \tag{2.39}$$

In the above relations, $U_0 = \varepsilon\ v_{\text{int}}$ = apparent saturated velocity of flow through the porous medium (same as conductivity $K$). The Reynolds number Re, apparent velocity $U_0$ and the *intrinsic permeability* $k$, in terms of average pore interface area $a_v$, are thus:

$$\text{Re} = \frac{4}{6}\frac{D_m U_0 \rho}{(1-\varepsilon)\mu} \tag{2.40}$$

$$U_0 = \varepsilon v_{\text{int}} = \varepsilon \frac{d_{pore}^2}{32\mu}\frac{\Delta P}{L} = \varepsilon \frac{1}{32\mu}\left(\frac{4\varepsilon}{a_v(1-\varepsilon)}\right)^2 \frac{\Delta P}{L}$$

$$= \varepsilon \frac{1}{2\mu}\left(\frac{\varepsilon^3}{(1-\varepsilon)^2}\frac{1}{a_v^2}\frac{\Delta P}{L}\right) \tag{2.41}$$

$$U_0 = \frac{1}{2}\frac{1}{36}\left(\frac{\varepsilon^3}{(1-\varepsilon)^2}\right) D_m^2 \frac{\Delta P}{\mu L} \tag{2.42}$$

$$k = \frac{Q}{A}\frac{\mu L}{\Delta P} = U_0 \frac{\mu L}{\Delta P} = \frac{1}{72}\frac{\varepsilon^3}{(1-\varepsilon)^2} D_m^2 \tag{2.43}$$

The above expressions all indicate a dependence upon the representative grain size and porosity of the medium, regardless of particle type. Thus, for a waste site as a porous medium, average grain size and average system porosity would be key influences over liquid or gas flow.

## PERMEABILITY (k) CORRECTION FOR PACKED BED FLOW

Experimental work has shown that, with the use of $L$ for pore length, the results of experiments did not agree with relations for *apparent saturated flow velocity* through the media $U_0$, *intrinsic permeability* $k$ and *mean interstitial velocity* $v_{\text{int}}(= U)$. It was recognized that one error source is that $L$, the depth of the porous bed, is not necessarily the actual pore length $L_p$, which is likely to be longer than $L$ because flow has to negotiate past surfaces of irregular particles and twisting pores. The effect of irregular pore pathways is recognized with use of a tortuosity factor $\zeta$, which is:

$$\zeta = \frac{L_{pore}}{L} \tag{2.44}$$

where $L = z$ = depth of bed material, $\zeta$ = average tortuosity of the packed bed and $L_{pore}$ is the actual (representative) pore length through the bed. The effect of packed bed pore tortuosity on the *mean interstitial velocity* $U$ for laminar flow through the pores of granular material in a soil bed can be defined in terms of the bed tortuosity $\zeta$:

$$U = v_{\text{int}} = \frac{U_0}{\varepsilon}\frac{L_{pore}}{L} = \frac{U_0}{\varepsilon}\frac{L_{pore}}{z} = \frac{U_0}{\varepsilon}\zeta \tag{2.45}$$

It was found that the expression for DP/H was also improved, compared with experimental results, when $U_0$ was multiplied by (25/15) (implying that $\zeta \approx 2.083$). This results in:

$$U_0 = \frac{1}{2.083}\frac{1}{2}\frac{1}{36}\left(\frac{\varepsilon^3}{(1-\varepsilon)^2}\right)D_m^2\frac{\Delta P}{\mu L} \simeq \frac{1}{150}\left(\frac{\varepsilon^3}{(1-\varepsilon)^2}\right)D_m^2\frac{\Delta P}{\mu L} \tag{2.46}$$

Note that this expression uses a fixed tortuosity value for all media types and was developed for soil, thus the tortuosity value may not be appropriate for flow through waste materials.

Various investigators have developed theoretical approaches to the influence of grain size, type and porosity on flow through a granular medium. Because one of the most important estimations is the pressure drop across a bed under flow, a focus has been the effect on pressure drop, including the kinetic energy loss, for beds of various types of particles. For instance, the effect of kinetic energy loss is included in the right-hand part of the expression for pressure gradient, $\Delta P/H$, for beds of spheres of different diameters and sand and coke particles, by Ergun (1952).

$$\frac{\Delta P}{H} = 150\frac{(1-\varepsilon)^2}{\varepsilon^3}\frac{\eta U_0}{D_p^2} + 1.75\frac{(1-\varepsilon)^2}{\varepsilon^3}\frac{\rho U_0}{D_p} \tag{2.47}$$

In this expression $D_p$ represents mean particle size. McDonald et al. (1979) developed a similar expression for pressure drop, $\Delta P/H$, for packed beds of spherical, cylindrical, fiber, sand and gravel particles and marbles for which the constants $A$ and $B$ were independently found:

$$\frac{\Delta P}{H} = A\frac{(1-\varepsilon)^2}{\varepsilon^3}\frac{\eta U_0}{D_p^2} + B\frac{(1-\varepsilon)^2}{\varepsilon^3}\frac{\rho U_0}{D_p} \tag{2.48}$$

The expressions by Ergun (1952) and McDonald et al. (1979) confirm that the effect of types of particles on flow resistance through a bed of granular material can be expressed in the form of these linear relationships.

Comiti and Renaud (1989) have, however, reported that these types of expressions for pressure drop $\Delta P/H$, for linear plus nonlinear laminar flow through a packed bed, had not addressed beds of platy particles such as wood chips that are used in the paper industry. In packed beds, the platy particles tend to be horizontally aligned, resulting in particle overlaps and layering, which lead to a longer fluid path (or higher value of tortuosity). Thus, the value of tortuosity $\zeta$ could be larger for packed beds of plate-like particles than for the Ergun (1952) and McDonald et al. (1979) expressions. Plate-type particles are of interest to characterization of waste sites such as landfills because wood, paper packaging and sheets, film and other types of flat particles are likely to be present and for organics composting operations where wood chips are used. The issue is how should waste site particles be represented in expressions for pressure drop or flow resistance in characterizing a site as a reactor.

The Comiti and Renaud (1989) reexamination of the expression for pressure drop $\Delta P/H$ for linear plus nonlinear laminar flow through a packed bed confirmed the expression of the form:

$$\frac{\Delta P}{H} = NU_0 + MU_0^2 \tag{2.49}$$

where $NU_0$ = the *viscous resistance* term ($0 < \text{Re} < 10$) and $MU_0^2$ = the *inertial resistance* or *kinetic energy loss term* ($10 < \text{Re} < 180$). As with the Ergun (1952) and the McDonald et al. (1979) relations for pressure loss $\Delta P/H$, these expressions have their roots in the Forcheimer (1901) hypothesis that departure of fluid flow in a porous medium from Darcy's Law might be due to added kinetic energy losses and proposal of the relation:

$$-\frac{dP}{dx} = \frac{\mu}{k}U + a(\rho U^2) \tag{2.50}$$

where $k$ = permeability of the medium, $U$ = pore flow velocity and the term $\rho U^2$ represents the kinetic energy of the fluid. Comiti and Renaud (1989) defined the viscous resistance in terms of the mean velocity $U = (U_0\zeta/\varepsilon)$, the permeability k and the tortuosity $\zeta$ as:

$$\frac{\Delta P}{H} = 2\gamma\zeta^2\eta a_{vd}^2\frac{(1-\varepsilon)^2}{\varepsilon^3}U_0 \tag{2.51}$$

where $\gamma$ is an *average pore shape factor* (Carman, 1956), which has the value $\gamma = 1$ for cylindrical pores, considered the mean value of $\gamma$ for many shapes of conduits and $a_{vd}$ is the dynamic specific surface area as before. For the kinetic energy loss or inertial resistance term $MU_0^2$, Comiti and Renaud (1989) used the Nikuradse friction factor $f$, for cylindrical pipes (assumed to be the general theoretical pore shape), where:

$$\frac{1}{\sqrt{f/2}} = 2.46\ \ln\left(\frac{d_{pore}}{2E}\right) + 4.92 \tag{2.52}$$

where $d_{pore}$ = equivalent pore diameter, as previously defined. $E$ represents a pore roughness factor accounting for the many changes of direction a fluid may take along its path through the representative pore, where $E$ varies as $d_{pore}$ and reaches the value $E = d_{pore}$ for a very rough pore. When $E = d_{pore}$, $f/2 = 0.0968$. Substitution of the definitions for superficial and interstitial velocity ($U_0$, $U$), tortuosity and friction factor $f$ for the Comiti and Renaud (1989) expression for pressure loss (viscous + inertial resistance or kinetic energy loss) through a packed bed of plate-like particles led to:

$$M = 0.0968\zeta^3\rho a_{vd}\frac{(1-\varepsilon)}{\varepsilon^3}$$

$$N = \left[2\gamma\zeta^2 a_{vd}^2\frac{(1-\varepsilon)^2}{\varepsilon^3}\right]$$

$$\zeta = \left[\frac{M^2}{N}\frac{2\gamma\eta\varepsilon^3}{(0.0968\rho)^2}\right]^{\frac{1}{4}}$$

$$a_{vd} = \left[\frac{N^3}{M^2}\frac{(0.0968\rho)^2}{(2\gamma\eta)^3}\frac{\varepsilon^3}{(1-\varepsilon)^4}\right]^{\frac{1}{4}} \quad (2.53)$$

where $\zeta$ = particle bed tortuosity, $\rho$ = fluid density, $\gamma$ = pore shape factor, $\varepsilon$ = packed bed void fraction or porosity and $a_{vd}$ = dynamic specific surface area or specific surface area of the packed bed actually exposed to the fluid flow. Thus, the Comiti and Renaud (1989) expressions contain the pore shape factor, tortuosity and, implicitly, the mean particle size and surface area.

## CORRECTION OF PACKED COLUMN PRESSURE DROP FOR WALL EFFECTS

Comiti and Renaud (1989) have noted that the actual surface area contributing *viscous resistance* to fluid flow through a packed bed includes the surface of particles in contact with the flow as well as the surface of any confining wall.

Using a circular column shell of internal diameter $D$ and height $H$ as the theoretical wall of the packed bed, Comiti and Renaud (1989) thus proposed that total wall surface area is $\pi DH$. This wall contributes to *viscous resistance* through contact with particles enclosed in the column. In the case of a waste site, the confining wall surface area becomes the bottom liner, soil embankment or bed surface in contact with waste and soil particles making up the site, with the distinction that confining surfaces for the waste site are of higher permeability or conductivity.

Total surface area $S^*$ contributing viscous resistance to the flow for the Comiti and Renaud (1989) representation of the packed bed as a cylindrical column would be:

$$S^* = S + \pi DH \quad (2.54)$$

where $S$ = total surface area of particles, $D$ = column diameter and $H$ = column height. The corrected specific surface area $a_{vd}$ for bed area contacted by the laminar flow was presented as total surface area contributing to viscous resistance/total volume $V$ of the column = $a^*_{vd}$

$$a^*_{vd} = \frac{S + \pi DH}{V(1-\varepsilon)} = a_{vd}\left[1 + \frac{4}{a_{vd}D(1-\varepsilon)}\right] \quad (2.55)$$

where $V = \pi D^2/4$, $d_{pore} = 4\varepsilon/[a_{vd}(1-\varepsilon)]$, $S = 4V\varepsilon/d_{pore} = [4V\varepsilon a_{vd}(1-\varepsilon)]/4\varepsilon = V/[a_{vd}(1-\varepsilon)]$. Note that, in the case of a waste site, the total surface area $S^*$ becomes the outer surface of theoretical volume containing the wastes and soil, or if the confining was geometric, as for a landfill cell, $S^*$ is the surface area of the cell in contact with the wastes and soil.

Similarly, Comiti and Renaud (1989) addressed kinetic energy loss due to packed bed particles near a confining wall by assuming an annular space of diameter $d_{particle}$ near the wall, where:

$$d_{particle} = \frac{6}{a_{vs}}$$

$$a_{vs} = \frac{\textit{mean surface area of the particles}}{\textit{mean volume of the particles}}$$

$$a_{vs} = \textit{geometric or static specific surface area of particle}$$

$$\frac{a_{vd}}{a_{vs}} = X \leq 1.0 \tag{2.56}$$

For this annular space the roughness factor E, pore roughness height, for the pore undergoing flow is assumed to be $d_{pore}/2$, as fluid flow in a conduit (circular) near a confining wall of a packed bed is considered likely to undergo less directional changes than flow inside the bed. Applying the Nikuradse relation for flow friction for a circular conduit, $f/2 = 0.0413$. The mean friction factor for the packed bed is then obtained from weighting between $f/2 = 0.0968$ (friction factor for capillary flow in a packed bed with pores that are tortuous) and $f/2 = 0.0413$ (friction factor for fluid flow next to a confining wall):

$$(f/2)_{mean} = \left[1 - 0.0413\left(\frac{D - d_{particle}}{D}\right) + 0.0968\left(\frac{D - d_{particle}}{D}\right)\right] \tag{2.57}$$

Note that this expression relates the friction factor correction to average wall length (diameter) to average particle length.

Thus, Comiti and Renaud (1989) corrected the factors $M$ and $N$ in their relation for $\Delta P/H$, by applying the above corrections for confining wall effect on viscous and inertial resistance (kinetic energy loss) to fluid flow through the porous medium. The corrected factors are:

$$M^* = \left[1 - 0.0413\left(\frac{D - d_{particle}}{D}\right) + 0.0968\left(\frac{D - d_{particle}}{D}\right)\right]\zeta^3 \rho a_{vd}\frac{(1-\varepsilon)}{\varepsilon^3}$$

$$N^* = \left[2\gamma\zeta^2 a_{vd}^2\left[1 + \frac{4}{a_{vd}D(1-\varepsilon)}\right]\frac{(1-\varepsilon)^2}{\varepsilon^3}\right]$$

$$\frac{\Delta P}{H} = M^* U_0^2 + N^* U_0$$

$$\zeta^* = \left[\frac{(M^*)^2}{N}\frac{2\gamma\eta\varepsilon^3}{(0.0968\rho)^2}\right]^{\frac{1}{4}}$$

$$a_{vd}^* = \left[\frac{N^3}{(M^*)^2}\frac{(0.0968\rho)^2}{(2\gamma n)^3}\frac{\varepsilon^3}{(1-\varepsilon)^4}\right]^{\frac{1}{4}} \tag{2.58}$$

For the tortuosity $\zeta$, Comiti and Renaud (1989) also showed that pore tortuosity increased as particles became flatter (thickness/length = e/a). Comparison of length/thickness ratios for various particle shapes and porosities ($\varepsilon$) from studies indicated agreement with Pech (1984):

$$\zeta - 1 = 1.6 \ln\left(\frac{1}{\varepsilon}\right) \tag{2.59}$$

for the porosity range $0.1 < \varepsilon < 0.6$. They proposed that the relation:

$$\zeta - 1 = P \ln(1/\varepsilon) \tag{2.60}$$

can be used for various types of packed beds of particles. Estimation of $P$-values for several particle shapes in packed beds further led to the correlation:

$$P = 0.577 \exp[0.18(a/e)] \tag{2.61}$$

where $a$ = (average) particle length and $e$ = average particle thickness. Tortuosity was thus:

$$\zeta = 1 + 0.577 \exp[0.18(a/e)] \ln(1/\varepsilon) \tag{2.62}$$

The issue of how flatter particles affect surface area exposed to fluid flow through a packed bed was also addressed by stating that the ratio $X$ of specific interface area to specific surface was:

$$X = \frac{a_{vd}}{a_{vs}} = 0.43 + 0.577(e/a) \tag{2.63}$$

where $a_{vd}$ = specific surface area actually exposed to fluid flow, $a_{vs}$ = geometric specific surface area, $e$ = average particle thickness and $a$ = average particle length.

For a mixture of wastes and soil, the above expression for the effect of particle flatness introduces four unknowns: average particle diameter, average particle thickness, average geometric and specific surface area. This highlights the value of other implicit approaches to defining waste site particle shape and geometric area.

Comiti and Renaud (1989) also noted that the above relationships for pressure drop of laminar fluid flow through a packed bed of irregularly shaped particles may be questionable with regard to applying the Nikuradse friction factor relation $[1/\sqrt{(f/2)}]$, applicable to turbulent flow in rough, circular pipes.

They argued, however, that the work of Dybbs and Edward (1984) indicates that flow in packed beds for the inertial regime of laminar flow ($10 < \text{Re} < 180$) differs from laminar flow in a tubular conduit, in that an inertial core develops outside (above) the boundary layer, resulting in a nonlinear relation between pressure drop and flow rate; and this suggests that flow will develop in a cell of length equal to the diameter of the (representative) pore. This would suggest agreement between the Dybbs and Edward (1984) study and the expression for $\Delta P/H$ developed by Comiti and Renaud (1989) for laminar flow in a packed bed. Thus, the modified Nikuradse friction factor for describing the effect of friction at a confining wall for a packed bed can be used with caution.

The above relationships represent various theoretical and empirical approaches to simple mathematical representation of the internal structure of a packed bed and how structure affects flow of a gas or liquid. These relationships improve physical representation of a packed bed as a reactor system; but they must still be modified for adaptation to waste site physical definitions, as particle properties are likely to be far more diverse. For instance, the particle shapes are not likely to be as amenable to simple representation and average geometric surface area is not likely to be explicitly determined.

## CORRECTIONS FOR PRESSURE DROP RELATIONS FOR FLUID FLOW THROUGH A WASTE SITE

Table 2.4 of this section shows that, while significant field or laboratory data is available for *hydraulic conductivity* at landfill sites, little or no information is available on basic physical properties — for example, *mean particle size, permeability, tortuosity* or *pressure gradient* vs. depth or *liquid holdup, water uptake or thermal characteristics*. This is a result of the black box nature of these environmental systems. The relative success of water controls in limiting groundwater impact of landfills reduces incentive to further exploration of these parameters. To address a waste site — of which the landfill is an appropriate example — as a packed bed flow or trickling filter, the physical properties listed above must be defined in terms of overall or statistical properties of the solid materials (wastes and soils) buried in the site.

Recent findings of investigators of packed bed and column hydrodynamics suggest that several uncertainties in the modeling of packed beds containing particles of irregular shapes, sizes and properties can be addressed through theoretical and empirical approaches.

For definition of waste sites as heterogeneous reactor systems made up of granular media, it appears that characterization of systems in terms of particle size distribution (or pore distribution) is of primary importance. Full characterization is likely to be relatively expensive, but the simplicity of soil sampling and analysis methods and the expanding body of data on physical properties of waste sites suggest that the use of existing field data could reduce the need for extensive measurement of basic waste site characteristics. Establishment of basic physical properties — and adaptation of theoretical approaches to waste sites defined as packed bed or column (or filter bed) systems — can then become the foundation for development of models suitable to predicting hydrodynamics, gas flow and biological kinetics of the system. Examination of the parameters discussed under hydraulic conductivity and permeability could further clarify how packed bed relationships might apply.

## WASTE SITE PARTICLE PROPERTIES: SIZE AND SHAPE

The particles used in packed beds for treatment of industrial effluents and gases are typically homogenous in type, regular in shape and uniformly graded. These characteristics are important to packed bed design for good performance, asbreak predictable flow rates and adequate surface area of the particulate bed are highly desirable for optimal treatment.

Waste sites, by contrast, contain a bewildering variety of particle sizes and shapes. Little can be surmised about these particle sizes and shapes without analysis of soil size distribution (by screening excavated samples) or other equally efficient methods. More is known, however, about types of waste and soil particles in the site and their densities and hygroscopic properties.

Past investigators of effects of particle properties on packed bed or soil column flow have chosen the approach that, when particles are of irregular shape, the other relations for flow behavior are applicable if an *equivalent spherical particle size* (EPS) is used. The basis of this approach is the Blake, Blake-Kozeny and Kozeny-Carmen theoretical models for flow in soil columns. In these models, fluid transport is represented as flow through a bundle of capillaries of irregular cross section. This bundle of capillaries theoretically has a characteristic hydraulic radius and conducts flow through a soil bed of known porosity. All particles present are redefined in terms of spheres of volumes equal to those of actual particles. This theoretical approach results in a characteristic diameter of the spherical particle, for each particle size. The variation between the size of the actual particle and the diameter of the equivalent spherical particle *(equivalent particle size or EPS)* can then be estimated as a ratio, called the *sphericity factor*.

The equivalent particle size (EPS) has been stated in various forms by investigators. For instance, Das (1998) presents the EPS for nonspherical particles as:

$$\textit{Equivalent particle diameter} = D_e = \left(\frac{6V}{\pi}\right)^{1/3} \tag{2.64}$$

$$\textit{Sphericity (bulky particles)} = S = \frac{D_e}{L} \tag{2.65}$$

$$\textit{Angularity} = A = \frac{\text{Average radius of corners and edges}}{\text{Radius of maximum inscribed sphere}} \tag{2.66}$$

where $D_e$ = diameter of equivalent spherical particle, $V$ = particle volume, $L$ = particle length, $S$ = particle sphericity and $A$ = particle angularity. Leva (1959) proposed a relation for the equivalent spherical particle to represent particles of arbitrary shape, as:

$$d_p = \frac{6V_p}{A_p\varphi_p} = \frac{6V_p}{A_{sp}} \tag{2.67}$$

where $d_p$ = particle diameter, $\varphi_p$ = sphericity or particle shape factor, $d_p\varphi_p$ = diameter of a sphere of equivalent volume, volume of a single nonspherical particle, $A_p$ = surface area of the nonspherical particle and $A_{sp}$ = surface area of a sphere of volume equivalent to the particle. The Das (1998) and Leva (1959) studies provide valuable approaches to characterization of unusual particle shapes.

Das (1998) also states that most (*soil*) particles can be classified as *bulky* (formed from weathering of rocks and minerals and angular, subangular, rounded and subrounded in shape and represented in quartz sands); *flaky* (particles of low sphericity, e.g., $S < 0.01$, represented by clays); and *needle-shaped* (thin, elongated particles, represented by coral deposits and attapulgite clays).

The investigations of Carman (1937), Ergun (1952), Wylie and Gregory (1955), Dullien (1975), McDonald et al. (1979), Saez and Carbonell (1985), Comiti and Renaud (1989), Mauret and Renaud (1997), Seguin et al. (1996), Cramer (1998), Seguin, Montillet and Comiti (1998), Comiti, Sabiri and Montillet (2000), Teng and Zhao (2000), Puncochar and Drahos (2000), Li et al. (2001), Chhabra, Comiti and Machac (2001) and Niven (2001) have shown that the capillary flow and related fluid flow behavior of soil columns and packed bed systems containing nonspherical particle shapes can be successfully modeled when particle shapes are represented as equivalent spherical particle sizes (diameters). The packed bed and granular media particle shapes covered by these investigators included sands, gravels, coke particles, wood chips, spherical particles, beads, marbles, plates, cylinders, high-porosity foams, sintered discs, foams, screens and mats.

Waste sites such as landfills will, however, contain a much wider variety of particle shapes. Classification of these shapes into manageable categories for modeling purposes is useful, even though this may not guarantee great modeling accuracy unless the classification is careful and thorough. Nevertheless, consideration of the shapes and types of particles in waste sites could significantly reduce black box errors, and thus advance the accuracy of system representation and increase sensitivity of the site model to materials present. Simplified schemes for use of particle properties can be devised. For instance, the following particle properties, listed in Table 2.6, can be devised for municipal solid wastes (MSW), solid waste materials and soils likely to be present in a waste site.

The *specific volume* ($m^3/gm$) values for the types of waste particles listed in Table 2.6 can be found from the *actual material mass-to-density ratio.* For many waste materials, the *as-received* or *dry* density is available from general engineering or applied science literature sources. Dry or *saturated density* values are more valuable, as these describe both the material's water uptake capacity and its specific gravity and can be the basis for particle and system porosity definitions.

Use of the specific volume of a material and particle as defined by the Das (1998) and Leay (1956) relations allows an approximation of the *equivalent diameter* $D_e$ of a spherical particle of the same volume ($V_p$) as the actual particle. The sphericity of a particle of estimated diameter $d_p = d_{particle}$ and volume $V_p$ can also be found if the material's specific surface area $A_p$ is known:

$$V_p = \frac{\text{Mass}}{\rho}$$

$$D_e = \left(\frac{6V}{\pi}\right)^{1/3}$$

$$\textit{Sphericity} = S = \varphi_p = \left(\frac{6V_p}{A_p d_p}\right) = \frac{6V_p}{A_{sp}} \tag{2.68}$$

**TABLE 2.6**
**Waste Particle Physical Properties**

| MSW | Particle Shape | IVU = Industrial Volume Unit | Thickness, m | Typical Density, $kg/m^3$ | Particle Volume Estimation | Dry Density | Specific Surface (BET) $m^2/gm$ |
|---|---|---|---|---|---|---|---|
| Food wastes | Mixed | $Lbs/feet^3$ | NA | 288.3 | Mass/density | | 120–150 |
| Paper | Flat, folds, regular, irregular edge | 156–244 $g/m^2$ | 1.9–3.0 E-3 | 81.7 | IVU × thickness | 0.950 | 0.92 |
| Cardboard | Flat, folds, regular, irregular edge | 293–381 $g/m^2$ | 6.0–8.0 E-3 | 49.7 | IVU × thickness | | 1.1 |
| Textiles | Flat, folds, regular, irregular edge | 110–130 $g/m^2$ | 0.1 – 0.5 E-3 | 64.1 | IVU × thickness | | 150–29 |
| Plastic film | Flat, regular, irregular edge, folds | $gm^2$ | 30–100 E-3 | 64.1 | IVU × thickness | 0.92 | 0.32–0.82 |
| Plastics | Geometric, subgeometric shapes | $g/m^2$ | NA | 64.1 | Mass/density | 0.91–0.95 | 0.32–0.82 |
| Rubber | Flat plate, and mixed | $g/m^2$ | NA | 128.2 | Mass/density | 1.64 | |
| Leather | Flat plate, irregular edge | 0.7–0.9 E-3 $g/m^2$ | | 160.2 | IVU × thickness | 0.85 | 540 |
| Garden | Mixed and combination | $Lbs/yd^3$ | NA | 104.1 | Mass/density | | |
| Leaves | Cycloid, fractal | $Lbs/yd^3$ | NA | 104 | Mass/density | | |
| Brush | Solid dendritic + planar fractals | $Lbs/yd^3$ | NA | 162 | Mass/density | | |
| Grass | Solid dendritic, elongated, spatular | $Lbs/yd^3$ | NA | 104 | Mass/density | | |

| | | | | | | | |
|---|---|---|---|---|---|---|---|
| Wood | Regular, irregular dendritic solid | Lbs/feet$^3$ | NA | 240.3 | Mass/density | 0.525* | |
| Glass | Flat/curved plate, fracture edge | Gms/cm$^3$ | NA | 193.8 | Mass/density | 2.5 | |
| Ferrous | Plate, cylindrical shell, angles, mix | Mass/density | NA | 320.4 | Mass/density | 7.8 | |
| Aluminum | Plate, cylindrical shell, angles, mix | Mass/density | MA | | Mass/density | 2.7 | |
| Dirt, Ashes | Bulky, porous spherical, flaky | Kg/m$^3$ | NA | 480.6 | Mass/Density | 1.6–1.8 | 2–10 |
| Sand | Bulky, flaky | Kg/m$^3$ | NA | 1600 | Mass/Density | 2.5 | 3.3–10 |
| Silt | Bulky, flaky | Kg/m$^3$ | NA | 1900 | Mass/Density | 46.6 | |
| Topsoil | Bulky, flaky | Kg/m$^3$ | NA | 1350 | Mass/Density | | 19–30 |
| Clay | Bulky, flaky | Kg/m$^3$ | NA | 1150–1450 | Mass/Density | | 15 |

The above values only provide a basis for the use of particle size estimation developed from soil sample screening. Wadewitz and Specht (2001) stated that, for nonspherical particles for which a diameter $d$ had been determined using sieve analysis, calculations of area and volume tend to be incorrect when only the sieve diameter ($d$) is known. Corrections proposed for particle volume and area include the use of a Nusselt number $N$, related to the heat properties of the material from which the particle originates:

$$\text{Characteristic particle diameter} = d_{ch}$$

$$\text{Particle volume/area} = \frac{d_{ch}}{6}$$

$$\text{Nusselt number} = Nu = \frac{\alpha}{\lambda}\frac{A}{6V}d_{ch}^2$$

$$d_s = diameter = \frac{6V}{A_{sa}}$$

$$d = sieve\ diameter$$

$$A_{sa} = \textit{Surface area, from BET, GAB or Caurie sorption method}$$

$$V = \frac{4}{3}\pi d^2 L$$

$$\alpha = thermal\ diffusivity$$

$$\lambda = thermal\ conductivity \tag{2.69}$$

Wadewitz and Specht (2001) also showed that for heat conduction in bodies of various shapes, the most *body-shape-independent* Nusselt number results from plotting the Nusselt number (*Nu*) vs. $d_s/d$, where $d_s$, the *Sauter diameter,* is found from the ratio of body volume to BET, etc., surface area for the particular material; $d_{ch} = d$, and $d$ is the body diameter from *sieve analysis.* The body-shape-independent Nusselt number was:

$$Nu = 1.80e^{-2.22[(d_s/d)-1]} \tag{2.70}$$

Note that to use this relation, thermal diffusivity and thermal conductivity of a material must be known. The *Sauter diameter* is defined as:

$$\text{Sauter diameter} = d_s = 6\frac{V}{S_a} \tag{2.71}$$

where $S_a$ = BET, *etc.,* specific surface area from absorption studies, and $V$ = volume/unit area for the material. It should be noted that the Sauter diameter could also be defined in terms of the total surface area $S$ and total mass $M$ of a material. For a material of density $\rho_i$, with n particles of spherical equivalent diameter $d_i$:

$$M_{tot,i} = \left(\frac{\pi \rho_i}{6}\right) \sum_{1}^{n} n_i d_i^3$$

$$S_{tot,i} = \pi \left(\sum_{1}^{n} n_i d_i^2\right)$$

$$Sauter\ diameter = d_{s,i} = \left(\frac{6}{\rho_i}\right) \frac{M_{tot,i}}{S_{tot,i}} \tag{2.72}$$

The above statistical expressions allow for the use of mass and size variations from sieving of samples to be related to the theoretical average grain size and, thus, to average grain surface area. Thus it is of relevance to waste site material characterization.

## Characterization of Surface Area and Related Physical Properties of Wastes

Information on the surface area of solid materials is useful to descriptions of industrial, manufacturing and treatment processes that involve reaction and partitioning of chemicals to solids. The uptake of liquids and vapors, surface-catalyzed or enzymic reactions, filtration of liquids and gases, the drying and preservation of products, wastewater or water treatment and catalysis are engineering applications where surface area and material properties data can be put to beneficial use.

In characterizing the materials in a waste site as degradable solid substrates, it is also useful to independently arrive at important physical properties likely to affect enzymic reactions and uptake of moisture and thus determine biodegradation rate. One of the most important is actual surface area involved in reaction processes. The overall surface area available to microorganism enzyme attack determines the decomposition or conversion rate for biodegradable or chemically reactive materials and the availability of moisture adsorbed by a system of degradable solids strongly affects (limits or enables) metabolic processes of microorganisms. Additionally, heat flow and moisture loss and gain are also mediated by (pore) surface area available in a physical system such as a waste site.

## Specific Surface Areas of Solid Materials From Liquid or Gas Sorption Isotherms

The theoretical approaches to establishing the surface area of a solid material typically arrive at a definition of the internal surface available to vapor adsorption vs. relative pressure of the surrounding environment, the value of which is stated in area units per unit weight of the material — for example, $cm^2/gm$ or $m^2/gm$. Porous, hygroscopic materials such as paper, plant residues, fruits, foods, grains and grain products, agricultural residues, wood products and soil are hygroscopic and are known to adsorb or evaporate water to the environment — for instance, depending upon the dryness or wetness of the surrounding atmosphere and temperature.

The quantitative relationship between vapor or liquid adsorption onto the surface of a solid material and the relative pressure of the surrounding environment can be represented in terms of adsorbate (liquid or vapor) content of the material vs. relative pressure and is called the *adsorption isotherm.* The shape of the plot or curve has various canonical forms (for example, Types I–V isotherms); but it has unique values for each solid material and thus can provide invaluable insight into the microstructure of the material. The chemical basis of the sorption isotherm approach was presented by Langmuir and is now referenced in terms of the mathematical expression developed as the Langmuir isotherm:

$$\frac{p}{x} = \frac{1}{B_{x_m}} + \frac{p}{x_m}$$

$$B = \frac{\kappa e^{(E/RT)}}{z_m v}$$

$$\kappa = \frac{N}{\sqrt{2\pi MRT}} \tag{2.73}$$

For the Langmuir isotherm, $p$ = pressure, $x$ = adsorption (grams per gram of material), $x_m$ = adsorption when the material's surface is covered by a layer of adsorbate one molecule thick (or the monolayer capacity). Also, $\kappa$ = kinetic theory constant and $N$ = Avogadro's number ($6.02252 \times 10^{23}$ particles of a substance per mole; for example, glucose with chemical formula $C_6H_{12}O_6$ has a molecular weight of 180.16, thus 180.16 grams of glucose should contain $6.02252 \times 10^{23}$ glucose molecules). $M$ is the molecular weight of the liquid or vapor, $M$ = gas constant and $T$ = absolute temperature (°$K$), $E$ = activation energy for desorption (or heat of desorption, $-\Delta H$), $z_m$ = number of molecules adsorbed per unit area at monolayer capacity $x_m$ and $v$ = vibration frequency of molecules of the material upon which the vapor or liquid is adsorbed.

Surface chemistry theories for liquid or vapor adsorption focus upon the amount of an adsorbate (gas or liquid) taken up by the surface of a material. They are generally based upon the theory that sorption on a solid material's surface should advance from a single layer of molecules of a vapor or liquid at very low pressures (or the pressure at which the most energy is required to remove molecules of the adsorbate) to multiple layers of vapor or liquid at saturation. If a single layer of adsorbate one molecule thick covers the internal surface of a solid material in equilibrium at a particular (low) relative pressure with its environment, the surface can be described by a simple relationship in terms of the cross-sectional area of the adsorbate molecule and the number of molecules per gram of adsorbate present.

Accurate determination of the minimum grams of adsorbate present at the point on the sorption curve representing a single layer of molecules of the liquid or vapor can be translated to number of molecules and, subsequently, to surface area. Thus, surface area can be deduced from basic chemical laws once the adsorption isotherm of a material has been determined and if the cross-sectional area of the vapor or liquid molecule is known (usually the case).

Isotherm information is thus important in estimating the exact amount of vapor or liquid adsorbed when the adsorbate (liquid or solid) layer is one molecule thick. This amount is also referred to as the *monolayer capacity*. Various theoretical approaches to adsorption of vapors or liquids onto surfaces provide mathematical relations whereby monolayer capacity of solid adsorbents can be determined. An abbreviated list of these mathematical models is shown below because they have been used in determining the sorption characteristics of many of the solid materials likely to be present in solid waste sites.

The sorption isotherm information developed by these models might or might not include monolayer capacity or surface area estimations.

An important note is that the area determined from monolayer capacity includes both external and internal surface area including the total area of all pores, thus it represents the whole surface of the solid material that could be exposed to sorption from the surrounding environment. Also, if water adsorption is involved for the study (for example, a study of the uptake of moisture by food, paper and packaging and construction materials), the surface area so determined would be larger than if nitrogen (molecular area = 16.2 sq.Å) adsorption was involved because the water molecule (molecular area = 10.6 sq.Å) is smaller and can penetrate more deeply into the pores of the material.

In the case of solid materials exposed to degradative enzymes expressed by microorganisms, even in the ideal case all of the surface area so determined cannot be affected because the fraction of surface involved in enzyme sorption depends upon both the size of the enzyme molecule (larger than the water or nitrogen molecule) and the size of the pore limiting access by the microorganism involved.

The distinction must thus be made between actual surface area and the area in contact with gas flow. The surface that can be involved in biological reaction would obviously be less than the surface area found from sorption because access to finer pores in materials is limited according to size of organism or size of enzyme molecule.

Investigators of sorption chemistry have introduced a wide variety of chemical uptake models involving specific surface area of materials. Some of these models are of interest to representation of waste site particles, as many of the materials investigated — and thus surface areas estimated — are the same as would be disposed in waste sites. Some of these models and the materials investigated are shown below.

$$\text{Langmuir (1918):} \qquad w = \frac{w_m a_w}{K + a_w}$$

(Theoretical: general use, applicable only to first convex portion of sorption isotherm and used for monolayer capacity determination and other sorption estimations)

$$\text{Brunauer, Emmett and Teller (BET) (1938):} \qquad \frac{w}{w_m} = \frac{C_B a_w}{(1 - a_w)[1 + (C_B - 1)a_w]}$$

(Theoretical: sorption on biological and other materials)

$$\text{Harkins-Jura (1944):} \qquad a_w = \exp\left[k_3 - \frac{k_4}{w^2}\right]$$

(Theoretical: liquid film adsorption in industrial materials)

$$\text{Oswin (1946):} \qquad w = K\left[\frac{a_w}{1 - a_w}\right]^m$$

(Empirical: sorption in biological materials)

$$\text{Halsey (1948):} \qquad w = \left[\frac{\exp[a + bT(^\circ C)]}{-\ln(a_w)}\right]^c$$

(Theoretical: sorption in starchy foods)

$$\text{Henderson (1952):} \qquad w = \left[\frac{\ln[1 - a_w]}{[T(^\circ C) + b]}\right]^c$$

(Partially theoretical)

$$\text{Chung-Pfost (1967):} \qquad w = \frac{1}{a}\left[\ln(a_w)\frac{b - T(^\circ C)}{c}\right]$$

(Partially theoretical: biological materials)

$$\text{Caurie (1981):} \qquad \ln\left[\frac{100 - [\frac{w}{100+w}]}{[\frac{w}{100+w}]}\right] = K_1 - K_2 a_w$$

(Empirical: sorption in foods, textiles and industrial materials)

$$\text{Chen (1971):} \qquad \ln(a_w) = K_1 + K_2 \exp(K_3 w)$$

(Partially theoretical: sorption in cereal grains, field crops)

$$\text{Iglesias, Chirife (1976):} \qquad \ln[w + (w^2 + w_{0.5})^{1/2}] = K_1 a_w + K_2$$

$$w_{0.5} = w, \text{ at } a_w = 0.5$$

(Empirical: sorption in fruits, high sugar foods)

Guggenheim, Anderson, De Boer (GAB)(1984):

$$\frac{w}{w_m} = \frac{C_T k a_w}{(1 - k a_w)(1 - k a_w + C_T k a_w)}$$

$$C_T = C_{T,0} \exp\left(\frac{\Delta H_1}{RT}\right)$$

$$k = k_0 \exp\left(\frac{\Delta H_2}{RT}\right)$$

$$\frac{a_w}{w} = A + B a_w + C a_w^2 \text{ (polynomial transformation)}$$

$$A = \frac{1}{w_m C_T k}$$

$$B = \frac{1}{w_m}\left(\frac{C_T - 2}{C_T}\right)$$

$$C = \frac{k}{w_m}\left(\frac{1 - C_T}{C_T}\right)$$

(Theoretical: general use for sorption in a variety of materials, including biological)

In the GAB adsorption model, $a_w$ = water activity = 0.0 – 1.0, $w$ = moisture content = $g/g$ dry solids, $w_m$ = monolayer adsorption capacity = $g/g$ dry solids; $C_T$ and $k$ are thermodynamic constants related to the heat of sorption of water for the monolayer ($H_m$) and of multilayers of moisture ($H_m$), where $\Delta H_1 = H_m - H_{mx}$, $H_m$ = monolayer moisture heat or sorption, $H_{mx}$ = multilayer heat of sorption, $\Delta H_2 = \lambda - H_{mx}$, where $\lambda$ = heat of condensation of pure water. The polynomial constants $A$, $B$ and $C$ can be found by various familiar statistical regression methods.

Ratti et al. (1989): $\ln (a_w) = -k_1 w^{k_2} + \left[k_3 e^{(-k_4 w)} w^{k_5}\right] \ln (p_{w0})$

(Use for sorption properties of foods and other materials)

Dubinin–Radushkevich (1947):

$$\frac{w}{w_m} = A \exp\left[B(RT \ln (a_w))^2\right]$$

(General use, multilayer model for condensation in pores of solid material)

Smirnov, Lysenko (1991):

$$a_w = a\,\mathrm{erf}(p(w - w_m)) + b$$

$$a = 0.5(a_{w,\mathit{final}} - a_{w,\mathit{lowest\ initial\ value}})$$

$$b = 0.5(a_{w,\mathit{final}} + a_{w,\mathit{lowest\ initial\ value}})$$

$$\mathrm{erf}(p(w - w_m)) = \frac{a_w - b}{a}$$

All of the above sorption models allow for linear or nonlinear regression determination of constants. As noted above, the constants for the GAB model in the polynomial form $\{a_w/w = A + Ba_w + Ca_w^2\}$ and other models can be readily determined, once (raw) adsorption (moisture content $w$ *vs.* water activity $a_w$) data are available by statistical regression methods.

In some cases this information is published as graphical plots or sorption isotherms, from which points along the isotherm may be taken with some caution for accuracy. Also, in many cases, solid material characterization from nitrogen vapor sorption, other adsorbates and from mercury intrusion are reported in the literature. In this case, *specific surface area* and other model corrections should be made based upon size or cross-sectional area of the water molecule and particular adsorbate sorption parameters.

One of the most important of the theoretical approaches to adsorption on solids is the Brunauer, Emmett and Teller (BET) relationship. The BET relation generalizes the Langmuir kinetic group theory for equilibrium between the quantity of a gas (vapor) adsorbed on a material and its environmental pressure to single and multiple successive layers of adsorbate, allowing analysis of a wider range of sorption behavior and materials than the earlier Langmuir approach. As with earlier and later sorption theory approaches, the BET relation presents an equation for the adsorption isotherm or relative sorption vs. relative pressure:

$$\frac{p}{x(p_0 - p)} = \frac{1}{x_m c} + \left(\frac{c-1}{x_m c}\right)\frac{p}{p_0}$$

$$c = e^{(E_1 - L)/RT} = \text{heat of adsorption constant}$$

$$x_m = \frac{\mathit{gram\ adsorbate}}{\mathit{gram\ solid}}, \text{ at monolayer capacity} \tag{2.74}$$

where $x$ is the gram adsorbate per gram of material, $p$ = vapor pressure and $p_0$ = vapor pressure at material saturation. Being linear (valid for $0.05 < (p/p_0) < 0.3$, (Gregg, 1961), a plot of $p/[x(p_0 - p)]$ vs. $(p/p_0)$ has an intercept of $1/(x_m c)$, and an intercept of $(c - 1)/(x_m c)$, the values of which allow $x_m$ and $c$ to be determined. The BET theory also assumes equivalence between the number of molecules adsorbed and the amount of vapor or liquid adsorbed:

$$\frac{x}{x_m} = \frac{Sz}{Sz_m} \tag{2.75}$$

where $S$ is surface area, $z$ = number of molecules covering the surface at adsorbate content $x$, and $z_m$ is the number of molecules on the surface at adsorbate concentration $x_m$, the monolayer capacity. For these relations, cross-sectional area of the vapor or liquid adsorbate molecule can be represented by $A_m$, in square angstroms (1.0 sq. angstrom or $(\text{Å})^2 = 10^{-20}\,\text{m}^2$). Adsorbates that have been used to determine the microstructure of porous solids include acetone ($A_m$ = 20 sq. Å = $20 \times 10^{-20}\,\text{m}^2$), ammonia ($A_m$ = 14.6 sq. Å), argon ($A_m$ = 15.5 sq. Å), benzene ($A_m$ = 25.0 sq. Å), carbon dioxide ($A_m$ = 19.5 sq. Å), carbon monoxide ($A_m$ = 16.3 sq. Å), methane ($A_m$ = 16.0 sq. Å), nitrogen ($A_m$ = 16.2 sq. Å) and water ($A_m$ = 10.6 sq. Å) ($A_m$ values reported by Gregg, 1961).

The BET single and multilayer equations and modifications thereof have been used widely and successfully for determination of the monolayer capacity of solid, granular and powdered materials. BET model determinations of monolayer capacity and other important thermodynamic or physical characteristics of solid materials have been found to be surprisingly accurate, despite the model's simplicity. A wide collection of adsorption studies includes BET model determinations of monolayer capacity or area; however, surface area is not always estimated, as monolayer capacity information is considered more critical. However, the BET adsorption model is known to apply only for lower adsorption pressures — for instance, to the drier range of an adsorption isotherm, e.g., below a relative pressure ($p/p_0$) of 0.35 (Gregg, 1961).

## Equivalence Between BET and GAB Water Adsorption Models

The BET model has been modified by various investigators interested in finding characteristics of materials such as surface area, porosity, water retention and other physical properties. One such modification is the Valsaraj (1995) version of the BET model:

$$\frac{1}{w}\left(\frac{a_w}{1-a_w}\right) = \frac{1}{B_v w_m} + \frac{B_v - 1}{B_v w_m} a_w \tag{2.76}$$

stated here in terms of the GAB model parameters $w$ and $a_w$. For this model, regression from adsorption data — when adsorption data or graph plots of $w$ and $a_w$ are available — allows ready determination of $B_v$ and $w_m$, from which water content variation with relative humidity and material surface area can be estimated. Values of $B_v$ and $w_m$ have been reported in literature in the case of soils and other materials. If these are available, soil water content vs. humidity and surface area can be easily determined.

## Areas from Nitrogen, Vapor Adsorption vs. Moisture Sorption

Specific surface area based upon sorption assumes single-layer coverage of the solid material by an adsorbate of known molecular cross-sectional area. However, many adsorption studies are reported for adsorbates other than water; for instance, many nitrogen adsorption isotherms for solid materials have been reported. Equivalence between specific surface areas can be found, according to the method suggested by Khalili et al. (2000). For example, if the specific surface area of a material from nitrogen adsorption is $SSA_{N_2}$ and the water adsorption surface area ($SSA_{H_2O}$) is desired:

$$SSA_{H_2O} = SSA_{N_2}\left(\frac{\sigma_{H_2O}}{\sigma_{N_2}}\right) \quad (2.77)$$

The term $\sigma$ refers the cross-section area of the molecule, which, in the case of water and nitrogen, is $10.8\text{Å}^2$ and $16.2\text{Å}^2$, respectively (1 square angstrom = $\text{Å}^2 = 10^{-20}\,\text{m}^2$).

## Relationship between Water Activity and Other Moisture Characteristic Terms

Water potential is a commonly used term for soil moisture uptake and for matric potential of soils and granular solids. Conversion between water activity and water potential is useful for cases in which moisture content vs. water or matric potential information is available from literature. The plot of this relationship is also called the suction or capillary curve of the material. The water potential $\psi$ or the matric potential $\psi_m$ is usually presented in terms of the relation between air and water pressure:

$$\psi = \frac{(p_{air}) - p_{water}}{\rho_{water}g} = \frac{RT}{g(MW)}\ln\left(\frac{p_v}{p_{v0}}\right) = \frac{RT}{g(MW)}\ln(a_w) \simeq \psi_m \quad (2.78)$$

where $P_{air}$ = air pressure ($\geq 0$), $p_{water}$ = pressure at the sorbed water phase, $p_v$ = vapor pressure of water in the gas phase at a curved meniscus, $p_{v0}$ = partial vapor pressure of pure water, $R$ = universal gas constant, $T$ = degrees Kelvin, $MW$ = molecular weight of the water phase, $g$ = gravitational constant and $\rho_w$ is the density of the water phase.

It is thus convenient to convert between the water potential of a material and its water activity, with the assumption that the temperature chosen is representative of the problem considered and if values of air and water pressure or water activity and molecular weight and density of the liquid absorbate or water are available. At 20°C, the relation between water potential and water activity, estimated for this section from regression of data developed by Papendick and Campbell (1981) for water in soil, is:

$$\begin{aligned}
&\psi = 0.036 + (1351)\ln(a_w)\ (\text{–bars})\\
&\text{range } -0.001 \text{ to } -5{,}000 \text{ bars}, a_w = 1.0 \text{ to } 0.025\\
&-1 \text{ bar} = 1020 \text{ cm water} = -0.978 \text{ atm} = -0.01 \text{ joules/kg}
\end{aligned} \quad (2.79)$$

## Range of Adsorption in Solid Materials and Water Availability to Organisms

Adsorption by solid materials ranges from absolute (theoretical) dryness (adsorbate content = 0.0) to saturation or maximum content. Between absolute dryness and the point where the material is at its maximum moisture content are points of significant importance to the role of biological organisms involved in decomposition. The first of these points is approximately the limit at which the adsorbed moisture is available to a biological cell and the second is that at which water is free or available to organisms present. The latter occurs near the saturation limit.

In terms of theoretical adsorption, sorption curves for moisture (and other adsorbates) are typically represented by second or higher order polynomials and can have two or three inflection points. These inflection points provide information about the adsorptive state of water in the material and are thus of some importance in establishing availability of the adsorbate (water). The adsorptive state provides a measure of how easy it is for microorganisms to obtain moisture from materials around them or the *water potential* of the material.

For example, in the S-type or Type II adsorption isotherm for water (vapor or liquid) common for biological and industrial materials and soils, the first inflection point on the plotted isotherm curve occurs near the end of the low-pressure range. This inflection point is also at or near the limit for the BET isotherms — for instance, it can occur at water activity ($a_w$) values of 0.4–0.6. Moisture adsorbed up to this (inflection) point represents *tightly bound* or *bound* water. This bound water is unlikely to be available to microorganisms because the energy required to extract it from the material (potential energy or matrix potential) could be too high for the microorganism involved. Microbial growth or metabolic activity below this water adsorption range — for instance, below $a_w = 0.5$ — would be, at best, limited.

Between the first and second inflection point of the adsorption isotherm, e.g., from $a_w = 0.5 - 0.8$, water adsorbed is referred to as *loosely bound.* Availability of this water depends upon the organism, the substrate and chemical conditions at the location. For example, some fungi are very xerophilic and can exist on very dry materials, at $a_w$ values of about 0.6.

Above the second inflection point is a water content range called *free water*, meaning it is available to many organisms. This range of water content is also the range most favorable to soil and decomposition bacteria. Figure 2.6 depicts adsorption in a typical Type II isotherm.

The state of adsorption by a material thus has an important and perhaps controlling effect on whether a material will be decomposed inside a waste site.

The relation between these ranges of adsorption and adsorbate content (for instance, moisture content) and water availability will be discussed further in sections addressing the development of kinetic models.

Uptake of water by solid materials is of great importance to many engineering and industrial fields because water as a solvent is often involved in the transport and adsorption of chemicals. For environmental and biological treatment, the uptake of water has added importance, as *water content* can control the efficiency of treatment processes, especially where solid materials are involved — for example, in dry fermentation systems and in biofiltration.

While water content is a directly measurable property of solid materials (oven heating or weight methods), water content does not correlate well with *water availability* to organisms and thus to biological activity. Water content is more strongly linked to and correlated with *water activity*, which is a thermodynamic property and

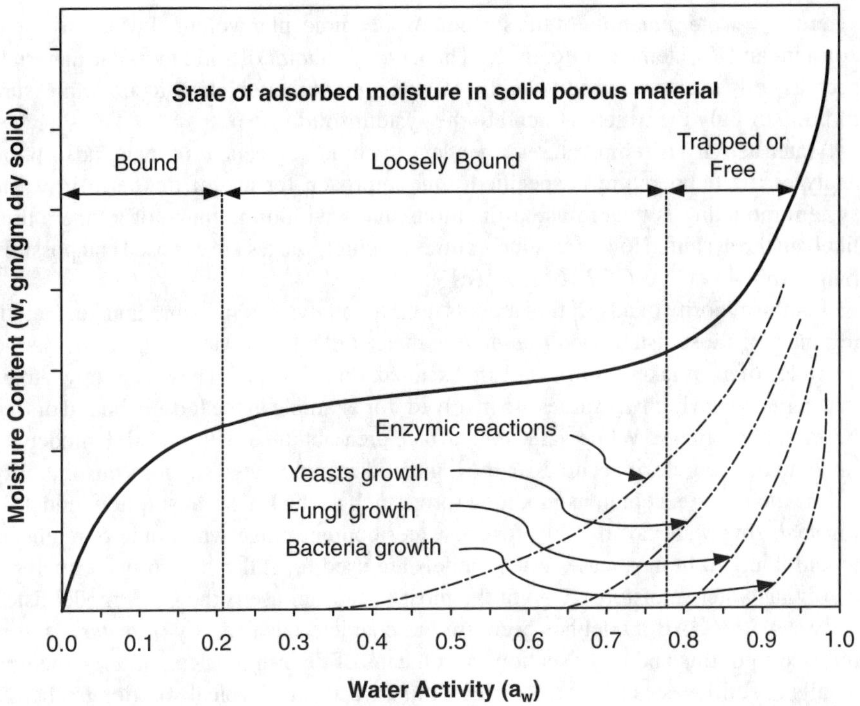


**FIGURE 2.6** Water sorption in porous solid materials (enzymic and soil biota patterns added)

value that quantifies the atmospheric availability of the water, and can be related to water availability to soil organisms.

Water activity relates directly to *equilibrium relative humidity* or the humidity of a solid material in equilibrium with the environment around it. It is stated in terms of the ratio of the equilibrium water vapor pressure exerted by a material to the vapor pressure of pure water:

$$a_w = \frac{p_w(\text{equilibrium water vapor})}{p_w(\text{pure water})} = \frac{\text{relative humidity}(\%)}{100} \tag{2.80}$$

As a thermodynamic term and a measure of the energy state of water adsorbed, the relation between water activity ($a_w$) and the (partial specific) Gibbs free energy of the material undergoing adsorption is:

$$a_w = \exp\left(\frac{\psi M_w}{RT}\right) = \frac{p}{p_0}$$

$$\ln(a_w) = \frac{\psi M_w}{RT}$$

$$\psi = \left(\frac{RT}{M_w}\right)\ln(a_w) \tag{2.81}$$

where $\psi$ = *water potential* of the system, $M_w$ = molecular weight of water, $R$ = gas constant and $T$ = temperature, in °K. The *water potential* of solid granular materials such as soils is also referred to as their *matric potential*, which is, again, a measure of how strongly the water is bound to the system solid matrix.

Water activity is temperature-dependent to some degree and in many adsorption analyses the temperature is specified. One approach for use of this term in waste system modeling is thus to use estimations made within the range of temperatures likely in the system. However, temperatures to which wastes are exposed can possibly range from 4 – 60°C (277.16–333.16°K).

For the majority of adsorption models used for analysis of moisture uptake, including most of those listed, *water content* is measured against water activity ($a_w$), and the range of $a_w$ is taken from 0.0–1.0. As noted, the BET model has been shown to be inaccurate when high $a_w$ values are involved, for example in the loosely bound or free water content range. While values of surface area obtained with the BET model are likely to be accurate, inaccuracy in the higher adsorption range suggests unsuitability of this model — except in its modified forms such as the Halsey adsorption model — for analyzing water availability for the water content range where microorganisms would flourish. In most cases other models are used for full-range sorption analyses involving water (moisture). One of the most general in use is the GAB model listed above. As the GAB model has been used to characterize a variety of materials, it is referred to in this and other sections as a means of determining important structural details of solid wastes likely to be present in a waste site. Typical sorption isotherms are shown in Figure 2.7.

## Determination of Solid Structure Characteristics from Adsorption Data

While *monolayer capacity*, in g/100 g of solids, is often provided in literature on studies of sorption by solid materials, in many cases (e.g., nitrogen or water) adsorption data/or plots may be published without estimated values of monolayer capacity, surface area in cm$^2$/gm or m$^2$/gm or other properties. It is thus useful to be able to calculate these properties from raw adsorption isotherm data, using the most convenient or applicable of the above adsorption models. For the (general use) GAB model, which is:

$$\frac{w}{w_m} = \frac{C_T K a_w}{(1 - K a_w)(1 - K a_w + C_T K a_w)} \tag{2.82}$$

the Type II adsorption isotherm can be represented by the quadratic polynomial:

$$\frac{a_w}{w} = A + B a_w + C a_w^2 \tag{2.83}$$

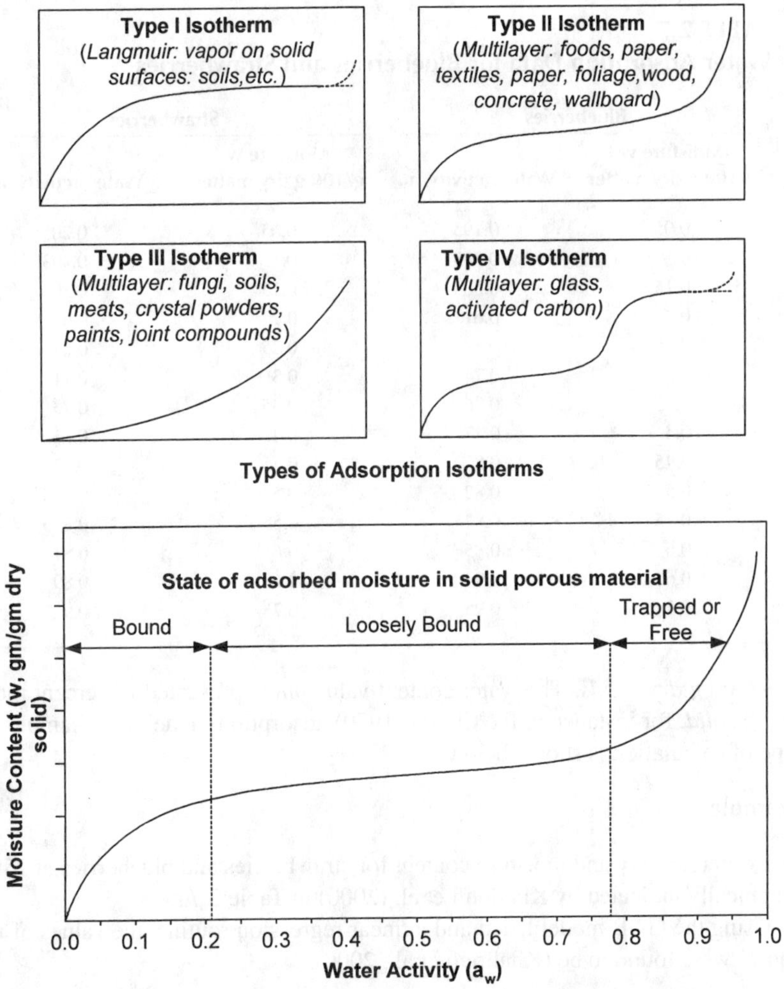


**FIGURE 2.7** Water sorption in porous solid materials (after Christen and Smith)

in which the constants $A$, $B$ and $C$ have the values:

$$A = \frac{1}{w_m C K}$$

$$B = \frac{1}{w_M}\left[\frac{K-2}{K}\right]$$

$$C = \frac{C_T}{w_m}\left[\frac{1-K}{K}\right] \tag{2.84}$$

The constants $A$, $B$ and $C$ can all be determined by regression analysis of $a_w$ (water activity) vs. water content ($w$ or $m_0$) data. The moisture content ($w$) is typically in units

**TABLE 2.7**
**Water Adsorption Data for Blueberries and Strawberries**

| Blueberries | | Strawberries | |
|---|---|---|---|
| Moisture w g/100 g dry matter | Water activity, $a_w$ | Moisture w g/100 g dry matter | Water activity, $a_w$ |
| 0.05 | 0.195 | 0.05 | 0.205 |
| 0.1 | 0.385 | 0.1 | 0.393 |
| 0.15 | 0.505 | 0.15 | 0.505 |
| 0.2 | 0.61 | 0.2 | 0.61 |
| 0.25 | 0.68 | 0.25 | 0.64 |
| 0.3 | 0.74 | 0.3 | 0.71 |
| 0.35 | 0.76 | 0.35 | 0.73 |
| 0.4 | 0.77 | 0.4 | 0.81 |
| 0.45 | 0.8 | 0.45 | 0.82 |
| 0.5 | 0.82 | 0.5 | 0.835 |
| 0.55 | 0.835 | 0.55 | 0.855 |
| 0.6 | 0.85 | 0.6 | 0.87 |
| 0.65 | 0.94 | 0.65 | 0.89 |
| 0.7 | 0.95 | 0.7 | 0.9 |

of *g*/100 *g dry solids*. The water content value $m_0$ is presented in percent moisture or *g/g solid*, for instance in the Caurie (1970) adsorption model. An example of this type of calculation is shown below.

## Example

The water activity and moisture content for strawberries and blueberries at 25°C are graphically indicated by Khalloufi et al. (2000) in Table 2.7.

Using the GAB model listed and a linear regression routine, the values of $w_m$, $C$ and $K$ were found to be (Khalloufi et al., 2000):

$w_m = 0.1065$ blueberries
$C = 5.374$ blueberries
$K = 0.983$ blueberries
$w_m = 0.096$ strawberries
$C = 0.384$ strawberries
$K = 1.099$ strawberries

The value $w_m$ is the monolayer adsorption capacity for the blueberry and strawberry. Gregg (1961) provides that the surface area of the adsorbent derived from the monolayer capacity can be expressed as:

$$S = \frac{w_m}{M} NA_m \tag{2.85}$$

where $w_m$ is stated in grams per 100 grams dry solids, $N$ = Avogadro's number = $6.02252 \times 10^{23}$, $A_m$ = cross-sectional area of the adsorbate molecule (for water this is $10.5 \times 10^{-20}\,m^2$) and $M$ = molecular weight of the adsorbate (water) = 18.0. This gives $S$ as 337.3 $m^2$/gm of internal surface for strawberry and 374.2 $m^2$/gm for blueberry.

A tabulation of estimated water adsorption model parameters and other properties for various solid materials that might be found in a waste site is presented in Table 2.8. Most of the GAB or BET values were calculated from graphs and data reported in literature.

## The Relation between Specific Surface Area and Sphericity of Waste Particles

To estimate the Sauter diameter $d_s$, which is a measure of the particle system pore area to particle volume, it is necessary to know the *specific surface area* $S$ and density $\rho$ of the material from which the particle is made. It is also useful to know the volume of the material in the system and its mass fraction. As noted, for granular materials, volume and mass fractions can be easily, though not always very accurately, determined from soil size analyses and material density information.

For a large number of natural or manufactured porous materials, absorption and porosity studies of liquid or gas sorption to food, agricultural products, stored products, articles needing preservation from decay, industrial products, water and effluents treatment units or air pollution filtration has required determination of specific surface area. The aforementioned Brunauer, Emmett and Teller (1938) or BET model has long been used to find specific properties (monolayer capacity or specific surface area) of wood, paper, foods, muscle tissue, cotton, wool and other solid materials and the GAB has been widely used in the food industry to determine water adsorption in foods and packaging materials. The Harkins-Jura (1944) method has been used to find and compile absorption factors for a variety of industrial materials. The Caurie (1981) method has also been used to determine moisture uptake properties of various materials. These and other analytical methods provide basic gas or liquid sorption parameters that can be used to estimate specific surface areas (SSA) as outlined. Some of these parameters have been calculated for this chapter and are tabulated in Table 2.8.

Particle characteristics of mixed solid materials can be determined by the following approaches.

For a particle of volume $v_p$, equivalent spherical volume and equivalent sphere area are given by:

$$\text{Equivalent spherical particle volume} = v_p = \pi \frac{D_e^3}{6} \tag{2.86}$$

$$\text{Equivalent spherical particle surface area} = S_p = \pi D_e^2 = (36\pi (v_p)^2)^{1/3} \tag{2.87}$$

**TABLE 2.8**
**Moisture Sorption Parameters and Specific Surface Areas of Various Waste Materials**

| | Waste Component | Water Absorbed % Dry Wt | Sorption Model Parameters $w_m$ g/100 g | C | k | Model | Specific Surface $m^2/gm$ | References |
|---|---|---|---|---|---|---|---|---|
| Foods | Pineapple pulp | 0–260 | 12.99 | 0.41 | 1.07 | GAB | 469* | AIChE Symposium Series, No. 297, Vol. 89 |
| | Strawberry/ blueberry | 0–70 | 12.1/11.9 | 2.0/2.2 | 0.95/0.93 | GAB | 428*/424* | Calculated from: Zalloufi et al., 2000. |
| | Apple | 0–40 | 4.5 | 10.5 | 1.29 | GAB | 161* | Data from plots by Iglesias, Chirife, 1982 |
| | Bread (white) | 0–21.5 | 9.2 | 0.55 | 0.58 | GAB | 331.5* | Wilson, Fuwa, *J. Ind. Eng. Chem.*, Oct., 1922, 913. |
| | Fruits | | | | | GAB | 134–685* | Calculated from sorption data** |
| | Citrus | 0–104 | 27.5 | 0.54 | 0.81 | GAB | 994* | Calculated from sorption data** |
| | Fruit skin (persimmon) | 0–70 | 11.43 | 1.6 | 1.001 | GAB | 413* | Telis et al., *Thermochimica Acta*, 343 |
| | Banana waste | | 8.3 | 1.44 | 1.04 | GAB | 299* | Calculated from sorption data** |
| | Vegetables -var. | | 18–25 | | | GAB | 151–249* | Calculated from sorption data** |
| | Cauliflower | 0–400 | 7.0 | 25.2 | 1.03 | GAB | 248.2* | Mulet et al., *J. Food Sci.*, 64, 1999 |
| | Sugar beet (raw) | 0–26.5 | 4.9 | 10.24 | 1.188 | GAB | 177.6* | ASAE Standards 1999, ASAE D245.5, October 1995. |
| | Potatoes, yams | | | | | GAB | 153–210* | Calculated from sorption data** |
| | Potatoes | 0–29.4 | 5.4 | 11.03 | 0.82 | GAB | 195.4* | ASAE Standards 1999, ASAE D245.5, October 1995. |
| | Meats (various) | | 14–32 | | | GAB | 116–253* | Calculated from sorption data** |
| | Meat (beef) | 0–64 | 5.9 | 17.7 | 0.91 | GAB | 212* | Calculated from sorption data** |
| | Meats (raw) | 0–58 | 7.44 | 13.95 | 0.87 | GAB | 268.6* | Data from plots by Iglesias, Chirife, 1982 |
| | Fish (various) | | | | | GAB | 200–274 | Calculated from sorption data** |
| | Rice (cooked) | 9–32 | 9.9 | 9.98 | 0.71 | GAB | 358* | Calculated from sorption data** |
| | Spaghetti | 0–36 | 5.0 | 457.5 | 0.86 | GAB | 181* | Lagoudaki et al., *Lebensm.-Wiss. U.-Technol.*, 26: 512–516, 1993. |

| | | | | | | | | |
|---|---|---|---|---|---|---|---|---|
| | Pasta, noodles | 3.5–21 | 6.1 | | | GAB | 52.7* | Calculated from sorption data** |
| | Seeds (beans, etc.) | | | | | GAB | 98 – 246* | Calculated from sorption data** |
| | Nut shell (pistachio) | 0–30 | 6.2 | 8.46 | 0.94 | GAB | 220* | Yanniotis et al., *Leb. Wiss. U.-Technol.*, 29, 1996. |
| | Peanut shells | 0–30 | 5.85 | 9.67 | 0.81 | GAB | 211* | Chianchung Chen, *J. Agric. Eng.*, 2000, 75: 401–408 |
| | Tea leaves | 0–40 | 6.71 | 0.403 | 0.878 | GAB | 242.4* | Temple, Boxtel, *J. Agric. Eng.*, 1999 |
| Paper | Newsprint | 0–11.3 | 3.6 | 10.9 | 0.69 | GAB | 131* | Calculated from sorption data** |
| | Bond/office | 0–13.8 | 5.15 | 9.32 | 0.49 | GAB | 186* | Wilson, Fuwa, *J. Ind. Eng. Chem.*, Oct., 1922, 913. |
| | Kraft (brown, etc.) | 0–25 | 2.8 | 60.5 | 0.77 | GAB | 99.1* | Richards et al., *ASHRAE Trans.*, BA-92-6-1 |
| | Corrugated cardboard | 0–17 | 4.8 | 20.1 | 0.73 | GAB | 172* | Calculated from sorption data** |
| | Milk carton | | | | | | | |
| | Boxboard | | 3.9 | | | GAB | 139.8* | Calculated from sorption data** |
| Plastics | HDPE | | | | | BET | 0.32–0.82 | Lee Favis, *Polymer*, 42, 2001 |
| | Expanded polystyrene | 0–4.5 | 2.9 | 273.8 | 0.41 | GAB | 72* | Data from Hansen, *Sorption in Industrial Materials* |
| | Polyurethane foam | 0–3.13 | 0.8 | 3.88 | 0.762 | GAB | 29.1* | Pissis et al., *J. Appl. Polymer Sci.*, 71: 1209+, 1999. |
| | PVC | 0–3.8 | 0.3 | 0.595 | 0.93 | GAB | 10.9* | A. M. Thomas, *J. Appl. Chem.*, 1, April 1951, 141–158 |
| | Nylon 6,6 | 0–8.94 | 6.5 | 0.839 | 0.597 | GAB | 235* | Lim et al., *J. Appl. Polymer Sci.*, 71: 197–206, 1999. |
| | EVOH package film | 0–7.66 | 7.18 | 0.815 | 0.541 | GAB | 259* | Zhang et al., *J. Appl. Polymer Sci.*, Part B, 37: 691. |
| | Biodegradable LDPE | 0–8.5 | 4.99 | 18.5 | 0.447 | GAB | 180.4* | Raj et al., *J. Appl. Polymer Sci.*, 84, 1193–1202, 2002. |
| Textiles | Wool | 0–36.3 | 12.04 | 7.71 | 0.686 | GAB | 434.7* | Wortman et al., *J. Appl. Polymer Sci.*, 79: 1054, 2001. |

(*Continued*)

**TABLE 2.8**
**Continued**

| Waste Component | | Water Absorbed % Dry Wt | Sorption Model Parameters $w_m$ g/100 g | C | k | Model | Specific Surface $m^2$/gm | References |
|---|---|---|---|---|---|---|---|---|
| | Wool (worsted) | 0–28.2 | 9.45 | 12.93 | 0.677 | GAB | 341.5* | Wilson, Fuwa, *J. Ind. Eng. Chem.*, Oct., 1922, 913. |
| | Cotton (clothing) | 0–17.6 | 5.2 | 8.69 | 0.718 | GAB | 188* | Wilson, Fuwa, *J. Ind. Eng. Chem.*, Oct., 1922, 913. |
| | Silk | 0–27.3 | 4.55 | 15.25 | 0.835 | GAB | 164.5* | Henry B. Bull, *J. Am. Chem. Soc.*, 66, 1944. |
| | Nylon | 0–12.8 | 2.23 | 5.84 | 0.832 | GAB | 80.6* | Henry B. Bull, *J. Am. Chem. Soc.*, 66, 1944. |
| | Linen (table) | 0–12.7 | 4.02 | 9.49 | 0.697 | GAB | 145* | Wilson, Fuwa, *J. Ind. Eng. Chem.*, Oct., 1922, 913. |
| | Jute | 0–14.6 | 6.56 | 22.37 | 0.565 | GAB | 237* | Wilson, Fuwa, *J. Ind. Eng. Chem.*, Oct., 1922, 913. |
| | Manila (rope) | 0–20.1 | 6.81 | 9.01 | 0.678 | GAB | 246* | Wilson, Fuwa, *J. Ind. Eng. Chem.*, Oct., 1922, 913. |
| | Sisal (rope) | 0–21.3 | 5.81 | 14.07 | 0.734 | GAB | 209.9* | Wilson, Fuwa, *J. Ind. Eng. Chem.*, Oct., 1922, 913. |
| Rubber | Auto tire (Goodyear) | 0–1.0 | 0.43 | 2.02 | 1.13 | GAB | 15.5* | Wilson, Fuwa, *J. Ind. Eng. Chem.*, Oct., 1922, 913. |
| Leather | Collagen | 0–59.3 | 11.25 | 18.84 | 0.813 | GAB | 406.4* | Henry B. Bull, *J. Am. Chem. Soc.*, 66, 1944. |
| | Hides (tanned) | 0–48.2 | 9.94 | 28.2 | 0.765 | GAB | 359.2* | Evans, Critchfield, *Bureau of Standards J. Rsch*, # 11 |
| | Leather (shoe sole) | 0–34.5 | 12.9 | 9.5 | 0.65 | GAB | 465* | Wilson, Fuwa, *J. Ind. Eng. Chem.*, Oct., 1922, 913. |
| Garden | Clover leaf (grass) | 0–15 | 3.3 | −55.3 | 0.774 | GAB | 115.2* | Stencl et al., *Grass Forage Sci.*, 2000 |
| Yard | Clover stem | 0–15 | 6.2 | −11.4 | 0.673 | GAB | 218.5* | Stencl et al., *Grass Forage Sci.*, 2000 |
| Waste | Grass (sedge) | 0–35 | 9.1 | 5.88 | 0.76 | GAB | 323.4* | Blackmarr, USDA SE-74, 1972 |
| | Grass (wire) | 0–35 | 7.04 | 9.88 | 0.92 | GAB | 249.5* | Blackmarr, USDA SE-74, 1972 |
| | Pine needles (1) | 0–40 | 8.7 | 28.6 | 0.77 | GAB | 310* | Blackmarr, USDA SE-74, 1972 |
| | Pine needles (1) | 0–40 | 7.5 | 28.6 | 0.82 | GAB | 265.4* | Blackmarr, USDA SE-74, 1972 |
| | Leaves-red oak | 0–43 | 7.0 | −25.2 | 0.88 | GAB | 248.5* | Blackmarr, USDA SE-74, 1972 |

| | | | | | | | |
|---|---|---|---|---|---|---|---|
| Leaves-pst oak | 0–45 | 9.0 | 31.2 | 0.81 | GAB | 319.6* | Blackmarr, USDA SE-74, 1972 |
| Leaves-hickory | 0–48 | 6.97 | −14.3 | 0.91 | GAB | 247.2* | Blackmarr, USDA SE-74, 1972 |
| Stems | | | | | | | |
| Alfalfa pellets | 0–24 | 5.6 | 25.7 | 0.77 | GAB | 201.3* | Fasina, Sokhansanj, *J. Agric. Eng.*, 1993, 56: 51–63. |
| Alfalfa hay (fresh) | 0–68.5 | 5.54 | −34.8 | 0.919 | GAB | 200.0* | ASAE Standards 1999, ASAE D245.5, October 1995. |
| Wheat straw (whole) | 0–100 | 5.3 | −23.4 | 0.947 | GAB | 191.4* | ASAE Standards 1999, ASAE D245.5 - October 1995. |
| Corn cobs | 0–49.9 | 5.5 | 16.57 | 0.895 | GAB | 189.7* | ASAE Standards 1999, ASAE D245.5, October 1995. |
| Corn cobs (whole) | 0–34.8 | 8.29 | 9.48 | 0.769 | GAB | 299.5* | ASAE Standards 1999, ASAE D245.5, October 1995. |
| Cottonseed hulls | 0–36.6 | 6.1 | −47.6 | 0.832 | GAB | 220.6* | ASAE Standards 1999, ASAE D245.5, October 1995. |
| Cocoa beans | 0–22.4 | 2.5 | −10.4 | 0.885 | GAB | 91.6 | ASAE Standards 1999, ASAE D245.5, October 1995. |
| Hops | 0–20.6 | 3.6 | 49.0 | 0.826 | GAB | 130.2* | ASAE Standards 1999, ASAE D245.5, October 1995. |
| Feathers (pillow) | 0–14.6 | 6.6 | 22.4 | 0.565 | GAB | 237* | Wilson, Fuwa, *J. Ind. Eng. Chem.*, Oct., 1922, 913. |
| Sawdust and shavings | 0–22 | 2.6 | 12.5 | 0.91 | GAB | 92.5* | Richards et al., *ASHRAE Trans.*, BA-92-6-1 |
| Wood chips (maple) | 0–40 | 4.4 | 1.8 | 0.9 | GAB | 153.4* | Baker et al., *Compost Sci. Utilization*, 7, 1999 |
| Wood chips (willow) | 0–131 | 5.2 | 913.4 | 0.96 | GAB | 188.2* | Gigler et al., *J. Agric. Eng.*, 2000, 77(4): 391–400. |
| Food composting material | 0–48 | 4.1 | 822.1 | 0.897 | GAB | 146.6* | Baker et al., Compost *Sci. Utilization*, 7, 1999 |
| Biosolids | 0–48 | 3.3 | −44.7 | 0.936 | GAB | 118.9* | Baker et al., Compost *Sci. Utilization*, 7, 1999 |
| Dried sludge | | | | | | 4.74 | Lu et al., *Fuel*, 74(3). 1995 |
| Twigs (ash, fresh, dead, with bark) | 0–130 | 18.6 | −65.6 | 0.826 | GAB | 670* | Griffith, Boddy, *New Phytologist*, 117(2), Feb. 1991. |
| Wood (pine) | 0–28 | 6.3 | 36.6 | 0.84 | GAB | 222.4* | Blackmarr, USDA SE-74, 1972 |
| Wood (southern pine) | 0–30 | 6.63 | 3.8 | 0.76 | GAB | 235.2* | Richards, ASHRAE, BA-92-6-1 |
| Plywood | 0–27 | 4.6 | 5.3 | 0.82 | GAB | 164.4* | Richards, ASHRAE, BA-92-6-1 |
| Particle board | 0–25 | 5.1 | 3.04 | 0.78 | GAB | 180.4* | Richards, ASHRAE, BA-92-6-1 |
| Gypsum board | 0–7 | 0.8 | 59.97 | 0.82 | GAB | 28.6* | Richards, ASHRAE, BA-92-6-1 |
| Brick | 0–0.5 | 0.17 | −62.2 | 0.67 | GAB | 6.0* | Calculated from sorption data** |
| Linoleum | 0–14 | 2.0 | 14.1 | 0.86 | GAB | 74* | Calculated from sorption data** |
| Glass | | | | | BET | 0.434 | Gregg, Adsorption . . . 1967 |
| Ferrous | | | | | BET | 0.57 | Gregg, Adsorption . . . 1967 |

*(Continued)*

**TABLE 2.8**
**Continued**

| Waste Component | Water Absorbed % Dry Wt | Sorption Model Parameters $w_m$ g/100 g | C | k | Model | Specific Surface $m^2$/gm | References |
|---|---|---|---|---|---|---|---|
| Topsoil | | | | | BET | 29.6 | Pennell, Rao, EST, 28, 1992 |
| Sand | | | | | BET | 1.4–4.2 | Weber et al., 1992, *Envir. Sci. Tech.* |
| Sand-gravel aquifer material | | | | | BET | 4.02 | Macintyre et al., *J. Contam. Hydrol.*, 1998 |
| Silt | | | | | BET | 46.6 | Pennell, Rao, *Envir. Sci. Tech*, 28, 1992 |
| Silty soil | 0–15 | 1.17 | 40.1 | 0.92 | GAB | 42.3* | Data from plots by De Seze et al., *Sci. Tot. Env.*, 253 |
| Soil | 0–7.5 | 0.6 | 18.3 | 0.82 | GAB | 21.7* | Data from plots by De Seze et al., *Sci. Tot. Env.*, 253 |
| Campus lake (type) sediment | 0–8.3 | 1.34 | 48.9 | 0.84 | GAB | 48.3* | Data from plots by De Seze et al., *Sci. Tot. Env.*, 253 |
| Clay | | | | | BET | 60.4 | Sharma et al., *Soil Sci.*, 107. |

* Values starred thus (42.3*) and GAB model values for wm, C and k and water sorption ranges were estimated for this work from isotherm plots or reported data, from the reference sources listed in the right-most column.

** These values were estimated from published water adsorption data for foods, construction and industrial materials and soils using MATHCAD and MATLAB routines developed for this purpose (Ph.D. Thesis, Percival A. Miller, Rensselaer Polytechnic Inst., 1998).

*Source*: Baker et al., 1999. Equilibrium moisture isotherms for synthetic food waste and biosolids composts. *Compost Sci. Utilization*, Vol. 7, No. 1, 6–13.

De Seze et al., 2000. Sediment-air equilibrium partitioning of semi-volatile hydrophobic organic compounds: Part I: method development and water vapor sorption isotherm. *The Science of the Total Environment*, 253: 15–26.

Khalloufi, S., J. Glasson and C. Ratti, 2000. Water activity of freeze dried mushrooms and berries. *Canadian Agric. Eng.*, 42(1): 7.1.

Yanniotis, S. and I. Zarmboutis. 1996. Research note: water sorption isotherms of pistachio nuts. *Lebensm.-Wiss. U.-Tech.*, 29: 372.

Richards, R. F. et al., 1992. Water vapor sorption measurements of common building materials. *ASHRAE Trans.*: Symp., BA-92-6-1.

Temple, S. J. and A. J. B. Boxtel. 1999. Equilibrium moisture content of tea. *J. Agric. Eng. Res.*, 74: 83–89.

Telis et al., 2000. Water sorption thermodynamic properties applied to persimmon skin and pulp. *Thermochimica Acta*, 343: 49–56.

Stencl, J. and P. Homola. 1998. Water sorption isotherm of leaves and stems of *Trifolium pratense*. *Grass and Forage Sci.*, 55: 159–165.

W. H. Blackmarr, Equilibrium Moisture Content of Common Fine Fuels Found in Southeastern Forests, USDA Forest Service Research Paper SE-74, 1971. (Loblolly, slash and longleaf pine needles; red oak, post oak and hickory leaves; wire grass, broomsedge and loblolly pine wood).

Wilson, Robert E. and Tyler Fuwa. October 1922. Humidity equilibria of various common substances. *J. Ind. Eng. Chem.*, 913–918.

Baldev, Raj, A. Eugene Raj, K. R. Kumar and Siddaramaiah. 2002. Moisture-sorption characteristics of starch/low-density polyethylene films. *J. Appl. Polymer Sci.*, 84: 1193–1202.

Zhongbin Zhang, Ian J. Britt and Marvin A. Tung. 1999. Water absorption in EVOH films and its influence on glass transition temperature. *J. Polymer Sci.*: Part B: 37: 691–699.

Kanapitsas, A., P. Pissis, J. L. Gomez Ribelles, M. Monleon Pradas and E. G. Privalko. 1999. Molecular mobility and hydration properties of segmented polyurethanes with varying structure of soft- and hard-chain segments. *J. Appl. Polymer Sci.*, 71: 1209–1221.

Lim, Loong-Tak, Ian M. Britt and Marvin A. Tung. 1999. Sorption and transport of water vapor in Nylon 6, 6 film. *J. Appl. Polymer Sci.*, 71: 197–206.

Wortman, F.-J., P. Augustin and C. Popescu. 2001. Temperature dependence of the water-sorption isotherms of wool. *J. Appl. Polymer Sci,*, 79: 1054–1061.

Evans, W. D. and C. L. Critchfield. July 1933. The Effects of Atmospheric Moisture on the Physical Properties of Vegetable and Chrome Tanned Leathers. Research Paper RP583, Bureau of Standards Journal of Research, US Department of Commerce. Vol. 11. 146–162

Bull, Henry B. 1944. Adsorption of water by proteins. *J. Am. Chem. Soc.,* 66: 1499–1507.

Moisture Relationships of Plant-Based Agricultural Products. ASAE Standards 1999. Document No. ASAE D245 OCT95.

For number $n$ particles, total particle volume $V_{tot}$ total surface $S_{tot}$ and the equivalent sphere diameter are given by:

$$\text{Total volume of particles} = V_{tot} = n\left(\frac{\pi D_e^3}{6}\right) \tag{2.88}$$

$$\begin{aligned}\text{Total surface area of equivalent spherical particles} &= S_{tot} = nS_p \\ &= n((36\pi V_p^2)^{1/3}) \end{aligned} \tag{2.89}$$

$$\frac{V_{tot}}{S_{tot}} = \frac{n(\frac{\pi D_e^3}{6})}{n((36\pi V_p^2)^{1/3})} = \frac{(\frac{\pi D_e^3}{6})}{((36\pi(\frac{\pi^2 D_e^6}{36}))^{1/3})} = \frac{D_e}{6} \tag{2.90}$$

The value of the equivalent sphere diameter applies to a material of a single-particle size, density and specific surface. It can thus be applied only to a single-particle size group at a time.

Another of several approaches is to find the mean size of a type of granular material from its particle size distribution (PSD), even in a mix of other particles. An essential requirement for this is prior knowledge of the size (and mass) variation of the material in a mix of granular materials. This, as outlined earlier, would require a separation method such as sampling and screening.

Screening of a representative or composite sample of a waste site allows careful separation and weighing of waste fractions such as paper, glass, soil and plastics for each total weight of waste $m_i$ materials caught between two standard screen sizes. For example, if waste paper particles of average screen size $d_j$ [= (smaller screen size + larger screen size)/2] are of total weight or mass $m_j$, where $i = 1 \ldots n =$ types of solid materials in the mixture, the relation between $m_i$ and $d_j$ represents the size distribution for this material:

$$\begin{aligned} j &= 1 \ldots m = \text{particle size categories from screen} \\ i &= 1 \ldots n = \text{number of material types in the mixed system} \\ m_{i,j} &= \text{mass of material of type } i \text{ between screen sizes} j - 1 \text{ and } j + 1 \\ f_{i,j} &= \text{mass fraction of } j\text{th material of screen size} \\ d_j &= \frac{m_{i,j}}{\text{mass of } i\text{th material } M_{tot,j}} \\ M_{i,total} &= \sum_{i=1}^{n} m_{i,j} = \text{total mass of material } i \text{and size range} j \\ V_i &= \frac{M_i}{\rho_i} = \text{volume of } i\text{th material} \end{aligned} \tag{2.91}$$

Statistical methods can then be applied to the PSD data for material type $i$ to arrive at other useful modeling properties (e.g., mean size and shape of the $i$th granular material). Where $v_{i,j}$ represents the collective volume of all particles of size $j$ and type $i$ of a material, of density $\rho_j$ and $\alpha_{v,j}$ represents the relative sphericity (how closely the particle represents a sphere) of particles from material type $j$:

$$j = 1 \ldots m = \text{particle size categories from screen}$$
$$i = 1 \ldots n = \text{number of material types in the mixed system}$$
$$m_{i,j} = \text{mass of material of type } i \text{ between screen sizes } j-1 \text{ and } j+1$$
$$v_{i,j} = \text{total volume of } n_i \text{ particles of size } j \text{ and type } i$$
$$v_{i,j} = \frac{m_{i,j}}{\rho_i} = \alpha_{v,j}\, n_{i,j}\, d_{i,j}^3$$
$$S_{i,j} = \alpha_{v,i}\, n_{i,j}\, d_{i,j}^2$$
$$(\textit{Volume-specific surface})_i = \frac{S_{i,j}}{V_j f_{i,j}}$$
$$\text{Spherical shape: } \alpha_{v,j} = 6$$
$$\textit{Volume-specific surface} = S_{v,i} = \alpha_{v,i} \sum_{i=1}^{n} \frac{f_{i,j}}{d_{i,j}}$$
$$\textit{Mass-specific surface} = \rho_i S_{v,i}$$
$$\textit{Surface volume shape coefficient} = \alpha_{v,i,A} = S_{v,i} \left( \sum_{i=1}^{n} \frac{f_{i,j}}{d_{i,j}} \right)^{-1} \tag{2.92}$$

Important particle properties such as *Sauter diameter* and *sphericity* can thus be estimated from soil analysis data, as illustrated.

## Example

A solid waste material makes up 20% by weight ($f = 0.20$) of all materials in a waste site. It has a BET specific surface area (*SSA*) of 125 m$^2$/gm, a dry density of 0.660 g/cm$^3$ and a saturated density of 0.785 g/m$^3$ (water content 19%). Wastes and soils buried in the site have a total estimated dry weight of 8000 kg. Particle size distribution analysis for this particular solid waste material has shown that 15% of its weight is in the particle size range 60 mm screen size ($d = 0.06$ m). There are also likely to be $n$ particles, of average screen size 60 mm, in this 15% size fraction. Determine the *equivalent spherical particle size* $D_e$, particle *sphericity* $\varphi$ and *Sauter diameter* $d_s$ for this material. Representative particle size distribution data for the solid waste material are shown below (Table 2.9).

Specific surface area = *SSA* = 125 m$^2$/g (for *wood*, using Caurie, GAB or BET sorption)
Density of particle = 0.660 g/cm$^3$

**TABLE 2.9**
**Sample Particle Size Distribution for *i*th Solid Waste Material**

| Sieve Size, mm | % weight | Sieve Mass *i*th Material % | Fraction Total mass, $w_i$ | Mass Fraction *i*th Waste. $q_i$ | Mean Sieve Size, $d$ mm | Mass-size Ratio, $q_i/d$ | $\sum(q_i/d)$ |
|---|---|---|---|---|---|---|---|
| +160 −320 | 5.4 | 6.9 | 0.004 | 0.069 | 240 | 0.288 | 0.288 |
| −160 + 80 | 11.3 | 25.9 | 0.0293 | 0.259 | 120 | 2.158 | 2.446 |
| −80 + 40 | 15.9 | 31 | 0.0493 | 0.31 | 60 | 4.493 | 6.939 |
| −40 + 20 | 17.3 | 21.9 | 0.038 | 0.219 | 30 | 7.300 | 14.239 |
| −20 + 10 | 14.4 | 8.8 | 0.013 | 0.088 | 15 | 5.867 | 20.106 |
| −10 | 35.7 | 5.5 | 0.02 | 0.055 | 5 | 11.0 | 31.106 |
| Total | 100 | 15.3 | 0.15294 | 1.0 | | | 31.106 |

Particle size category $= j =$ size between two screens, e.g., $d = (20 + 10)/2 = 15$ mm

$Q_j =$ mass fraction of $j$th size particle between screens $= m_i/M_{j(total)}$

*Volume-specific surface* = VSS = 6 × 31.106 = 186.64 $m^2/m^3$

Sum (mass fraction/mean screen size) $= \{\sum(q_i/d_i)\} = (31.106) \times m^{-1}$

For material $\rho = 660\,kg/m^3$, mass-specific surface = (VSS/660) $(m^2/m^3)(m^3/kg)(1000\,g/kg)$

*Mass-specific surface* = MSS = 282.78 $m^2/g$

Surface volume shape coefficient = SVSC $= SSA/\{\sum(q_i/d)\} = 125/31.106 = 4.02$

Mass-specific shape coefficient = MSSC $= MSS/\{\sum(q_i/d)\} = 282.78/31.106 = 9.09$

Surface volume mean diameter $= \{\sum(q_i/d)\}^{-1} = 0.03215\,m = 32.15\,mm = d_{sv} =$ *Sauter diameter* $= d_s$

$d_a = (SVSC \times d_s)/6 = 21.53$ mm (mean projected area particle diameter)

$$\text{Equivalent spherical diameter} = ESD = 6/(SSA \times \rho) = 6/[(125\,m^2/g) \times (0.66\,g/cm^3)] = 72{,}727.27\,\mu m = 0.07272\,m = 72.72\,mm\ (ESD) = D_e$$

Sphericity $= \varphi = d_s/d_p = 32.15/72.72 = 0.442$

## Particle Shape Considerations

A significant difficulty in characterizing packed beds of mixed particles such as waste sites is that the grains or particles present are decidedly heterogeneous in physical properties; irregular in shape, compared to a thickness of soil or sand or gravel;

**TABLE 2.10**
**Particle Types and Sphericities Used in Industrial Packed Bed Reactors**

| Particle Type | Sphericity Factor $\varphi$ |
|---|---|
| Sphere | 1.0 |
| Cylinder (d = length) | 0.87 |
| Cube | 0.81 |
| Sand | 0.75 |
| Coal dust (pulverized) | 0.73 |
| Crushed glass | 0.65 |
| Berl saddles | 0.3 |
| Raschig rings | 0.3 |

and arbitrarily compact, when compactness means the degree of curvature along the main surface plane. For instance, a particle of brown paper or plastic may be thin enough to qualify as a lamina in shape; its volume may be found by dividing its mass by its density; and its specific surface and porosity may be found from adsorption isotherm considerations, yet disposal and normal compaction conditions may induce considerable crinkling and folding. While the particle is self-homogenous in that its physical properties should not change due to crinkling and folding (sorption and density properties remain the same), the final (compact) geometry of the particle affects bed parameters such as pore connectivity and liquid holdup, as crinkling and folding increases the number of system seams, pockets and holes that can act to hold liquid in pendular or bridge arrangements under surface tension or act as dead pores trapping gas or air. While it may not be possible to know the final shape of all particles in a waste system, sufficient information should be known about other particle properties to define overall volume, possible packing arrangement, pore surface contribution and liquid uptake. Individual shape properties are less important, as definitions of bulk flow tend to derive shape properties from overall statistical distributions.

However, particle shapes are of some importance because available surface area is a determinant in heat and mass transfer and reaction kinetics, which depends upon both particle shape and arrangement. The following particle shape factors or sphericities have been reported (Geankopolis, 1978).

In this section, it was shown how the average diameter of a distribution of particles of a specific material can be obtained from sieve screening information. If a thorough sample of paper grains in a waste site is defined as varying between 180 and 2 mm, with mass fractions assigned to each screen size, the mean size of the paper grains in the waste site can be estimated from particle size distribution statistics. From material density and specific surface area, the *sphericity* (how closely the grain approximates a sphere) can then be estimated. A useful form of the average particle diameter for the ongoing consideration of packed bed dynamics is the *surface volume mean diameter* or *Sauter diameter*. Stated in terms of the particle size distribution, the Sauter diameter is:

$$q_j = \frac{m_j = \text{mass, } j\text{th particle size group, type } i}{M_i = \text{total mass of } i\text{th particle type}}$$

$$d_j = \text{screen size of } j\text{th particle group}$$

$$\text{sum}\left(\frac{\text{mass fraction}}{\text{total mass}}\right) = \sum\left(\frac{q_j}{d_j}\right)$$

$$\text{Surface volume mean diameter} = \left[\sum\left(\frac{q_j}{d_j}\right)\right]^{-1} = d_{sv} = \text{Sauter diameter} \tag{2.93}$$

In this case, the total mass of a type of waste particle, say paper ($i = 1$) or plastic ($i = 2$), is grouped by the average between two screen sizes and the mass of that type of grain ($m_i$) between these two screens. Summation provides the mean diameter. Because this diameter ($d_{sv}$) is based upon an exponent of screen size, for example volume $= Ad_j^3$, where $A$ is a constant, it does not describe shape or surface properties of the particle. However, in soil science, the *equivalent spherical diameter* (ESD) of a particle relates to its *surface volume mean diameter* by the expression:

$$\text{Equivalent spherical diameter} = \text{ESD} = \frac{6}{(\text{SSA}) \times \rho} \tag{2.94}$$

where SSA = specific surface area and $\rho$ = particle density. Particle sphericity $\varphi$ is then the ratio of the surface volume mean diameter to the equivalent spherical diameter:

$$\varphi = \frac{d_{sv}}{ESD} \tag{2.95}$$

As the surface area of a sphere is $\pi D^2$, the volume of the equivalent sphere for a particle is $\pi(ESD)^2$. The actual surface area of the particle $S_p$ can be derived from the relation between volume and surface area:

$$\frac{\text{Surface area}}{\text{volume}} = \frac{S_p}{v_p} = \frac{\pi(ESD)^2/\varphi}{\pi(ESD)^3/6} = \frac{6}{\varphi(ESD)} = \frac{6}{d_{sv}}$$

$$S_p = \frac{\pi(ESD)^3}{d_{sv}} = \text{actual particle surface area} \tag{2.96}$$

Note that this approach is a convenient way to relate intrinsic particle properties of waste types, such as specific surface area (SSA) and density ($\rho$), to their apparent (sieve) diameter, with the proviso that particle diameters found from sieve diameters are prone to some degree of error in the value of $d_i$. If the approach is applied to various granular materials present inside the waste sites, approximations of specific particle sphericities can be made, noting that the specific surface areas and densities of various materials must be known.

Assuming that the mixture of waste particles and soils in a waste site are amenable to characterization through size distribution analyses and the previously discussed parameters, the mean specific surface of the mix of particles (or packed bed) can be derived from the mean surface fraction for individual particle types. As noted, each waste type would constitute a total solid volume fraction $V_i$, in the waste site, where $V_i$ is found from the total mass $M_i$ of waste site solid type $i$, of density $\rho_i$:

$$V_i = \frac{M_i}{\rho_i} = \frac{\sum m_{i,j}}{\rho_i} \tag{2.97}$$

and the total solid volume of particles in the waste site is:

$$V_{TOT} = \sum_i V_i = \sum_i \left(\frac{M_i}{\rho_i}\right) \tag{2.98}$$

The mean surface fraction ($MSF_i$) of waste site particle type $j$ is then:

$$MSF_i = S_{p,i} \times V_i = \frac{\pi (ESD(i))^3}{d_{sv}(i)} \times V_i \tag{2.99}$$

where $S_{p,i}$ represents the mean particle surface area of waste particle type $i$. The (weighted) mean specific particle surface (*MSPS*) for the whole mix of waste site particles is then the sum of the mean particle surfaces (Geankopolis, 1978):

$$MSPS = \sum_i MSF_i = \sum_i \left(\frac{\pi (ESD(i))^3}{d_{sv}(i)} \times V_i\right) \tag{2.100}$$

The mean specific particle surface (*MSPS*) has the same relation to the mean (system) particle diameter as for a single particle, thus for the particle mix in the waste site the *mean particle diameter* (*MPD* or $d_{MS}$) can be stated as:

$$MPD = \frac{6}{MSPS} \tag{2.101}$$

The value of MPD does not indicate the *sphericity* of the mean particle representing the site. However, the calculated parameters incorporate individual waste or soil type sphericity and surface area, and thus are likely to represent particle characteristics in a mixed waste site.

Packed bed flow modeling traditionally characterizes particle mixes in a bed system in terms of specific surface and effective particle diameter (or MSD). In these models heterogeneity of particles is rarely assumed and any overall particle sphericity value is based upon the assumption that all particles should have the same shape. Obviously this assumption would lead to errors in modeling a waste site and is only suitable for a homogenous mix of soils.

It should be noted that waste composting sites do not quite qualify as packed beds because they are not *packed* in the sense of compacted mixes of granular materials. In any case, the packing of composting materials would restrict airflow and thus impair the rates of aerobic decomposition.

## APPLICATION TO MIXTURES OF GRANULAR MATERIALS

Note that the above approach applies to one type of material. The approach can, however, be used for a mixture of granular materials if the system property desired does not depend upon the specific density of any particular material.

An unseparated waste site sample screening can only provide size distribution data for the whole mix of materials present and does not distinguish between types of materials. Yet these data can be used to estimate a system property such as overall grain size or porosity and can be related to the system's macroscopic air or water permeability. For instance, the volume of a particle $v_p$ combines its specific surface $a_v$ and its actual surface area $S$, where:

$$v_p = \frac{m_p}{\rho_p} = \frac{\text{particle mass}}{\text{particle density}}$$

$$v_p = S_p/a_v$$

$$a_v = \frac{6}{D_e} = \frac{6}{\varphi_s D_p}$$

Here $D_e$ represents the equivalent spherical diameter, as can be estimated from the particle size distribution of a material, $D_p$ represents the mean particle size and $\varphi_s$ represents the particle sphericity, as previously described. The volume of the $i$th type of material, say plastics, is its total mass in the mixture or in the site divided by its density:

$$V_i = \frac{\sum_{i=1}^{i=n} m_{i,j}}{\rho_i} \tag{2.102}$$

The volume fraction $x_i$ of the $i$th material is the ratio of its volume $V_i$ to the total volume of other solid materials present in the system:

$$x_i = \frac{V_i}{\sum_i^n (V_i)} \tag{2.103}$$

For the $i$th material in a mix of solid materials, its average specific surface $a_{v,ji}$ can be stated, in terms of its mean particle size(s), as:

$$a_{v,i} = \frac{6}{D_{e,i}} = \frac{6}{\varphi_i D_{p,i}} \tag{2.104}$$

The *mean specific surface area* of the mixed solid materials in the system can then be stated as:

$$a_{v,mean} = \sum_{i=1}^{j=n} (x_i a_{v,i}) \tag{2.105}$$

From this, the *effective mean particle size* $D_{p,mean}$ can be simply stated as:

$$D_{p,mean} = \frac{6}{a_{v,mean}} \tag{2.106}$$

For models of fluid flow through packed particle beds, estimations of the *effective mean particle size* $D_{p,mean}$ and *mean specific surface area* or $a_{v,mean}$ are important to modeling of laminar fluid flow (gas or liquid), and pressure drop or liquid holdup or retention. This approach, among others as efficient, may thus be applied for estimation of these system properties.

## Application of Particle-Based Properties to the Kinetic Modeling of Reactors

For kinetic studies and modeling of reactors, the above approach is limited since it can introduce errors. The factors affecting reactor kinetics also include water or chemical uptake, chemical composition and reactivity of materials, specific pore volume and specific thermal properties, as well as pore accessibility to biological agents. Approaches inclusive of specific material properties must be used, provided these properties are known.

For waste sites, most important properties of waste components can be found with some diligence, or approximated or determined from laboratory investigations. However, physical properties found from size distribution factors such as *mean particle size* and *Sauter diameter* allow material surface area effects to be modeled, and permit development of kinetic models that integrate specific material reactivity with that of the whole system. The goal of assessment of particle properties, for the modeler, is to provide basic information that can be used to develop simplified mathematical models for waste sites as bioreactive but particle-based reactor systems, amenable to representation with careful examination of likely natural characteristics and processes.

# 3 Characterization of Disposed Wastes: Physical and Chemical Properties and Biodegradation Factors

One of the more important parameters for waste treatment studies is the characterization of the substrates. In the case of a waste site, inactive or not, substrates for treatment processes can be the solid wastes buried, the leachate or the gas if the treatment approach proposes to examine amelioration of natural effects. Review of these substrates in order to select representative characteristics is thus useful to analyses. In many waste decomposition studies, the solid wastes are treated as a single substrate, with bulk kinetic factors determined from laboratory or pilot scale studies. Recent solid waste composition studies provide a more accurate description of the types of materials likely to be disposed and amounts — thus, disposed solid wastes can possibly be addressed as more than a single substrate once individual waste types are sufficiently defined.

## DETERMINATION OF PHYSICAL AND CHEMICAL CHARACTERISTICS OF WASTES

To determine which municipal solid waste (MSW) types would be present in waste sites, in what quantities and which were biodegradable, the literature on waste stream materials was reviewed. Details such as density, moisture-as-received, volatile solids content, chemical composition, carbohydrates, fats and protein content, specific surface, heat (and water) properties and estimated fraction per ton or kilogram of MSW were collected.

Chemical composition, moisture content, ash content, volatile solids content and biodegradable volatile solids content are particularly useful properties for stoichiometric approximation of likely waste decomposition products, while thermal and water properties are particularly useful in the development of mass and heat transfer approaches for modeling. The results of the review of various literature sources and of estimations of solid waste (water sorption and surface area) properties are included in referenced tables.

Handling, burial and compaction significantly change available surface and density properties of some solid wastes. Review of literature for information on *in situ* waste volumes and surface areas was generally unfruitful. This was not considered critical, because it is believed inaccuracy in this area of studies should not greatly affect the accuracy of decomposition analyses. Also, surface area approximations can be made from hydraulic and water sorption considerations.

Loss from the input due to macroorganisms (rodents, birds, marsupials, worms and insects) before burial was also reviewed. The review indicated that the level of solid waste removal, directly due to the intervention of larger organisms, the waste's *history* should be relatively small. Some larger organisms, for example, some bird species and some soil-dwelling fauna, could have a small role in waste disappearance. Quantitative impacts of these organisms are discussed in detail in Chapter 4.

Freshly disposed MSW has similar composition to MSW discard streams as estimated and sampled for municipal waste management. In recent years there have been many thorough studies of this kind, including the well-known Franklin-EPA 1988, 1990 and 1995 studies of MSW composition in the United States. Original landfilled MSW composition was assumed to be the same as from the EPA (1990) study or from the recent solid waste stream characterization study done for the City of New York (1992) for simplicity.

While local and regional weight or volume fractions of MSW may differ, it is believed variations of this type should not adversely affect decomposition analyses because the major fractions of the material in waste sites and dumps is cellulosic under most known circumstances.

## MSW COMPOSITION VS. LANDFILL LAYER DEPTH OR AGE: DATA FOR INITIALIZATION

It is unlikely that MSW composition will not change with age, depth and burial. However, there are few studies that have addressed the content and volume of wastes as correlated with site depth. Two of the most relevant have been by Attal et al. (1992), which studied variations of wastes' volatile solids content with depth and thus with the correlated age (from buried newspapers) and that of Bogner (1990), which studied the biodegradability of samples (with descriptions of solid wastes found) taken at different depths at landfills. These studies were used as bases for rough approximation of initial waste contents and type variation with depth at the hypothetical landfill used for modeling.

The approximation of initial landfill waste composition vs. depth or age is likely to be prone to error. Possible approximations include linearly or exponentially decreasing MSW *percentage* vs. depth. The data and plots provided by Attal et al. (1992) suggest this pattern, with number plots suggesting exponentially decreasing volatile solids content. This approach reduces the difficulty of modeling initial conditions, thus the usefulness of statistical analysis of waste content vs. depth and age. However, development of an interactive database permits hydrolysis rate vs. landfill layer vs. age approximations, so that approximate mass distributions of MSW types vs. layer age could be directly made. This has the obvious advantage of reducing assumptions of waste types vs. depth, an important (heterogeniety) issue for landfill biodegradation models. Approaches to depth vs. wastes age can thus be:

(1) Linear or exponential models initialized with partial depth sampling of the site correlated with age markers (newspaper dates, etc.).
(2) Setting initial MSW fraction soil contents vs. depth or age, based upon sampling from local landfills. While this approach reduces errors associated with wastes decomposition rates, there is usually very little reliable or detailed information on MSW fraction vs. landfill soil depth or age.
(3) Assuming an age vs. depth, then initializing layers' MSW fraction distributions based upon likely waste fraction status at the age of the layer. This approach is particularly suited to modeling studies. Error could arise from incorrect assumption of layer age, but field studies have indicated age can be approximated based on discovered unrotted printed material. Thus, this approach can possibly be verified in the field.

A combination of (1) through (3) could give the most reasonable MSW distribution vs. depth from surface for analysis.

## INDIVIDUAL WASTES AND CHARACTERISTICS

Because of the wealth of solid waste information contained in literature, albeit limited relevance of some data to modeling, it is useful to define each MSW waste class (paper, yard wastes, food wastes, etc.) separately to limit the number of variables considered. For this reason, separate discussions of wastes and properties are included below.

## CHARACTERISTICS OF PAPER WASTES

Paper wastes comprise the largest single (weight and volume) fraction of solid wastes entering landfills and dumps, ranging from 35 to 50% by weight. Typically, the MSW paper stream includes the following:

- Newsprint, discarded newspapers or colored inserts
- Corrugated cardboard, paper packaging, cartons and wrapping paper
- Office paper, printed and computer paper
- Brown paper bags and sacks
- Cardboard
- Milkboard (milk cartons) and foodboard (coated cups and plates)
- Books, magazines and commercial varieties of paper including mailing forms

As a major fraction, mainly cellulosic in nature, paper content of MSW highly influences the overall rates of landfill decomposition. The nature of cellulose decomposition should thus be of interest to landfill modeling. Cellulose decomposition involves solubilization by cellulase enzymes expressed onto the surface of cellulose materials and absorbed thereupon by hydrolytic organisms, including numerous species of aerobic or anaerobic bacteria and fungi.

Extensive research has been conducted on the enzymic or acid hydrolysis of cellulosic materials and pure cellulose, including for the production of ethanol, gas and other products of commercial value protein. An important point from many of these studies is that the controlling step in the decomposition of cellulosic materials is hydrolysis and that hydrolysis in the main is controlled by available surface area of the material. It has also been shown that the actual rate of hydrolysis is controlled by accessibility of the material pore structure to the relatively large cellulose enzyme, which is on the order of 51 angstroms.

The extensive research into aerobic or anaerobic hydrolysis provides information on relative rates of decomposition or hydrolysis rate constants, of various paper or cellulosic materials that may enter the landfill stream. These constants have to be applied with some care; in many cases they have been derived under more favorable or controlled conditions than in the field or at landfills. Many of these constants are tabulated elsewhere in this book.

Important properties of paper wastes for landfill analyses also include chemical and physical composition. Chemical composition of papers, as for food, yard and other degradable wastes, is often presented in proximate, ultimate and biochemical forms. The most useful chemical composition form depends upon the type of analysis to be attempted. For example, both aerobic and anaerobic biological decomposition process simulation could benefit from chemical characterization of wastes, because this could roughly allow estimations of degradation pathways and likely products if reasonable stoichiometric representation decomposition could be developed.

It is useful to conduct a literature search for chemical compositions of the various solid materials. Of all MSW stream fractions, the paper waste stream is perhaps the best characterized. However, except for the Niessen (1978) and the various EPA (1990) waste stream characterization (and related) studies, there is often little consistency between MSW characterizations, particularly for paper. More recent characterizations of the MSW stream have tended to include more detailed information on fractions according to source, i.e., percentages of newspapers and magazines in the medical waste stream, but little added information on types of paper.

While this increases the difficulty of characterizing paper fractions according to material type, the wide range of literature on paper types provides sufficient background to supply missing data for full chemical and physical characterization of paper wastes. The results of literature research indicate that most of the paper waste stream in MSW could be adequately characterized. This makes it possible to use substrate composition approaches when useful or necessary, i.e., in calculating water balances or output gas component balances. A similar approach allows characterization of other MSW fractions for analytical consistency.

Solid waste digestion proceeds with hydrolysis by cellulolytic organisms, primarily by the facultative (aerobic/anaerobic), aerobic bacteria and the fungi (Stearns and Ross, 1973).

Decomposition process simulation would thus require determination or collection of aerobic and anaerobic hydrolysis constants for the paper constituents of MSW and for other solid organic materials likely to be present in a waste site.

## CHARACTERISTICS OF FOOD WASTES

Food wastes include leavings from residential food consumption and from commercial (or industrial) sources. Approximations of food content of the landfilled waste stream have ranged from 8.5% (1988 U.S. national estimate, EPA, 1990) to 10.22% (New York City Department of Environmental Protection [NYCDEP], March 1992). Specific food waste characteristics were developed from a variety of sources as listed. The food waste fraction should be considered important to decomposition analyses of waste sites including landfills and composting operations because of its significant biodegradability. Thus, full characterization is useful.

New York Department of Agriculture statistics on estimated per capita food consumption vs. food type include fruit and vegetable types as well as meats. U.S. Department of Agriculture publications also include food consumption and type statistics, food and agricultural materials compositions and estimates of edible and nonedible fractions for the same. These sources can be used to develop fractional percentages of food wastes, weight percentages of total MSW stream and physical and chemical characteristics. Several other sources reviewed corroborated information on food waste types and compositions. A representative waste characterization, based on synthesis of data from these information sources, is provided in Table 3.1.

**TABLE 3.1**
**Food Waste Characteristics**

| MSW Type | % Weight in MSW | % $H_2O$ | % Ash |
|---|---|---|---|
| **FOOD:** Fish (refuse = 44.1%) | 0.04 | 71.34 | 1.66 |
| Beef (refuse = 19.3%) | 0.614 | 53 | –7 |
| Pork (refuse = 12.2%, cooked values) | 0.723 | 42.9 | 1 |
| Lamb (refuse = 17.7%, cooked values) | 0.015 | 46.1 | 1.2 |
| Chicken (refuse = 33.8%) | 2.46 | 57.3 | 0.88 |
| Eggshells | 0.119 | 0 | |
| Coffee grounds | 0.217 | 62.9 | |
| Tea leaves (refuse = 50%) | 0.024 | 86 | 4.92 |
| Bread | 0.45 | 35.3 | 1.1 |
| Fats and oils | 0.21 | 0 | |

*(Continued)*

**TABLE 3.1**
**Continued**

| MSW Type | % Weight in MSW | % $H_2O$ | % Ash |
|---|---|---|---|
| Orange peel (waste = 26% fresh wt., skin, seed in para.) | 0.273 | 72.5 | |
| Tangerine and tangelo (peel) | 0.31 | 73.5 (7.9) | 0.68 (4.3) |
| Grapefruit (peel values and peel = 23%) | 0.047 | 72.5 | 0.6 |
| Apple | 0.266 | 25.2 | 1.46 |
| Avocado (seed values and seed = 34% of fresh weight) | 0.376 | 52 | |
| Banana (skin values and skin = 16–38%) | 0.079 | 88 | 1.24 |
| Cherry and grape | 0.52 | 82 | 88 |
| Peach and nectarine (pits = 7.5%) | 0.134 | 86.9 | |
| Pear | 0.076 | 85.8 | |
| Pineapple (refuse = 23–50% of fresh) | 0.023 | 83.9 | 0.54 |
| Plum | 0.058 | 85.8 | 0.4 |
| Strawberry | | 89.9 | |
| Prune (pits = 5.5%) | 0.01 | 85.7 | |
| Watermelon (rind = 56% and seeds = 3.8%) | | 92 | |
| Cantaloupe and honeydew (rind = 40.5%, fruit values in parenthesis) | 0.512 | 92 | 49.3 (91.4) |
| Vegetables: potato wastes | 0.352 | 91.15 (90.5) | 0.68 (0.5) |
| Broccoli | 0.929 | 11 | 1 |
| Cabbage (refuse = 15%) | 0.118 | 91.5 | 1.22 |
| Carrots (refuse = 20%) | 0.118 | 77.7 | 0.9 |
| Cauliflower | 0.1 | 70.6 | 0.9 |
| Celery (refuse = 20%) | 0.125 | 92.3 | 0.7 |
| Corn (refuse = 61.0%) (cob*) | 0.061 | 75.6 | 0.8 |
| Cucumber and eggplant (refuse = 15%) | 0.276 | 81.1 | 75.4% edible part |
| Garlic and onions (refuse = 10%) | 0.016 | 79 | 0.5 |
| Green beans (cooked & refuse) | 0.14 | 95.3 | 0.5 |
| Green pepper | 0.012 | 92.4 | 0.9 |
| Lettuce and spinach (refuse = 15%) | 0.046 | 94 | |
| Tomato | 0.53 | 94.1 | 92.7 |
| Cooked rice | 0.185 | 72.5 | |
| Other | 0.108 | 94.1 | 0.5 |
| Total | 10.1 | | |

## CHARACTERISTICS OF YARD WASTES

Yard wastes compose a significant fraction of the MSW stream. Percentages found include 2.29% weight (82% grass, 18% brush) for New York City (NYCDEP, March 1992), 20% (uncharacterized) (USEPA, 1990), 17% (ANSWERS [NYS

Capital Region Wasteshed] Solid Waste Plan, 1990), 11.8% (Brookhaven Solid Waste Management Plan, 1989) and 14.5% (uncharacterized, combination average of several states, including NYS; CalRecovery, Inc., 1993). Of the above percentages, the least likely is the low 2.29% for New York City, which might reflect a more urbanized environment and the national estimate of 20%, which is likely to be too high. The percentages reported for the local region (17%), including the New York cities of Troy, Schenectady and Albany, are not statistically different from the value reported by CalRecovery, Inc. (1993). Thus, a 14.5% weight yard waste content of MSW could be considered representative.

While an important biodegradable fraction of MSW, the yard waste fraction is often poorly characterized, if at all, as noted above. One of the few detailed characterizations of yard wastes, widely used in subsequent literature, was included in a complete MSW stream characterization by Niessen (1978). Included was the following information from proximate and ultimate analyses of the trees, wood, brush and plants content of the MSW stream.

As is obvious, percentages of the yard waste fraction were not provided by the Niessen (1978) study. However, the more recent Brookhaven (NY) Solid Waste Management study (March 1989) indicated that, of the 11.8% weight yard waste fraction delivered during the waste stream analysis period, 2.2% was leaves, 2.4% was grass and 7.2% was brush. For analytical purposes the percentage distribution is considered sufficient, because this state region could be conservatively representative of the yard waste stream. Percentage distribution of the yard waste fraction was estimated as shown below:

| Yard Waste Type | % Wt. in MSW | % Total | If 14.5% of MSW |
|---|---|---|---|
| Leaves | 2.2 | 18.64 | 2.70 |
| Grass | 2.4 | 20.3 | 2.95 |
| Brush | 7.2 | 61 | 8.85 |
| **Total** | **11.8** | **100** | **100% of 14.5%** |

The yard waste percentages above, as fractions of the 14.5% average content of MSW, were used in the analyses for this book. Other proximate and chemical analyses values for typical yard wastes and reference sources are listed below (ASME, 1987). It should be further noted that the yard waste values are based on U.S. waste disposal statistics and require modification for both overall waste stream percentage and makeup for non-USA locations.

## CHARACTERISTICS OF PLASTICS WASTES

Plastics are a minor but important weight fraction of landfilled solid wastes. Importance to bioreactor processes include (1) they protect other wastes against wetting, (2) they trap air pockets and occupy a larger volume than their mass would indicate, (3) they are the most hydrophobic waste fraction and (4) subsequently are often the slowest to degrade.

**TABLE 3.2**
**Proximate Analyses and Physical Characteristics of Yard Waste Fractions**

| MSW Type | % $H_2O$ as Received | % Ash | Ultimate Analysis: % Element in Dry Wt. of Yard Wastes | | | | |
|---|---|---|---|---|---|---|---|
| | | | C | H | O | N | S |
| **YARD**: Evergreen shrubs | 69.0 | 2.61 | 48.51 | 6.54 | 40.44 | 1.71 | 0.19 |
| Balsam spruce | 74.35 | 3.18 | 53.3 | 6.66 | 35.17 | 1.49 | 0.2 |
| Flowering plants | 53.94 | 5.09 | 46.65 | 6.61 | 40.18 | 1.21 | 0.26 |
| Lawn grass | 75.24 | 6.55 | 46.18 | 5.96 | 36.43 | 4.46 | 0.42 |
| Lawn grass | 65.0 | 6.75 | 43.33 | 6.04 | 41.68 | 2.15 | 0.05 |
| Leaves | 9.97 | 4.25 | 52.15 | 6.11 | 30.34 | 6.99 | 0.16 |
| Leaves | 50.0 | 8.2 | 40.5 | 5.95 | 45.10 | 0.20 | 0.05 |
| Wood and bark | 20.0 | 1.0 | 50.46 | 5.97 | 42.37 | 0.15 | 0.05 |
| Brush | 40.0 | 8.33 | 42.52 | 5.9 | 41.2 | 2.0 | 0.05 |
| Mixed green waste | 62.0 | 13.0 | 40.31 | 5.64 | 39.0 | 2.0 | 0.05 |
| Grass, dirt and leaves | 21–62 | 30.08 | 36.20 | 4.75 | 26.61 | 2.10 | 0.26 |
| TOTAL % MSW WEIGHT | 14.5 | | | | | | |

**TABLE 3.3**
**Yard Waste Analyses — Ultimate Analyses**

| Waste | % $H_2O$ as Received | % Ash | Ultimate Analysis: % Element in Dry Wt. of Wastes (ASME, 1987) | | | | |
|---|---|---|---|---|---|---|---|
| | | | C | H | O | N | S |
| Grass, lawn clippings | 75.24 | 6.55 | 48.18 | 5.96 | 36.43 | 4.46 | 0.42 |
| Plants, flower garden | 53.94 | 2.34 | 46.65 | 6.61 | 40.18 | 1.21 | 0.26 |
| Shrubs, evergreen cuttings | 69.00 | 0.81 | 48.51 | 6.54 | 40.44 | 1.71 | 0.19 |
| Lawn grass | 65.0 | 6.75 | 43.33 | 6.04 | 41.68 | 2.15 | 0.05 |
| Leaves | 9.97 | 4.25 | 52.15 | 6.11 | 30.34 | 6.99 | 0.16 |
| Leaves | 50.0 | 8.2 | 40.5 | 5.95 | 45.10 | 0.20 | 0.05 |
| Wood and bark | 20.0 | 1.0 | 50.46 | 5.97 | 42.37 | 0.15 | 0.05 |
| Brush | 40.0 | 8.33 | 42.52 | 5.9 | 41.2 | 2.0 | 0.05 |
| Mixed green waste | 62.0 | 13.0 | 40.31 | 5.64 | 39.0 | 2.0 | 0.05 |
| Grass, dirt and leaves | 21–62 | 30.08 | 36.20 | 4.75 | 26.61 | 2.10 | 0.26 |

The range of plastics found disposed in the waste stream include low-density polyethylene (LDPE) used for grocery and garbage bags and film, high-density polyethylene (HDPE), polypropylene (PP), polystyrene (PS), polyvinyl chloride (PVC), polyethylene terephthalate (PET) used for beverage containers, polyurethanes used in sheets and cushioning foams, nylons, polyesters, acrylonitrile butadiene styrene (ABS), polyester used in films and textiles, nylon, phenolic resins and acrylics. Typical weight percentages of plastics in the United States waste stream (Franklin Associates, Inc., 1988) are:

| | |
|---|---|
| Low-density polyethylene (LDPE) | 2.21% |
| High-density polyethylene (HDPE) | 1.73% |
| Polyproplyene (PP) | 1.32% |
| Polystyrene (PS) | 1.32% |
| Polyvinylchloride (PVC) | 0.55% |
| Polyethylene therephthalate (PET) | 0.4 % |
| Polyurethane, foams (PU) | 0.41% |
| Acrylonitrily butadiene styrene (ABS) | 0.22% |
| Polyester (P) | 0.04% |
| Nylon | 0.04% |
| Phenolic resin | 0.03% |
| Acrylic resin | 0.01% |
| Total, wt. fraction: | 8.7 % |

Plastics in MSW represent more space than their weight indicates because they are mostly light, ranging in specific gravity from 0.01 to 0.04 for urethane foams to 1.35 for polyvinyl chloride (Modern Plastics Encylopaedia, 1988). Plastics not normally listed in residential waste stream composition studies, but which appear in landfills and dumps, include:

- Auto tires
- Shoe sole rubbers
- Natural rubbers
- Synthetic rubber articles

Rugs and carpets, except for the more expensive types (treated wool, cotton or other fibers), are likely to contain or be made of the plastics used for manufacturing textiles, i.e., the polyesters, nylons, polyethylenes and blends of the same.

The above definitions of the plastic waste stream complicate characterization according to plastic type. Though plastics can generally be characterized by type, from the various proximate and ultimate analyses published in literature, *the difficulty is determining which plastic corresponds to a waste source*. For some waste sources, this was possible to a reasonable extent. For example, milk or soft drink bottles are indicated in literature to be mainly polyethylene terephthalate (PET); bags, sacks and plastic film indicated to be low-density polyethylene (LDPE); and rubber tire material indicated to be acrylonitrile-butadiene-styrene (ABS) or vulcanized rubber.

The modeler can make the decision to use a *most likely* assumption of the type of plastic making up the MSW plastic fraction when direct information is not available. For example, it could be assumed that plastic water bottles in MSW could be polypropylene. For decomposition, heat and moisture behavior analyses, the similarities of chemical and physical properties among plastics, with the exception of the nylons, polyesters and urethanes, make it unlikely that this approach to plastics waste stream characterization should incur great error.

Characterizations of plastic types are listed in the Table 3.4.

## PLASTICS DETERIORATION IN WASTE SITES

Plastics disappearance is often too poorly characterized in experimental studies to be analyzed as substrates in a hypothetical reactor. There is a lack of research of plastics decomposition under the conditions typical of waste sites. For this reason, plastics degradation in waste sites, with the landfill as representative, merits a detailed discussion.

In waste decomposition studies the deterioration of ordinary plastics is often considered insignificant because of their resistance to biological and chemical attack. In systems such as landfills the influence exerted on overall decomposition might be larger than expected from their relatively minor weight fraction, because disposed plastics have a relatively large volume and surface area compared to weight. Also resistant to water uptake, plastics might limit biodeterioration of other solid wastes. Though likely to be degraded after quite extended soil burial, rates of plastics disappearance in landfills are likely to be slower than most other wastes. Little information on relative rates of deterioration in landfill soils is available as noted.

Degradation of plastics or synthetic polymers in the landfill waste soil environment would be affected by environmental factors causing solubilization or embrittlement. The rates of deterioration would thus depend both upon the likelihood of such factors and the associated lengths of exposure. Environmental factors that could lead to plastics deterioration in landfills include microbial enzyme attack or hydrolysis, temperature cycles, corrosion by chemicals or trace solvents in landfill gas or leachate and deterioration effects of water absorption. All of these pathways involve breakage of secondary or primary chemical bonds between molecules. These bonds vary in strength and accessibility between materials, thus soil deterioration of plastics could be quite material-specific. Changes that occur include reduction (or increase) in molecular weight as a result of random molecular chain scission (or cross-linking). For example, deterioration of condensation polymers such as PE and PP involves random chain scission, resulting in hydrolysis (Hawkins, 1972). Ratings of various plastics for stability upon exposure to environmental factors have shown that nearly all of the plastics listed above have good to excellent stability in the presence of moisture, ozone and temperatures normal to a landfill. Exceptions to ozone susceptibility are some rubbers; and the stability of some phenol-formaldehyde thermosetting resins are susceptible to moisture (Hawkins, 1972).

**TABLE 3.4**
**MSW Plastic Waste Characterization**

| Plastics | % MSW | % WTR | % V | % ASH | S.G. | Density, KG/CU.M | Ultimate Analysis And Dry Weight | | | | |
|---|---|---|---|---|---|---|---|---|---|---|---|
| | | | | | | | C | H | O | N | S |
| Low-density polyethylene (LDPE) | 2.21 | 0.2 | | 1.19 | 0.91–0.925 | 915 | 84.54 | 14.18 | 0.00 | 0.06 | 0.03 |
| High-density polyethylene (HDPE) | 1.73 | 0.2 | | 1.19 | 0.95 | 950 | 84.54 | 14.18 | 0.00 | 0.06 | 0.03 |
| Polypropylene (PPE) | 1.32 | 0.2 | | | 0.90 | 905 | | | | | |
| Polystyrene (PS) | 1.316 | 0.2 | | 0.45 | 1.055 | 160 | 87.10 | 8.45 | 3.96 | 0.21 | 0.02 |
| Polyvinyl Chloride (PVC) | 0.55 | 0.2 | | 2.06 | 1.35 | 1350 | 45.14 | 5.61 | 1.56 | 0.08 | 0.14 |
| Polyethylene terephthalate (PET) | 0.203 | 0.2 | | | 1.33 | 1330 | | | | | |
| PBT | 0.20 | 0.2 | | | 1.33 | 1330 | | | | | |
| Polyurethane | 0.40 | 0.2 | | 4.38 | 0.95 | 200 | 63.27 | 6.26 | 17.65 | 5.99 | 0.02 |
| Acrylonitrile-butadiene-styrene (ABS) | 0.22 | 1.2 | | | 1.035 | 1080 (70 foam) | | | | | |
| Unsaturated polyester | 0.04 | | | | 1.34 | 1150 | | | | | |
| Nylon | 0.04 | | | | 1.16 | 1120 | | | | | |
| Urethane foam | 0.04 | | | 4.38 | 1.11 | 150 | 63.27 | 6.26 | 17.65 | 5.99 | 0.02 |
| Phenolics | 0.03 | | | | 1.41 | 1310 | | | | | |
| Acrylics | 0.01 | | | | 1.145 | 1180 | | | | | |

## CHEMICAL DETERIORATION OF PLASTICS

Chemical reactants normally present in the landfill environment that can attack polymers include oxygen, water, acids, bases, solvents and nonsolvents. For optimal deterioration these reactants must penetrate or diffuse into the material. This is obviously more likely with reactants that are more available, such as the trace gases and leachate compounds and with exposure of thinner plastic materials. Water immersion at ambient temperatures has been shown to cause serious swelling and softening of plastics such as cellulosic esters and nylons. The presence of acids or acidic products can promote hydrolysis under these conditions, and polyurethane foams have also been found to seriously deteriorate after 16 months of immersion at ambient temperatures (Hawkins, 1972).

Hawkins (1972) notes that hydrolysis of polymers and their low molecular weight analogs can be initiated by trace acids or bases; for example, some reactants found in industrial smogs can cause plastics deterioration. Nitrogen oxide, nitric acid, sulfuric acid and alkalis have been known to erode polyethylene, polystyrene, polybutenes and phenolic resins. These reactants are unlikely in the landfill or waste site environment.

## BIOLOGICAL DETERIORATION OF PLASTICS

Biological deterioration of plastics, of importance to deterioration considerations, is likely to be quite material-specific. Although most synthetic plastics are considered organic, they are quite resistant to biological attack. The polyethylenes, polyamides, polyethylene terephthalates, polypropylenes and polystyrenes have been shown to be resistant to microbial attack under soil burial (Hawkins, 1972). Plasticized polyvinyl chlorides were all attacked by soil bacteria in soil burial studies (Booth and Robb, 1968), although PVC resins were not (Coscarelli, 1964). However polyurethanes, depending upon formulation, were attacked by soil organisms, with the flexible forms more susceptible (Hawkins, 1972).

There is wide variation in susceptibility to deterioration between different types of plastics. Colin et al. (1981) concluded, based upon studies of 32-month garden soil burial of films of 25 $\mu$m thick polypropylene (PP), 110 m low-density polyethylene (LDPE), 25 $\mu$m high-density polyethylene (HDPE), 25 $\mu$m nylon 6.6 and 25 $\mu$m polyethylene terephthalate (PET), under tropical conditions (29°C, 85% humidity), that the buried plastics can be ranked as follows according to sensitivity to degradation (in soil): polyester $=$ propylene $<$ LDPE $=$ HDPE $<$ nylon 6.6. Obviously from the list, nylon 6.6 would be the most degradable in soil and PP the least degradable. The authors tested for the possible sources of degradation and concluded that there was little evidence of the surface deterioration, decomposition products or surface characteristics that would be associated with microbial attack and the deterioration that occurred could be due to oxidative attack in the soil. PE and PP were concluded to be highly resistant to deterioration in soil. Pirt (1980) also noted that, although water uptake varies between plastics, a characteristic of most is that they are insoluble in water, which presents a formidable barrier to their biodegradation. Wessel's (1964) review of a variety of plastic textiles and other physical forms in shake flasks, soil burial and other methods indicated that a wide range of plastics are resistant to biodegradation under soil burial conditions. It can be shown, however, that plastics in soil take extensive time to show conclusive evidence of biodeterioration.

## Effect of Physical Structure of Plastic on Degradability

In terms of physical structure of synthetic plastics, degree of microbial attack may be an effect of particle or macromolecule size rather than intrinsic resistance (Clesceri, 1996). This is confirmed in some respect by Nickerson's (1969) study of pitting and penetration of the surface of natural rubber and synthetic polyisoprene rubber by a "black, yeast-like fungus and pink bacterium," which indicated that particle size was the limiting factor in pitting. Pirt's (1980) review of the biodegradability of plastics and the results of other investigators concluded that various aerobic soil bacteria could degrade synthetic plastics especially if maximum size division was possible (small particle size and using the chemostat approach organisms could develop both anaerobically and aerobically, which could degrade certain plastics after 1000 generations of acclimation to plastic substrates). The capacity of the methane-oxidizing bacteria (methylotrophs) was suggested as a possibility for co-oxidation of plastics. The work of Colin et al. (1981) — concluding slow aerobic degradation of most common plastics in film form was likely but could be either chemical (oxidative) or biological — used thicker films than would be found in the MSW stream, which might reduce biological deterioration evidence. Karlsson et al. (1987) showed that soil biodeterioration increases as plastics are increased in area (i.e., reduction in particle size).

## Organisms Involved in Plastics Biodegradation

For plastics buried in soil, some information exists on rates of weight loss (hydrolysis) because this has been one method of testing the biodegradability of synthetic polymer formulations for some time. The organisms involved are typically bacteria and fungi. The former require liquid water and the latter humidity for chemical reactions involved in enzyme expression, growth and reproduction; thus, in tests these conditions are typically optimized. Growth is typically on the surface of the material, thus loss can be measured in weight loss per unit surface area.

Though fungi can grow on the surface of plastics, this growth is often superficial and associated with the presence of nonplastic matter. Fungi, however, have been shown to attack plasticizers in plastics, as is the case for bacteria. Fungal hyphae have also been shown to be capable of penetrating various plastics and elastomers used in electric sheathing (Blake, 1950). While equipped to grow in more adverse conditions than bacteria — for example, the higher fungi can tolerate drier conditions — the filamentous fungi need an atmosphere with sufficient oxygen and are most effective under warm to mild temperature conditions. Given the low water absorption of plastics in soils, it is likely that most bacterial attacks of the kinds important in landfills would only occur in plastics fractions in intimate contact with moist soil or leachate.

Soil-burrowing rodents and soil insects are also known to be agents of plastic particulization, with mice, gophers, squirrels, rats, termites and beetles associated with these activities — soft, malleable plastics are most vulnerable and rigid plastics are attacked only at the edges. Because these activities would have the effect of increasing the plastics surface and reducing particle size, it could be argued that other degradative processes in landfills could be hastened thereby.

## VARIATION OF DEGRADATION WITH PLASTIC TYPE

The chemical structure of plastics is known to be a factor in relative microbial deterioration. Molecular weight, for instance, correlates with microbial attack resistance and lower weights of plastics are more susceptible. This has been shown to be the case for polyurethanes, polyesters and polyethylenes.

The polyurethanes have for some time been shown to be susceptible to microbial attack in soil or in culture, with both commercial foams and linings attacked and with the degree dependent upon chemical formulation. Work by Darby and Kaplan (1968) on the susceptibility of over 100 polyurethanes to biodegradation in culture by mixed fungi *Aspergillus niger, A. flavus, A. versicolor, Penicillium funiculosum, Pullularia pullulans, Trichoderma* sp. and *Chaetomium globusum* showed that polyurethanes moderately or slightly attacked were those formulated from low molecular weight alkane diols and higher molecular weight polypropylene glycols. Crum et al. (1967) reported deterioration of polyurethane linings in culture when exposed singly or in combination to the bacterium *Pseudomonas aeruginosa* and to the fungus *Cladosporium resinae*, with the linings losing integrity in 3 to 4 weeks.

Fields et al. (1974) reported culture dish losses in $mg/cm^2$ for polyesters of 16 mg for the lowest molecular weight and 4 mg for the highest after 42 days of incubation. It was also found that higher molecular weight polyester polymer containers buried in soil had lost 95% of weight after one year of soil burial. Diamond et al. (1975) found that, after one month of soil burial, in five fungal *(Aspergillus)* species the weight losses of ten polyester films averaged 18%; and these losses did not correlate with fungal growth or molecular weight. It was also found that laboratory sample showed weight losses 2.5 times that of the sample buried in soil.

Studies of the polyethylenes, which are olefin polymers and can be the largest single MSW plastics fraction, also indicated microorganism attack variation with molecular weight. For instance, Watkins (1950) found the ability of polyethylenes to support fungal *(Aspergillus, Penicillium and Trichoderma* spp.) growth, with molecular weights above 10,000 not supporting growth. Jen-Hao and Schwartz (1961) found only molecular weights below 4800 stimulated significant bacterial growth *(Pseudomonas, Nocardia and Brevibacterium* spp.) on polyethylene, with growth at the higher molecular weights due to the presence of low molecular weight oligomers present in the plastics. Potts et al. (1972) showed that both high-density and low-density polyethylene (HDPE, LDPE) supported mixed fungal *(Aspergillus niger, A. flavus, Chaetomium globusum and Penicillium funiculosum* species) growth, at lower molecular weight, with the susceptible molecular weights approximately the same for both PE types. Commercial LDPE (household) wrap supported medium growth of mixed fungi of 30 to 60% surface coverage. One of the few discussions of actual rates of PE disappearance in soil indicated relatively slow decomposition. Albertsson and Ranby

(1975) found that incubation of LDPE film in composted garbage for 2 years resulted in a 0.1% weight loss (2000 years for disappearance); and under optimum conditions weight loss was 0.001 to 0.1% per month (83.3 to 8330 years for disappearance).

Both vulcanized and natural rubbers have been known to be susceptible to attack by fungi *(Aspergillus, Penicillium)* and bacteria (actinomyces, *Bacillus, Pseudomonas* and *Thiobacillus*) (Hawkins, 1972). Tsuchii et al. (1985) found that the bacterium *Nocardia* sp. was able to degrade natural rubber (latexes) up to 100% after eight weeks of incubation (0.154 years); and Nette et al. (1959) found that the actinomycete *Proactinomyes ruber* could degrade natural rubber by 25.8% in 45 days (2.093/year).

The synthetic rubbers are usually more resistant to microbial attack. These rubbers can contain emulsifiers, stabilizers, coagulants, oil, or carbon black (Zyska, 1988). Of compounds contained in synthetic and vulcanized rubbers, Zyska (1988) reports that carbon blacks, nonblack fillers and coloring materials are resistant to fungal attack except when in contact with edible materials; the vulcanizing agent sulfur appears slightly fungitoxic; and most rubber plasticizers, softeners and extenders are fungicidal or semiresistant. Rates reported by Zyska (1988) include soil burial weight losses of 40% after 91 days for natural vulcanized rubber without carbon black (1.6/year) and 15% weight loss for natural vulcanized rubbers with a carbon black content of 45 ppm (0.60/year). Carbon black content protected against fungal growth. Williams (1986) reported that vulcanized natural rubber lost 0.5 to 4.5 of its weight after six months of soil burial (44.8 to 99.3 years for disappearance). Kwiatkowska et al. (1980) also reported a decrease of 90% of weight in a three-month soil burial test for thin sheets (0.07 mm) of material (3.65/year). This could also indicate relative surface-to-weight effects — i.e., the higher the exposed surface-to-weight ratio, the more rapidly such synthetic rubber formulations would be degraded in landfill soils. Under anaerobic conditions, only butyl rubber, of neoprene, butyl and silicone rubbers, was affected by sulfate-reducing bacteria (Coscarelli, 1964).

Many natural polymers (rubbers, cellulosics, lignin, chitin, keratin, starch, pectin and polysaccharides), however, are attacked by biological organisms, especially under aerobic conditions (Steinberg, 1961). These include bacteria, fungi, insects and rodents. Keratin of hair, feathers, claws and horns is decomposed by fungi, which decompose chitin as well.

## Effect of Plastics Biodegradability Test Method on Published Results

Because of the long times required for evidence of degradation of most commercial plastics, some investigators have concluded that plastics biodegradation could be insignificant, even when based upon sophisticated tests for evidence. Others, notably Albertsson (1990), have indicated that quite long times, on the order of several years, are required to show breakdown in soil for plastics in commercial form (ground plastics are usually more susceptible). Nearly all studies in soil are thus relatively brief compared with the slow rates of biodegradation.

Jones et al. (1974) developed methods for testing plastics (PP, PE and PS) based upon a variation of the Warburg biodegradability test. Prepared plastics (milled to under 200 mesh, air oxidation or irradiated) were placed in samples of garden, forest and landfill soils kept at optimum moisture content and 20°C for 6 months. It was shown that forest soil was most bioactive, while the landfill soil was least bioactive, corresponding to relative carbon content of the soil. In contrast to other investigators, it was found that polypropylene (PP) was degradable, more so in landfill soils than in forest or garden soils. It was concluded this was due to focus of some other studies upon degradation by fungi rather than bacteria. Testing with sewage sludge samples showed the plastics degraded more rapidly than in other soils, confirming biodegradation, with polystyrene (PS) showing the slowest response or no response. The expected rate for complete biodegradation of photodegraded (fragmented) PP in natural soils was estimated at about four years. From the plots published by Jones et al. (1974), it can be shown that the rates for biodegradation of the PE and PP samples in garden soils might actually be on the order of 11.3 and 8.0 years, respectively (net oxygen uptake vs. time [days] taken from the curve produces statistical values of $a$ and $b$ for the characteristic curve $y = a + b \ln x$, which has first and second derivatives $f' = b/x$ and $f'' = -b/x^2$; and if a value of $f''$ is taken such that it is very small, i.e., 0.002, the time $t$ can be shown to be 11.3 and 8.0 years, where a, b for PP = 25.57 and 4.67; and a, b, for PE = 50.64 and 9.39).

PS degradation was further tested using sewage sludge and garden soil (Guillet et al., 1974) and it was shown that biodegradation occurred; but rates were too slow for validation of the testing methods developed. These rates showed a decline with time, toward a constant value. Guillet et al. estimated the rates for complete biodegradation of PS as 20 to 80 years in garden soil and 11 to 24 years in activated sludge.

The issue of length of testing is thus quite important to consideration of rates of landfill conversion of plastics. The validity of conclusions reported on plastics tested by soil burial often depends upon the length of test. Albertsson (1988) conducted a 10-year aerated, humidified soil burial test showing conclusively, based upon analyses of relative and cumulative $CO_2$ production over that time, that film plastics were biodegraded. It was also concluded that even 10 years is a short time to be shown because of the relative inertness of the material; and the rate is increased by relative exposure to ultraviolet (UV) radiation. PE exposed for 7 days to UV light lost 0.3% of weight as $CO_2$ over 10 years (implying 90 years for conversion). Maximum rates of loss averaged 0.02 to 0.05% per year (indicating 2000 to 5000 years for disappearance). Albertsson and Karlsson (1990) concluded in a later paper that all of the plastics biodegradation curves were similar, showing three distinct stages of behavior; and at 10 years it is only possible to see small indications of a coming mineralization point.

Measurement of $CO_2$ evolved from the plastic material undergoing degradation has been indicated to be a relatively reliable method of quantifying both the occurrence of biodegradation and the amount, as discussed in the work of Albertsson (1988), Albertsson and Karlsson (1990) and Yabannavar and Bartha (1994). Yabannavar and Bartha (1994) showed — by measurement of evolved $CO_2$ in a test with 3-month (160-day) laboratory soil cultures with buried plastic film samples kept at optimum pH, moisture and aerobic conditions for biodegradation — that $CO_2$ readings were all above the "5% above soil background" limit considered evidence of biodegradation.

Weight losses of the plastics ranged from 1.2% to 13.1% for three types of PE plastic garbage bag materials, one PE shopping bag plastic and two PP and PVC food wrap plastics. These weight losses could not all be concluded to be associated with PE, PP and PVC degradations, however, since degradation of plasticizers and fillers in the plastics, shown to be degraded, could have been the source of weight loss. Molecular weights of most plastics, including PE and PVC, did not change significantly. Such a change is considered a truer measure of biodegradability. It was concluded that, while actual weight loss was significant, under the American Society for Testing of Materials (ASTM) new testing procedures for plastics biodegradability, the films would not be considered biodegradable because the molecular weights of the plastics were not proven to have significantly changed. It was noted, however, that burial of these plastic films for an additional 12 weeks resulted in additional decreases in PE garbage bag film molecular weights.

One of the few studies validating the long term-nature of biodegradation by Otake et al. (1995) concluded that even the work of Albertsson et al. (1988) was too short to estimate the practical biodegradation life of LDPE and that accelerated biodegradability tests are impossible to apply in such circumstances. These authors used excavated PE, PS, PVC and UF resin samples that had been buried for more than 32 years. LDPE films showed whitening, pitting and gouging where in contact with soil, which was conclusively identified as characteristic of metabolic action of fungi and bacteria. PS, PVC and UF resin, however, showed little or no evidence of degradation. These authors did not attempt to quantify the weight loss of the material.

## Effect of Air or Oxygen Content on Plastics Degradation

Many studies of plastics exposures and rates reported, however, have been for aerobic or oxygen-available conditions. Oxidation of plastics can occur at temperatures below combustion — for example, the attack of ozone on unsaturated polymers. Limited oxygen conditions may limit or enhance the degree of attack by various reactants, thus discussion of landfill deterioration has to be cognizant of the importance of anaerobic conditions to relative deterioration rates. Gould's (1988) discussion of control of microbial growth through the exclusion of air states that, while air contains about 21% oxygen ($O_2$), its solubility in water (which would control oxygen-dependent growth rates) is only about 5 $\mu$g/ml; and the half-saturation constant for various microorganisms ranges from 500 to 25 times (0.01 to 0.2 $\mu$g/ml) less than the levels in air-saturated water. Thus, inhibition of $O_2$-dependent growth would occur only when oxygen levels are reduced to a few percent or less of saturation levels. However, Shaw and Nicol (1969) reported that: "many molds and *Pseudomonas* spp. can still grow in levels of $O_2$ as low as 1%."

Hawkins (1972) notes, however, that for bacteria the amount of oxygen present only serves to limit the kinds of bacteria that can develop, with nitrate and sulfate becoming the electron acceptors for nitrate- and sulfate-reducing bacteria, respectively, rather than oxygen. Gould (1988) also notes that for facultative anaerobes, cell yield can be 2 to 6 times less when oxygen is absent; the presence of alternate electron acceptors can allow strict aerobes to grow and increase the yield, thus the deterioration potential of facultative anaerobes. Gould's discussion of the food-preservative effect of raised levels of carbon dioxide ($CO_2$) can have some relevance to the conditions inside landfills, because the elevated levels of $CO_2$ compared to air may have reduction effects on the plastic deterioration rates found for normal soil burials. Sulfate-reducing bacteria can grow on complex organic materials and abet steel corrosion, thus it is possible they may have a role in plastics deterioration in landfills. Scott (1992), however, states that plastics are essentially inert in landfills.

### Plastics Deterioration Rates

It is useful to summarize the rates reported in the literature reviewed or estimated for this work as a background to simulation of plastics disappearance rates in landfills. While the accuracy of application to waste site conditions cannot be determined at this point, these rates are useful to full consideration of the reactor concept. The rates are listed in Table 3.5.

## LANDFILL LEACHATE AND LANDFILL GAS CHARACTERISTICS

The chemical quality of landfill gas and leachate can influence whether biological treatment can be effective *in situ* or whether a combination of treatments, at the site or off-site, is needed. Data on gas and leachate characteristics is thus basic to establishing controls. Regulatory standards for landfill leachate release off-site or to potable groundwater have been established throughout the U.S. and in many other countries. Often effluent collection and treatment at wastewater plants for leachate and landfill gas collection and combustion permits achievement of these standards. For inactive landfills and dumps where there is no leachate collection system, the potential groundwater or surface water impacts of seepage, runoff and air pollutant impacts are not less than at controlled landfills because the same decomposition processes apply — although for uncontrolled sites there may be less funding of treatment or control. Large composting sites where seepage and runoff have been collected have also shown the capacity to generate leachates rich in decomposition products, including phenol, for which water quality standards have long been established. The growing experience with landfill gas collection for energy recovery for heating and power has been often focused toward sites generating enough (anaerobic) methane to make collection economically viable — for instance, sites with an estimated over one million tons of wastes in place. U.S. air quality regulations for landfill gas control limits requirements for gas flaring or energy recovery to sites generating over 50 Mg of volatile organic compounds (VOCs) per year, based upon the statistically developed argument that these relatively large landfills are the major source of air quality impacts. However, the vast number of dumps and landfills throughout the world often comprise a larger number of medium-to-small than large sites. All of these sites pose management problems, particularly at a local level, when the need for environmental rehabilitation arises.

Modeling of these sites would require adequate representation of typical waste site effluents such as gas and leachate. This might require original testing; or if sufficient

**TABLE 3.5**
**Plastics Degradation Rates**

| Material | Rate as Reported | Degradation/Year | Test Conditions | Reference |
|---|---|---|---|---|
| Polyurethane linings | 3–4 week deterioration | NA | Lab culture, fungi, bacteria | Crum et al., 1967 |
| Polyurethane foam | 16-month serious deterioration | NA | Water immersion | Hawkins, 1972 |
| Urea formaldehyde resin | No significant deterioration in over 32 years | NA | Garden soil burial | Otake et al., 1995 |
| Polyesters | 4–16 mg/sq cm in 42 days | NA | Lab culture | Fields et al., 1974 |
| Polyesters | 95% wt. loss in one year | 0.95/year | Aerobic soil burial | Fields et al., 1974 |
| Polyester films | 18% average loss after 1 month | 2.16/year | Aerobic soil burial | Diamond et al., 1975 |
| LDPE film | 0.1% wt. loss/2 years | 0.0005/year | Composted garbage | Albertsson and Ranby, 1975 |
| LDPE film | Optimum 0.001–0.01/month | 0.00012–0.012/year | Composted garbage | Albertsson and Ranby, 1975 |
| PE films | 0.3% wt./10 years | 0.0003/year | Aerated soil burial | Albertsson and Karlsson, 1988 |
| PE films | Max. wt. loss 0.02–0.05%/year | 0.0002–0.0005/year | Humidified, aerated soil burial | Albertsson and Karlsson, 1988 |
| PE (milled) | NA | 0.0884/year (est. from curve) | Milled PE buried in soil | Jones et al., 1974 |
| Polypropylene (PP) | NA | 0.125/year (est. from curve) | Milled PP buried in soil | Jones et al., 1974 |

| | | | | |
|---|---|---|---|---|
| Polypropylene (PP) | Degrade in 4 years | 0.25/year | Milled PP buried in natural soils | Jones et al., 1974 |
| Polystyrene (PS) film | 20–80 years in garden soil (est.) | 0.0125–0.05/year | Garden soil | Guillet et al., 1974 |
| Polystyrene (PS) film | 11–24 years in activated sludge (est.) | 0.042–0.091/year | Activated sludge | Guillet et al., 1974 |
| Vulcanized natural rubber | 40% in 91 days | 1.604/year | Rubber without carbon black, test conditions NA | Zyska, 1987 (reported) |
| Vulcanized natural rubber | 15% in 91 days | 0.602/year | Rubber with 45 ppm carbon black, test conditions NA | Zyska, 1987 (reported) |
| Vulcanized natural rubber | 95% in 8 weeks | 6.18/year | Culture, conditions not available | Tsuchii et al., 1985 |
| Vulcanized natural rubber | 0.5–4.5% in 6 months | 0.01–0.09/year | Thin 0.07 mm sheet, soil burial | Williams, 1986 |
| Vulcanized natural rubber | 90% loss in 3 months | 3.6/year | Thin sheets, soil burial | Kwiatkowska et al., 1980 |
| Natural rubber films | 25.8% in 45 days | 2.09/year | Soil burial | Nette et al., 1959 |
| Natural rubber films | 100% in 8 weeks | 6.5/year | Fungi, soil burial | Tsuchii et al., 1985 |

data are available, representative chemical component levels — for example, in leachate — may be used. From the biological or chemical treatment point of view, the chemical components of gas or leachates become substrates, the conversion of which can be modeled using kinetic relationships. It is thus useful to establish descriptions of concentrations, properties and biodegradability characteristics of chemical substances found in waste site leachates and gas.

## LANDFILL LEACHATE

Landfill leachate forms through the percolation of water through landfilled wastes. During percolation the water picks up soluble organics from biological and chemical decomposition, dissolves gases and metal salts. Leachate quality varies widely and is very site specific. Concentrations of organics — for instance, volatile acids — are usually high, as are ammonia, suspended solids and several chemical compounds typically elevated above water quality standards. Organic acid content is measured in biological or chemical oxygen demand (BOD or COD), i.e., the amount of oxygen that would be used by microorganisms in converting the substance or the amount of oxygen needed to completely convert organic carbon to $CO_2$ and water, respectively. For wastewaters with organics, a familiar relationship between BOD and COD level is:

$$COD = 1.087\ (BOD) \tag{3.1}$$

For *young* sites — meaning leachates from waste sites of relatively recent operation — leachate CODs have been reported up to 90,000 mg/l and total suspended solids (TSS) up to 1000 mg/l, as compared to ranges of 250–1000 mg/l COD and 250–500 mg/l TSS for municipal (sewage treatment plant) wastewaters (Harris and Gaspar, 1989). Trends detected in leachates indicate that younger leachates are richer in the organic acids typical of the early stages of solid waste decomposition. For waste sites of longer standing (inactive or operating), COD and BOD levels in leachate may be less than the values for young leachates; but they generally exceed levels found in wastewaters, thus cannot be discharged without ground or surface water impact. Below, some typical leachate effluent limits required for granting landfill operation permits, in the left column and in mg/l units, are compared with values from leachate from a large, partially active landfill (Kusterer et al., 1996):

| Allowed Limit | Location | Levels in Landfill Leachate |
|---|---|---|
| BOD: 300 | Montgomery County, MD Landfill | BOD: 1100 to 300 |
| COD: 500 | Montgomery County, MD Landfill | COD: 3500 to 442 |
| TSS: 400 | Montgomery County, MD Landfill | TSS: 230 to 27 |
| TDS: 8000 | Montgomery County, MD Landfill | TDS: 10,000 to 850 |

The COD value of the above leachate indicates quality is dominated by COD output from older rather than younger wastes. Previous works by Pohland, 1975; Leckie et al., 1979; and Barber and Maris, 1984 actually indicate a COD range of 18,000 to 90,000 could be expected from MSW decomposition in simulated or actual landfills. Johansen and Carlson (1976) have also noted that landfill leachates with higher COD values can be typical of landfills where the waste receives little water from the outside, so is undiluted; and that leachate COD can vary with wetness of weather. More concentrated leachates could thus be expected of capped, relatively deep inactive landfills. They might also be expected in areas of the world with low rainfall. Higher concentration of COD in leachates from deeper landfills was confirmed by the studies of Raveh and Avnimelech (1979), Ham and Bookter (1982) and Shibani (1987), who reported that COD concentration doubles with each doubling of MSW depth in a landfill. Poorly capped or uncapped inactive landfills, inversely, could have lower range COD values, as more open conditions could both increase the opportunity for dilution of throughflow and for oxidation of organic substances in leachate. In choosing leachate data to represent a waste site for the purpose of modeling, particular site conditions and climatic regime should thus influence which range of COD or BOD or other chemicals is appropriate.

If the purpose of waste site effluent assessment is for establishing whether the leachate can be managed at the locally available publicly owned treatment works (POTWs), leachate testing is obviously useful, as it allows planning for pretreatment or on-site treatment as alternatives to management at the POTW. While POTWs can handle leachates with lower CODs without great difficulty, high BOD or COD leachate of quantity might stress the treatment capacity. However, a variety of studies have been conducted on POTW treatments of leachates and show insight into leachate quality vs. treatability.

## Leachate Organics

Ghassemi et al. (1983) showed, based upon tests from 11 U.S. landfills, that the organics in highest concentrations in landfill leachates are acetic acid, butyric acid, methylene chloride, 1-1 dichloroethane and trichlorofluoromethane; and that inorganics with highest leachate concentration were iron (Fe), calcium (Ca), magnesium (Mg), cadmium (Cd) and arsenic (As). Johansen and Carlson (1976) reported from studies of leachates from several Norwegian and U.S. landfills that ranges of concentrations of leachate organic acids were:

| | |
|---|---|
| Acetic acids: | 2570–100 mg/l |
| Propionic acids: | 4375–37 mg/l |
| Butyric acids: | 5875–12 mg/l |
| Valeric acids: | 550–112 mg/l |
| Caproic acids: | 600–56 mg/l |

Ghassemi et al. stated that, under dry conditions or with high-strength leachates, organic acids make up most of organics in leachate, with acetic, butyric and propionic acids contributing nearly 90% of all organics (of concern since butyric acid contributes a particularly bad odor). It was also stated that low organics concentration indicates a slow biodegradation rate or a late stage of anaerobic decomposition.

## LEACHATE BOD/COD RATIO AS AN INDICATOR OF BIOLOGICAL TREATABILITY

The organic content of waste site leachate affects the degree of biological treatability. Leachates with higher BOD/COD ratios or higher organics contents should be more treatable (Johansen and Carlson, 1976). This suggests that biological treatment could be more effective for leachates from compacted or relatively dry inactive landfills as compared to leachates from open or ineffectively capped landfills. Johansen and Carlson (1976) have reported that the BOD/COD ratio of landfill leachates varies from 0.6 to 0.2, compared with about 0.50 for municipal wastewater, with the lower ratio associated with decreased degradability. Harris and Gaspar's (1989) study of leachate from 24 landfills indicated an average BOD/COD ratio of 0.317. Ariati et al. (1989) (reported by Avezzu et al., 1995) noted that leachate BOD/COD ratio declines with age of landfills from 0.61 to 0.33. This generally suggests leachates from inactive landfills or leachates from older sites would not be as responsive to wastewater treatment methods; and leachate treatments *in situ* (for instance, via leachate recycle as in the landfill bioreactor or on-site collection and treatment) should consider methods to increase the BOD/COD (or BOD/TOC) ratio. Methods that enhance organic nutrient or biological mass content (leachate recycle, sludge or nutrient addition or aerobic treatment) of leachate should improve BOD/COD ratio and, thus, leachate biodegradability prospects. Avezzu et al. (1995) studied landfill leachates for effect upon activated sludge treatment processes. It was shown that leachate BOD and COD concentration, volatile acids and most leachate metals steadily declined with age. For example, COD declined from 25,000 mg/l at the beginning to 150 – 210 mg/l after 10 to 11 years; $BOD_{20}$ declined from 11,500 mg/l to 120 to 180 mg/l; and $BOD_5$ declined from 7,660 mg/l to 0.0 after 10 to 11 years. It was concluded that, although treatment always resulted in high BOD and COD removals, the leachate metals tended to accumulate in the solid phase of the sludge.

It has also been shown that as landfills age, the organic nitrogen content of leachate increases (Avezzu et al., 1995) and ammonia content decreases. High ammonia content (over 500 ppm), as well as high volatile organic compounds concentration, can inhibit biological growth. As indicated in the discussion of inhibition kinetics, inhibitory substances do not improve decomposition or organic substance conversion rates, as they adversely impact microorganism growth and substrate consumption. There is no indication, however, that older leachates would have inhibitory concentrations of ammonia or volatile organic compounds (though levels may be well above those acceptable for liquid effluent discharges to ground or surface waters). In many cases of leachate treatment as wastewater, ammonia is air-stripped using shallow retention ponds or packed tower scrubbers (Harris and Gaspar, 1989). Wu et al. (1988) also concluded that, while anaerobic process biofilters are effective for treatment of concentrated leachates, without effluent recycle, external pH adjustment or pretreatment; and for metals removal, these biofilters are ineffective in removing ammonia from leachates. These investigators recommended chemical processes or biological nitrification to remove residual BOD and ammonia from leachate. The waste site modeler with the goal of leachate management thus has to consider how to simulate the effect on waste site decomposition and organic chemicals conversion of process modifications such as induced biological nitrification and pH adjustment of recycled leachate.

One important point made by Harmsen (1983), from the view of difficulties of biological treatment of older waste site leachates, was that the older leachates were associated with methanogenic activity; and even though they might contain lower concentrations of fatty organic acids overall, they can have higher concentrations of the higher molecular weight organic compounds. These include humic or fulvic acids, which have relatively low biological treatability. Collection and removal from leachates from these sites might thus involve the need for chemical reduction of levels of these acids.

## Hazardous or Toxic Compounds in Waste Site Leachates

A wide range of organic contaminants considered toxic or priority pollutants have long been routinely identified by a number of leachate studies. Venkataramani and Ahlert (1984) showed that, in addition to raw leachate that had COD levels of 23,000–28,000 mg/l, TOC levels of 8000–10,000 mg/l and pH of 7.5–9.0, landfill leachates could have the following levels of priority pollutant organic compounds:

| | |
|---|---|
| Benzene | 0–1930 $\mu$g/l |
| Chlorobenzenes | 0–4620$\mu$g/l |
| Dichloroethane | 180 $\mu$g/l |
| Methyl chloride | 3.1 $\mu$g/l |
| Tetrachloroethylene | 0–590 $\mu$g/l |
| Toluene | 0–16, 200 $\mu$g/l |
| Trichloroethylene | 0–7700 $\mu$g/l |
| Xylene | 0–3300 $\mu$g/l |

These investigators found that for leachates, mixed aerobic microbial populations had a maximum specific growth rate ($\mu_{max}$) of 0.06 to 0.07 /hr and a half-saturation constant ($K_s$) of 67 mg/l. This was attributed to the toxic nature of the leachate. It was reported by this study that the acclimated organisms could degrade the organic compounds present in the leachate (under aerobic conditions) up to 80%, in the presence or absence of glucose or other nutrients, with the degradation pathway suspected to be cometabolism. The authors indicated that biological treatment could be economical, simple and a primary leachate treatment step, with high biological oxidation potential for hazardous organics; and a stable DOC content of the leachate can indicate the presence of acclimated biological species. The implication of these findings includes that on-site aerobic treatment may be able to address even levels of toxic or hazardous contaminants in the liquid phase of landfill sites, once aerobic conditions and organism acclimatization are established. In this case, modeling of site processes provides opportunity for feedback control.

However, leachate from waste disposal sites can contain significant levels of many other organic substances of health impact concern. Harris and Gaspar's (1989) review of the composition of 24 landfill leachates included average ranges for water pollutants, heavy metals and volatile and semivolatile organic compounds. Results of this review are shown below:

**TABLE 3.6**
**Leachate Composition**

| Chemical Parameter Milligrams/Liter | Average, Mg/l (Harris and Gaspar, 1989) | Range (Venkataramani and Ahlert, 1984) |
|---|---|---|
| pH | 6.65 (5.3–8.2) | 7.5–9.0 |
| Conductivity | | 13–18000 μmhos/cm |
| COD | 7698 (80–36000) | 23–28000 |
| BOD | 2442 (6–24200) | |
| TDS | 6222 (180–28000) | 15–17000 |
| TSS | 248 (74–860) | |
| TOC | 2816 (48–16000) | 8–1000 |
| Sulfate | 416 (15–980) | |
| Ammonia | 322 (13–910) | |
| Potassium | 220 (6–726) | |
| Sodium | 603 (17–2000) | |
| Calcium | 575 (54–2700) | |
| Chloride | 763 (23–2700) | |
| Cyanide | 0.11 (0.005–1.30) | |
| **HEAVY METALS, in mg/l:** | | |
| **Antimony (Sb) (Heavy metals)** | 0.02 | |
| Arsenic (Sb) | 0.17 | |
| Barium (Ba) | 0.52 | |
| Beryllium (Be) | 0.02 | |
| Cadmium (Cd) | 0.08 | |
| Chromium (Cr) | 0.10 | |
| Copper (Cu) | 0.04 | |
| Iron (Fe) | 284 | 4 mg/l |
| Lead (Pb) | 0.15 | |
| Magnesium (Mg) | 241 | 0.5 mg/l |
| Mercury (Hg) | 0.00 | |
| Nickel (Ni) | 0.31 | |
| Selenium (Se) | 0.02 | |
| Silver (Ag) | 0.01 | |
| Thallium (Tl) | 0.04 | |
| Zinc (Zn) | 3.89 | |

(*Continued*)

**TABLE 3.6**
**Continued**

| Chemical Parameter Milligrams/Liter | Average, Mg/l (Harris and Gaspar, 1989) | Range (Venkataramani and Ahlert, 1984) |
|---|---|---|
| **VOLATILES, in mg/l: Acetone ($\mu$g/l)** | 2131 | |
| Benzene ($\mu$g/l) | 15.23 | 0–1930 |
| 2-Butanone (MEK) | 2452 | |
| Carbon disulfide | 3.08 | |
| Chlorobenzene | 6.20 | |
| Chloroethene | 54.7 | |
| Chloroform | 2.70 | |
| 1,1-Dichloroethane | 66.8 | 180 |
| 1,2-Dichloroethane | 3.19 | |
| 1,4-Dichlorobenzene | 3.24 | |
| trans-1,2-Dichloroethene | 63.87 | |
| Ethylbenzene | 63.5 | |
| 2-Hexanone | 61.25 | |
| Methylene chlorode | 552 | |
| 4-Methyl-2-Pentanone | 67.80 | |
| Tetrachloroethene | 8.80 | 0–590 |
| Toluene | 327 | 0–16200 |
| Trichloroethene | 28.10 | 0–490 |
| Trochloroethylene | | 0–7700 |
| Trichlorofluoromethane | 3.80 | |
| 1,1,1-Trichloroethane | 59.70 | |
| Vinyl chloride | 14.87 | |
| Xylenes | 392 | 0–3300 |
| **SEMIVOLATILES, in $\mu g$/l:** | | |
| **Benzoic acid** | 289 | |
| bis (2-Ethylhexylphthalate) | 17.54 | |
| Diethyl phthalate | 59.58 | |
| 2,4-Dimethylphenol | 4.52 | |
| Napthalene | 4.28 | |
| Phenanthrene | 129 | |
| Phenol | 107 | |

# 4 Waste Site Ecology

The rich variety of refuse entering a landfill, waste dump (a term still used in many of the less-industrialized regions of the world) or compost pile can provide an energy-rich environment that is attractive and valuable as a food source to a host of organisms that range in size from large animals to fungi and bacteria. At a compost site, organisms present may be limited to leaf-eating insects (detritovores) with easy access from the background area to the compost heap or that are naturally present in the leaf and vegetation wastes as delivered — flying insects, litter-dwelling arthropods, ubiquitous vegetation litter fungi, bacteria, worms and other soil-dwelling fauna. Organisms at a waste site largely free of solid organic and decomposable wastes such as food or agricultural residuals — for example, a site contaminated by oily or petroleum wastes — are likely to be soil bacteria and fungi that have undergone varying degrees of adaptation to contaminants present, possibly including a capacity to co-metabolize hydrocarbons. The organism population of a biofiltration unit made up of aerobic soil is also likely to be populated by aerobic bacteria and fungi.

Sanitary landfills or refuse dumps can, however, because of their size and the variety of materials present, support the widest variety of living organisms at a nutritional level likely to be competitive with other sources of food for these organisms. Access to food wastes at dumps or landfills can reduce energy expenditures for animals or birds that might otherwise forage elsewhere over wide areas to obtain a similar daily food intake.

The soil surface and subsurface environment of dumps and landfills additionally support a unique ecology, likely to be dominated by species that can thrive in and adapt to this relatively hostile and competitive environment. Adverse factors are likely to include unfavorably high temperatures, unfavorable or fluctuating moisture content, a lack of protection from predation (in soil or above the soil) and unpredictability of access to or quantity of the most desirable food sources. Animals favored by this environment would include those that are aggressive or large in size or numbers, can consume a variety of materials or are opportunistic. Soil microflora (fungi, actinomycetes, bacteria) are less likely to be limited by the adverse factors mentioned, but species present are likely to conform to a particular microenvironment — for instance, methanotrophic bacteria in the upper aerobic soil layer of a landfill and methanogens in anaerobic regions of the same landfill.

For the waste site envisioned as a type of reactor, consideration of waste site ecology is of importance as a means of assessing the potential effect of the above-mentioned organisms. The cumulative effect of these organisms on the reactive content of the waste site includes minor removals and conversion of wastes by animals and birds, soil fauna and microflora. For soil and above-ground animals, the wastes may be ingested and occasional fecal deposits by these animals colonize the site with bacteria; the soil microflora provide initial attack, through hydrolysis processes, on cellulosic and food wastes by fungi and bacteria present in the soil and wastes, setting in motion the initial stage of decomposition. Soil organisms such as larvae of flying insects, beetles, mites and other small animal species are also likely to provide a degree of size reduction of cellulosic or plant wastes and addition of organic matter and fungal or bacterial inoculants to the waste-soil matrix on death.

The overall effect of organism numbers and types at waste sites thus represents an important though under-recognized factor in mass balance of the waste site. Their interaction can substantially influence the rate of waste conversion. Discussion of likely organisms and possible effects can thus improve waste site analyses.

Accurate representation of the effect of all biological organisms at the site would be an extremely laborious task, mainly because of the limited studies of this topic or of the range of species that may be present. Additionally, the ecology, especially in the case of refuse dumps or landfills, is likely to be site-specific. However, most sites within a geographic region could have the same range of species represented, with variation in species numbers between sites influenced by the mix of materials disposed, length of access to food sources and opportunity for invasive organisms. Climate may be more of an influence on species present if sites in different regions are compared.

Open dumps with higher food waste content should naturally attract a wider range of animals and birds than landfills designed and operated with modern standards — for example, the use of soil as daily cover for wastes. However, the regionalization of the modern landfill, which increases the quantity of food materials present at a modern disposal site, can subsequently provide alternative food support for a larger concentration of opportunistic animals.

## INFLUENCE OF THE WASTE SITE ENVIRONMENT ON TYPES OF ORGANISMS PRESENT

The mix of physical factors, materials and human practices at a landfill or dump is likely to present a relatively hostile environment for foraging or soil-dwelling animals. Unfavorable physical factors, especially for small or slow-moving animals, include increased likelihood of predation, perhaps lower oxygen concentration near ground level, the experience of disagreeable odors or volatile chemicals in the air, noise and disturbance from human (site-operation) activities and diurnal or access limitations during foraging or feeding periods.

Soil-dwelling fauna such as worms, mites and beetles additionally prefer moist but aerobic environments and are thus likely to be discouraged by elevated carbon dioxide, methane or temperature, which may occur in the upper horizon of landfill soils. Soil fauna feeding at inactive landfills or dumps are likely to be those that can burrow to nutritive materials and are small enough to survive in near-surface soil macrospores under typical waste site conditions, which can include compaction and atypical moisture inputs and gas composition. Nevertheless, the topmost horizon of a landfill or dump may provide a limited soil environment that is sufficiently aerobic and wet or cool enough to support many of these small animals. Additionally, some rodents may simply set up burrows or nests at the fringe of or near the site, avoiding unpleasant site soil conditions yet remaining near enough to exploit the foraging opportunity.

Larger and faster-moving animals (such as rodents, large birds or animals of size) could have improved advantage at modern waste disposal sites where access to materials is additionally limited by regular soil coverage of wastes.

Modern landfill operation practices, which usually involve the introduction, mixing and heavy compaction of soil covers onto freshly disposed wastes, may substantially reduce the numbers of worms, mites and beetles previously present in the upper soil layer. However, these species are likely to persist, in diminished or fluctuating numbers and in patchy distribution, if access to food sources remains and does not become detrimental.

The availability of food to a species and a sufficiently hospitable microenvironment are thus likely to be key determinants of which organisms are represented in the ecology of a particular waste site.

## SPECIES COMPETITION FOR FOOD AT A WASTE SITE

While there is obviously a wide variety of materials at a typical landfill or dump that animals or microorganisms can use for food, the kinds of materials that can be used by larger animals are relatively limited. Larger organisms prefer to feed on easily accessible, carbon-rich and nutritive organic wastes such as protein-rich meat, seeds, digestible fresh plant material, vegetable and fruit residuals and human foodstuffs derived from these materials. For these organisms, the materials accessed provide nutritional components that are a normal fraction of diet. As noted, the soil fauna such as worms, mites, beetles and arthropods are likely to be attracted to and consume the decaying plant litter. Under conditions of soil decay, these small soil animals also benefit nutritionally from protein provided by the precolonization of plant litter by fungi and bacteria, which, in turn, benefit from the organic matter provided by larger organisms as well as from enzyme decomposition of the wastes normally present. More recalcitrant but degradable organic wastes such as plant leaves and stems, leather and cellulosic wastes are eventually decomposed by soil microorganisms such as fungi and bacteria through hydrolysis-to-methanogenesis, to gases, organic liquids and salts.

Organism competition for available food at a waste site is additionally influenced by diurnal patterns (for instance, foraging when predation is less likely), climatic (rainfall vs. drought, cold vs. warm weather) and other environmental factors.

The specificity of these requirements per organism suggests food and opportunity niches. Interdependence among organisms would thus be expected. How strongly organism interdependence affects the disappearance or conversion of wastes at a site is thus likely to be more simply correlated with total numbers and mass of organisms present than the types or sizes of species present. Relative size of an organism could be less important than overall species mass and metabolic efficiency. In terms of analysis of this effect at a waste site, a straightforward though inexact approach would be to represent species according to their capacity to remove or convert wastes and to limit influence to the likely period of organism involvement in and attraction to the site.

## THE RANGE OF ORGANISMS AT WASTE SITES

The lack of thorough studies of this topic limits information available. Nevertheless, the presence of various biological species and organisms at waste sites and their preferences have been noted by various published sources.

For waste sites where daily disposal has ended or that have been undisturbed for some time, decreasing species size is favored; and smaller organisms, ranging from beetles and worms to soil bacteria, are likely to dominate. Access to food supply in this case will be limited by organism size and soil depth or compaction to be negotiated. For active landfills or open dumps, smaller and soil-dwelling organisms are still favored in terms of waste removal impact. Thus, overall long-term impact of organisms on decomposition and removal of waste should thus vary inversely with organism size.

Most investigators and biological treatment experts, in any case, consider smaller organisms such as soil bacteria and microorganisms to be the most important group in the decomposition process. The waste site is no different in this regard, as suggested by the previous discussion. However, the assumption that larger organisms have no effect is likely to be inaccurate, hence, the presence and activity of all organisms present should be of interest to the waste site analyst.

What organisms are likely to be present at a landfill or dump? At the upper size limits, for active modern landfills or city or rural dumps, species reported range from primates (scavenging humans and apes); black and grizzly bears; foxes; raccoons; rats; birds such as crows, gulls and other marine birds; starlings and other seed-eating birds; various insects and their larvae; worms; mites; ants; soil arthropods and the very small but very important organisms such as protozoa, fungi, yeast and bacteria.

## ORGANISMS FOUND IN COMPOST PILES

Many of these smaller organisms have been reported by studies of waste composting piles. Composting, which is essentially aerobic, contains organic wastes that have not been compacted. Compost piles may thus harbor organism species also present in landfills — though landfills are likely to support a wider range of organisms. Dindal (1990) reported the stratification of organisms found in solid waste composting piles made up of vegetable and fruit peelings, coffee grounds, egg shells, clam and oyster shells, peanut and nut shells, leaves, grass clippings, weeds, garden residues, sawdust, manure, pet wastes, newspapers, meal (soy, cottonseed, bone) and sewage sludge.

The types and trophic or consumer levels of the compost pile organisms reported by Dindal (1990) are listed in Table 4.1. Columns added at right commenting on possible conditions favoring these organisms in waste materials or in landfill soils. Trophic levels additionally describe food supply or predation stratification among the organisms present — e.g., those at the 1-trophic level are preys of organisms at the 2-trophic level and are, in turn, prey of organisms at the 2 to 3 trophic level. Organism size only roughly correlates with trophic level.

**TABLE 4.1**
**Soil Organisms vs. Consumer or Trophic Levels**

| Organism | Size Range, mm | Consumer Level | Conditions for Presence in Wastes | Likelihood in Inactive Waste Sites at Depth |
|---|---|---|---|---|
| Centipedes | 30 | 2–3 | Aerobic, protected | Very Limited |
| Ground (Carabid) Beetles | 8–20 | 2–3 | Aerobic, protected | Limited |
| Rove Beetles | 10 | 2–3 | Litter porosity, aerobic | Limited |
| Ants (Formicid) | 5–10 | 2–3 | Shallow, aerobic | Limited |
| Pseudoscorpions | 1–2 | 2–3 | Litter environment | Unlikely |
| Predatory Mites | 0.5–1.0 | 2–3 | Porous, aerobic soil | Limited |
| Soil Flatworms (Turbellarians) | 70–150 | 2 | Moist, aerobic | Limited |
| Protozoa | 0.01–0.5 | 2 | Unknown | NA |
| Rotifera | 0.1–0.5 | 2 | Unknown | NA |
| Feather-Winged Beetles (Ptilids) | 1–2 | 2 | Upper soil horizon, ground surface litter | Limited |
| Mold Mites (Acarina) | 1.0 | 2 | Upper soil horizon | Limited |
| Beetle Mites | 1.0 | 2 | Upper soil horizon | Limited |
| Springtails | 0.5–3 | 2 | Upper soil horizon | Limited |
| Fly and Larvae (Diptera) | 1–2 | 1 | Shallow soil | Limited |
| Sow bugs | 10 | 1 | Aerobic conditions | Limited |
| Roundworms (Nematodes) | 1.0 | 1 | Mixed, moist conditions | Depth dependent |
| Millipedes | 20–80 | 1 | Aerobic, litter conditions | Limited |
| Earthworms | 50–150 | 1 | Moist, aerobic soil | Limited |
| Land Snails and Slugs | 2–25 | 1 | Aerobic, litter conditions | Limited |
| Beetle Mites | 1.0 | 1 | Aerobic shallow soil | Limited |
| White worms and Pot worms (Enchytraeids) | 10–25 | 1 | Aerobic moist conditions | Limited |
| Actinomycetes | | 1 | Mixed conditions | Present |
| Molds | | 1 | Aerobic conditions, mainly | Present |
| Bacteria | | 1 | All conditions | Present |

## TROPHIC RELATIONS AND ENVIRONMENTAL FACTORS DETERMINING ORGANISMS AT WASTE SITES

In a perfectly functional food supply or food web system, the stratification of organisms should reflect basic food supply and the functional relationships among the organisms. The presence and growth of 1-trophic level organisms attract organisms from the higher trophic levels. Strictly speaking, a low mass gain in the lowest trophic level should also affect mass gain at the higher trophic levels; thus, variation in overall mass of higher trophic organisms should reflect mass variation at the lowest trophic or consumer levels. These functional relationships among organisms develop from their feeding habits or niches and require that (1) the basic food supply is such that feeding habits of the lowest trophic or consumer levels is well supported and (2) access between the trophic levels is established (meaning the various organisms do not encounter severe restrictions on their feeding habits). These two simple requirements can guide consideration of whether a particular type of waste site should attract various types of organisms.

For example, sites with contaminated soils or biofilters are likely to contain enough carbon-nitrogen sources in the form of solid organic debris or dissolved substances to support a sizeable bacteria population; however, the physical characteristics of these two types of sites might discourage the presence of organisms that prey on bacteria. The biofilter might have enough fine organic plant debris to support the turnover of a healthy population of bacteria and fungi that, in turn, can support worms, protozoa and rotifers. However, the latter organisms are aerobic and need a certain level of moisture, oxygen content and porosity to survive. How the biofilter supplies these needs would determine whether a functional food web exists among organisms in this system.

For contaminated soils, the mass of bacterial organisms present depends on the levels of organic materials (plant and vegetation matter) still present after many years of soil development. While the presence of certain organic contaminants might add to the food supply and improve numbers and mass, soil organic content is likely to be the main source. Access to this source, as well as additional organic input, is likely to be restricted to very small organisms that can enter and live in soil pores — this group is likely to be dominated by bacteria. These conditions make for a relatively stable population. However, bacterial numbers are the main determinant of rate of contaminant removal; thus population increase has to be induced by applying nutrients and soil conditions conducive to larger bacterial populations.

In Dindal's (1990) trophic (food web) scheme for the compost pile, energy from food sources flows upward from the lowest trophic level to the highest. At the lowest level, the bacteria, actinomycetes and fungi feed on organic residuals including plant and animal matter. The first trophic level is thus the bacteria and fungi. The fungi are eaten by mites, springtails and feather-winged beetles, while the protozoa and rotifers feed on the bacteria. The mites, springtails, beetles, protozoa and rotifers thus make up the second trophic level. These are in turn eaten by beetles, mites, ants, scorpions and centipedes (with the exception of the protozoa and rotifers), which form the third trophic level. However, another group of lower-level organisms — the worms, bugs, beetle mites, millipedes and flies — do not feed directly on fungi or bacteria. This group feeds instead on the same materials as the bacteria and fungi, i.e., plant and animal matter, and thus, are also members of the first trophic level in a compost pile. The worms, bugs, beetle mites, millipedes and flies (larval) are predated by the second trophic level (beetles, mites, ants, scorpions, centipedes and flatworms). The functionality of the compost system ecology is dependent on both a rich supply of the basic food material (carbon and nitrogen-rich plant and animal matter) and favorable environmental factors, including sufficient water content for survival of the various organisms, sufficient oxygen supply for this aerobic waste decomposition system and sufficient porosity for physical access among organisms.

For the waste site or dump, interactions among organisms are established early and persist, especially through the active stage of waste disposal. For organisms other than the soil microflora, trophic levels among the organisms at landfills, where they exist, are likely to be patchy or incidental. While the heterogeneous food sources might attract a host of organisms, the opportunity to establish fully functional trophic relationships is limited by physical access to these sources. Organism access to food sources in landfills would be affected by limited foraging time for above-soil organisms and environmentally harsh conditions for soil-dwelling fauna, including fluctuating or limited moisture and oxygen content, occasionally unfavorable temperature and soil porosity and a food supply that varies unpredictably with distance and soil depth.

A simplified description of the likely distribution of wastes at the site can suggest why trophic relationships are unlikely to be as developed as at compost sites. The collection and truck disposal of food, paper, plastics and other wastes at landfills initializes the activity of the various organisms. Chemical processes subsequent to the death of plants, vegetables, animal wastes and remains and other materials produce volatiles that naturally attract various insects, both flying and crawling species.

The forbs-chewing insects (moths and their larvae, grasshoppers, worms) are naturally attracted to the high nitrogen water content (1.5 to 9.7% dry weight), (65 to 95% fresh weight) of the leaves of forbs plants (Tabashnik, Slansky, 1987). Relevant examples are cabbages, carrots, melons, onions, potatoes, tomatoes, beans and alfalfa. These insects would be able to identify and be attracted to these types of plant materials where they are accumulated through disposal practices (both leaves and other plant parts, although forbs fruits may be of lower nutritional value). The presence of these types of food wastes at landfills is likely to be limited in industrialized countries. The forbs-chewing insects mentioned are also vulnerable to predation by larger organisms. These animals are also temperature-sensitive and display heat avoidance (Kingsolver, Watt, 1983) and are thus likely to be discouraged by unfavorable temperatures and lack of shade at landfills and dumps. Hence, the presence and effect on disposed wastes of forbs-chewing insects at landfills, despite the presence of an attractive food source, is likely to be minimal, particularly in countries where disposed wastes are covered each day.

Tree leaves and grasses disposed with wastes at landfills, dumps and composting operations can vary from plant species to species, and with leaf age in nitrogen, water and fiber content, but they can attract leaf-feeding insects (foliovores). Young leaves, with more nitrogen and water and less fiber are likely to be more nutritionally valuable to leaf-eating insects, which include worms, butterfly and moth larvae and adults, and ants. Older leaves, such as from fall yard sweepings in temperate climates, are thus likely be a food source for foliovores that can cope with nutritionally poor and fiber-rich leaf wastes (through low growth rates and low food conversion efficiency (Scriber, 1978)). Larger foliovore insects are also better able to extract nutrients from older or more fibrous leaf and grass materials, while smaller foliovores might prefer younger leaf and grass waste materials. Several species of foliovores may be present because leaf and grass wastes disposed and composted can vary from fresh to aged through the year and the foliovore sizes can range from 10 mg to 10 gm (Stevenson, 1985) in a waste accumulation such as yard waste or other composting operation. However, foliovores are, in the main, arboreal in feeding habits, which might suggest that forest-floor-feeding herbivorous insects are more favored at waste sites with disposed leaves and grass. Local temperature at a waste site can also be a complicating factor for foliovores. This group is not the most capable of regulating body temperature in an environment where the temperature is subject to fluctuation (Stevenson, 1985).

Grass typically is nutrient-rich for insects, with young grass leaves having more nitrogen and water and less fiber, thus making young grass more desirable as food for grass-eating insects (leaf beetles, worms, larvae of moths, locusts and grasshoppers), but grass can be used as a food supply for a large variety of insects, some of which specialize in eating seeds of a wide variety of grasses. Because these insects graze, they are prey to bird and other species; thus, their presence associated with grass wastes would be tempered by predation by other animals.

Food and plant wastes, given the right conditions, attract various mite species. Mites have been studied because of their agricultural importance, and *Tetranychus* spp., *Panonychus* spp. and *Oligonchyus* spp. have been associated with plants such as citrus, apple, beans, strawberry, cotton, chestnut, fir, grass, various vegetable plants and flowering plants (Rodriguez and Rodriguez, 1987). Conditions favorable to plant-eating mites include sufficient environmental moisture (but not saturated conditions), as dry conditions of food materials reduce nutrient availability. Mites have the advantage of rapid and successful colonization of wastes through adaptation to fluctuating environmental factors such as temperature and moisture content. Mite species have a short generation time (8 to 30 days) and high reproduction rates in the elevated moisture content (relative humidity) range of 40 to 95% (Rodriguez and Rodriguez, 1987). However, high and low temperatures do not favor these patterns.

Dust mites feed on materials such as cereal-related foods, seeds, fruits, vegetables, legumes, proteinaceous and high-fat foods, yeasts and fungi. Factors favoring their presence at waste sites include inoculation, colonization from the surrounding environment, the ubiquitous presence of fungi in solid wastes, small size, the capacity to perceive favorable water vapor and volatile compounds, and adequate oxygen content. Unfavorable conditions are below-optimum oxygen content and dry, hot conditions. Although some mites (dust and seed mites) thrive in relatively dry conditions, those partially dependent on fungi would be affected when the environment is dry enough to cause fungi numbers to be reduced. This suggests that this group of animals may be more represented at compost sites than at landfills, even though landfills may have a richer mix of the typical mite foods mentioned.

The presence of wood-eating insects — buprestids (tree borers of sapwood and heartwood), cerambycids (inhabiters of dead wood), beetles (bark and wood-boring and their larvae), siricids (wood fungi eaters), anobiids (lumber infesters) and xiphydriids (wood fungi eaters) — in waste sites is more likely to be the result of inoculation with wood and bark already infested rather than from colonization from the surrounding environment. These animals infest living, dying and dead woody materials. Types of woody wastes disposed at landfills or dumps are typically lumber and dead or living parts of tree limbs, trunks, roots and wood bark. Of these materials, disposed lumber is least likely to be a food supply for wood-eating insects (beetles, beetle larvae, larvae of wasps) because of its low nutrient value and low water content (Hickin, 1975).

Of course, lumber in contact with the soil is likely to become colonized by fungi, cellulose-degrading bacteria and termites (*Isoptera* spp.) over time, which makes lumber more nutritious for wood-eating insects. Termites (*Isoptera* spp.) consume living as well as dead wood and plant materials, forest litter, grass, fungi, dung of other animals and soils, and harbor bacteria and protozoa as symbiotic intestinal organisms (Waller, La Fage, 1987). Their presence in landfills might be disadvantaged by the low nutritional value of dead wood (without fungal or bacterial colonization, increased moisture content and decreased hardness associated with decayed wood) and by the temperature and moisture fluctuations of landfills. In addition, termites are preyed on by ants, invertebrates and birds. Soil-dwelling termites, for instance, are known to prefer high-moisture environments and respond poorly to environmental disturbance (Wood et al., 1982), while wood-dwelling and -feeding termites prefer dry conditions; some die if humidity is high (Collins, 1969). Nevertheless, tree parts and stumps in some stage of rotting, if disposed in dumps or landfills, are likely to have already been infected by wood-eating insects or by fungi. The poor nutritional quality of commercially based wood, as disposed, can affect the representation of wood-eating insects at waste sites. Poor nutritional quality of commercial wood results in a longer development period for wood-eating insects (Fisher, 1940). Bark beetles (*Scolytidae* spp.), followed by sapwood feeders, are known to rapidly inhabit dying and recently deceased tree material (Haack et al., 1983), where nutritious but ephemeral substances are released and wood fiber moisture content is relatively high. Neither the typical residence time of uninfected lumber particles in a compost site nor soil burial within a landfill favors colonization attack by aerobic, slow-developing, wood-eating insects, suggesting that wood-eating insects without capacity to survive in either environment would not be present. However, many of these organisms use more than one food type, and, if undisturbed and environmentally favorable conditions exist at waste sites, numbers of individual species are likely to increase. Well-decayed wood with sufficient moisture content, well-developed hyphal fungi colonies, other alternative food sources and a sufficiently aerobic microenvironment may abet the presence of members of this group in some parts of a landfill and more so at a composting operation.

The ubiquitous presence of fungi at sites with solid organic waste provides a food supply for various arthropods, animals or small organisms that feed on fungi. These include mites, springtails, millipedes, thrips, beetles, fly larvae, nematodes and even protozoa.

The protozoa group, microscopic in size (10 to 50 $\mu$m) and large in number (1000 to 60,000 in natural soils), live in soil water films or tiny pores filled with water (down to pore size 8 $\mu$m) and adhere to soil particles. Dominant species are the ciliates, flagellates, amoebae and testaceae. This group feeds on bacteria, protozoa and fungal spores. Their location in soil water films or pores is typically richer in biological matter than the nearby environment and they are usually in symbiotic relationships with nematodes. The protozoa are known to be good indicators of soil environments under constraints because, though usually the largest in number for soil fauna where water or food supply is limited; some species exhibit dormancy or shell formation as protection against dehydration or limited food supply.

The microscopic (size 0.5 to 2 mm) nematodes, one of the largest groups of all soil fauna ($10^6$ to $10^7$ individuals/m$^2$ or 100 to 200 mg/m$^2$), primarily feed on fungi (fungivores), bacteria (bacterivores), herbivores (plant parasites), predatory types and omnivores (fungi, bacteria, invertebrates, litter) (McSorley, 2000). Their natural predators are mites (e.g., *Mesostigmata, Oribatei*) and other invertebrates. As with the protozoa, they are very important to organic materials' decomposition because of their high food consumption rates and nutrient recycling rates. They also go into dormancy or form protective shells under food- or water-limited conditions.

The microscopic (0.16 to 2.6 mm) soil mites (Acarina), found in natural soils in numbers up to 200, 000/m$^2$, are also mainly fungivores; but they also eat rotting leaf litter or rotting wood (Crossley, Coleman, 1999). This group includes the orabatid mites, long associated in large numbers with forest litter and found in considerable numbers in compost piles. The orabatid mites, 10,000 to 100,000/m$^2$ in soils (Crossley, Coleman, 1999), have the effect on plant litter of fragmentation and tunneling (in rotting wood), which colonizes the material with smaller organisms.

The minute (0.16 to 7 mm) collembolans (*Collembola*) or *springtails*, which are fungivores or eat plant litter, are very important soil fauna because of their large numbers in soil and litter (10, 000 to 100, 000 m$^2$) (Crossley, Coleman, 1999). They are preyed on by millipedes and other invertebrates. Their effects on materials are similar to that of the orabatid mites, but they can inhabit perhaps deeper soil horizons than the mites. While important soil fauna such as the mites and collembolans live at or near the soil surface, they prefer moist but unsaturated conditions, and, if soil-dwelling, they need access to the soil surface (for nutrient variation). The requirement of aerobic, porous conditions means that most would be limited at soil depth, especially so in landfills, where temperature or moisture content might not be favorable on a long-term basis.

The enchytraeids (*Enchytraeideae*) are soil-dwelling, pale-colored earthworm-like animals of 1 to 60 mm lengths — the largest species are in arctic environments — and are found in most plant-litter-rich environments in substantial numbers, where they eat soil mineral particles, leaf litter (about 50% of the enchytraeid diet) and fungi and bacteria (also about 50% of the enchytraeid diet). They have a preference for plant litter already partially rotted and colonized by bacteria and fungi (van Vliet, 2000). Their importance in the organic matter decomposition process is due to (1) fragmenting the material and thus enhancing bacterial decomposition; (2) their grazing results in younger microorganism populations with higher metabolic activity; and (3) they incorporate organic matter into the soil (van Vliet, 2000). Their numbers range from a few thousand per $m^2$ to $100{,}000/m^2$; but they appear to be in higher numbers in moist, acidic, temperate-climate soil environments rich in organic matter. While the enchytraeids must live in a moist soil environment and their locations and numbers in soil are influenced by soil moisture content and temperature, they have developed adaptation strategies to cope with harsh conditions such as cocooning, diapauses or delayed reproduction.

The earthworms (*Lumbricidae*) are well known for their role in agriculture, soil tilth improvement, land reclamation and waste management. They are mostly aquatic or semi-aquatic, meaning that, in the soil environment, moisture content must be high enough for their survival. Their abundance in natural soils can range from about 10 to $2{,}000/m^2$ or a mass range of 1 to 300 $g/m^2$. Their feeding and life habits (they eat soil and surface plant litter) often result in physical changes in soil structure. They create fixed or cast burrows of various sizes (1 to more than 10 mm diameter), depths (up to 1 meter) and varying direction. As they burrow, they consume the soil and cast and cement it behind them. These burrows, the walls of which are often rich in plant-derived nutrients, can become networks or the largest pores in finer soils through which water and air can infiltrate.

The millipedes, which are part of the group (centipedes, millipedes, sow bugs, pseudoscorpions, flies and fly larvae, spiders and beetles), are considered macro arthropods by size definition (2.6 to 70 mm). The macro arthropods can benefit by eating plant or organic matter nutritionally enriched by the colonies of fungi or bacteria, or they may eat animals that feed on bacteria and fungi. Many, if not all, members of this group have been identified in organic waste composting environments. The millipedes (*Diplopoda*) mainly feed on rotting organic matter and can live in environments from arid to wet (Crawford, 1979). Their presence in litter and soil environments can range from 70 to 8000 individuals/$m^2$ or 200 to 1250 $mg/m^2$. Sow bug (*Isopoda*) numbers and mass in soil and litter typically range from 0 to 8000 individuals/$m^2$ or 5 to 1600 $mg/m^2$. The centipedes (*Chilopoda*), on the other hand, are soil and litter predators of sizes 30 to 300 mm; the larger sizes are tropical species. The beetles (carabid, tenebrionid and staphylinid) eat plant litter, plants and other small animals. The pseudoscorpions are known to prey on *Collembola* (Weygoldt, 1969). Spiders (*Aranae*) are carnivorous, feeding on insects and other soil animals. Some of the smaller species live within the soil environment.

The presence of various types of insects or soil-dwelling fauna at a waste site such as those described previously can obviously attract predatory insects or soil animals, which occur in nearly all insect orders. For instance:

- Ants (*Hymenoptera: Formicidae*) eat soil arthropods and their larvae.
- Flies (*Diptera: Alsidae, Syripidae, Tachinidae* spp.) eat aphids, leaf hoppers, mealy bugs and other insects.
- Praying mantises (*Mantodea*) eat many insects and mites, grasshoppers and locusts; (*Ortheoptera*) also eat a variety of insects.

- Carabid beetles (*Pterostichus*) and coccinellid beetles (*Coeloptera: Coccinellidae*) eat aphids, mealy bugs and scale bugs.
- Lice (*Psocoptera*) eat various insects and their eggs.
- Wasps (*Chalcicoidea, Braconidae*) eat aphids, scale, whiteflies, other wasps and mealy bugs.
- Lacewings (*Chrysopidae*) eat mites, aphids, thrips, whiteflies, bugs and the eggs or larvae of these insects.
- Spiders feed on insects and invertebrates.
- Centipedes feed on invertebrates and worms.
- Dragon files (*Odnata*) eat other small flies.
- Predatory mites (*Phytoseiidae*) eat all types of mites.
- Bee flies (*Bombyliidae*) eat the larvae of flies, wasps, beetles and ants.
- Rove beetles (*Staphylinidae*) eat aphids, springtails (*Collembola* spp.), nematodes and flies.
- Springtails (*Collembola*) are known to eat fungi, molds and worms.

Note that many of these are seen at various types of composting operations, dumps or landfills. They generally benefit from the growth of fungi on rotting wood, plant and animal (remains and fecal product) materials.

The numbers of many of these predators at waste sites would be affected by such factors as prey abundance, access and cover and limited environmental disturbance. This again favors the composting operation over the landfill site.

Many of the species described are also likely to be present within the landfill environment or at its periphery. Access by small animals to fungi-populated materials at waste sites may be limited to the surface or aerobic upper portions of landfills or dumps, if these locations are undisturbed for a sufficient time period. Compost sites containing various types of plant wastes and fungi are known to show association between fungi and animals present.

A diverse web and interdependent web of soil fauna and microflora are thus likely to be present at many waste sites, as waste site surface soils would contain many of the food sources used by these animals in other soil and litter environments. Feeding by soil arthropods such as the orabatid mites (*Acarina*), springtails (*Collembola*), pseudoscorpions, isopods, diplopods, arachnids and earthworms, on dead and decomposing plant and animal matter, fecal material, carrion, plant leaves, wood or roots, fungi, bacteria and algae, would promote the lower trophic levels as well as higher trophic levels.

The waste decomposition role of small soil insects is of high importance in the breakdown of solid organic materials in the upper horizon of waste sites. They consume large quantities of materials to compensate for poor food quality and produce relatively large amounts of wastes, while their movement repopulates the soil with microorganism from their fecal matter or body structure. Werner and Dindal (1987) report that fungi-feeding soil fauna (60% of all) can consume 24 mg of fungus per square meter per day and can ingest 7% of their dry body weight per day, while soil fauna predators can consume 2.5% of their body weight per day.

## INFLUENCE OF SITE ENVIRONMENTAL FACTORS ON ORGANISM TYPES

As a system, the waste site, albeit disordered, provides an environment for living organisms that may be less predictable than the background terrain. Microenvironmental factors believed to be especially influential include the local temperature, local soil or wastewater content. Available water content of the organic waste affects rate of colonization by the first trophic level of waste consumers; thus, drier conditions should slow colonization by bacteria — and fungi to a lesser extent — at any type of waste site. Some solid organic wastes in compost piles, landfills or dumps can contain enough water to support initial phase bacterial and fungal colonization; but site moisture content variation due to rapid evaporation of water, heat or exposure can reduce moisture availability to living organisms, slowing biological processes. System porosity can benefit nearly all organisms. For anaerobic organisms, increased soil pore area provides greater opportunity for attached organisms. For aerobic organisms, increased porosity could also mean higher oxygen availability.

For waste sites such as solid waste landfills or dumps, however, elevated levels of carbon dioxide or methane in soil pores, as well as unpredictable or elevated local temperature, may determine which bacteria or fungi species might thrive. For smaller organisms in landfills and dumps, decreased porosity or randomly advantageous pore size distribution can additionally limit access to food sources. These factors, which will directly affect growth and mass turnover rates of all of the organisms of site treatment (remediation) interest, also make for more patchy organism distribution patterns in landfill soils than in compost piles, contaminated soils or biofilters. In developing descriptions of the landfill or dump site in terms of an overall biological process, i.e., as a reactor, one has to be cognizant that these patchy processes cannot be directly represented with any satisfactory degree of accuracy. Rather, they should be indirectly assessed with careful selection of the factors used for indirect assessment. The landfilled waste site cannot, because of its unpredictable and frequently hostile nature, provide the perfect environment for prey-predator relationships between larger and smaller organisms. Organism size thus becomes less important than capacity to compete for wastes in a specialized environment.

The fact that landfilled wastes in a site may be buried in soil for the longest period of its decomposition history naturally favors a long-term impact from smaller soil-dwelling organisms, and the best equipped to survive under landfill soil conditions are the fungi and bacteria. The *waste decomposer* microorganisms or waste/soil microflora, i.e., the protozoa, fungi, actinomycetes, bacteria (Coleman, 1978) and in some case insect and worm adults, larvae and eggs, should exert a greater waste effect than larger organisms because of natural adaptation to soil life. Factors favoring landfill site survival—though not at optimal levels compared with compost sites — are small size, rapid growth and reproduction and capacity for reproductive adaptation to environmental extremes. These factors ensure a role for soil microflora in the landfilled solid waste decomposition process — even under the harshest conditions — and significant and rapid response to any favorable conditions at a landfill site.

In their competition for the mix of food materials present, the larger organisms also aid smaller organisms, particularly in the earlier stages of landfilled wastes burial, by facilitating access to wastes through shredding solid materials such as leaves, food materials or paper. Smaller soil animals such as mites and springtails, often important to decomposition of litter in compost piles, forest litter or soil, can by their grazing habits fragment leaf and cellulosic material, increasing the rate of bacterial decomposition (Coleman and St. John, 1987). Any degree of comminution of the solid organic wastes increases the surface area accessible to bacteria and fungi. Additionally, larger organisms burrowing for food not long after wastes are placed could increase the waste site system permeability to air or water, while the mixing of waste and soil particles involved could improve system inoculation by the smaller organisms. Egestates of and mortalities from larger organisms also supply organic matter to bacteria.

## THE WASTE SITE AS AN ENVIRONMENT FOR ORGANISMS

The landfill or dump is a less than ideal environment for the presence of a wide representation of the organisms mentioned earlier.

However, the presence of a wide variety of food or plant wastes and the presence of prey organisms are known to attract a multitude of insects and animals to dumps and landfills. Surface-feeding organisms such as insects, worms and flies would feed on landfilled wastes and on excretory products from larger organisms; yet feeding opportunity can depend on length of time before the wastes are buried by a layer of soil, opportunity and capacity for competition with other organisms at landfills, burrowing and excavating ability and agility.

For landfilled wastes, relationships among the lower-level organisms are considered key to simple models of the decomposition process. Bacteria are usually found to be largest in numbers and relative mass in waste-soil profiles, followed by fungi — suggesting a correlated level of waste removal or transformation. Fungi are considered to have a primary and vital role in soil waste decay as this group contains the universal decomposers of cellulose, which is usually the major chemical mass fraction of modern landfilled refuse. Fungal decomposition of solid organic material is primarily enzymatic and is referred to as hydrolysis or the solubilization of solid particles. The soluble products can be used by both fungi and bacteria to produce acids (acidogenesis), which can be used by other bacterial groups in metabolism and growth. This results in the production of gases such as methane, sulfur dioxide and carbon dioxide. Thus the fungi would play a key role in mutualism among the lower trophic levels of landfills and hence in all types of biological degradation of landfilled wastes.

The physical characteristics and nature of the basic food source at a particular type of waste site determines which organisms will be present. Environmental factors — water supply and food supply — and physical factors at the soil surface or in pores are key controls of interactive biological processes. The three examples — compost piles, contaminated soils and biofilters — illustrate this and suggest similar constraints would apply for landfills and dumps. It also points out that an important consideration for any modeling of the more complex type of waste site is not only which organisms dominate biological processes but whether the model recreates a physical system where the effects of less dominant organisms are not considered merely because of inconvenience.

Interaction between these organisms is thus an important subject in assessment of the effect of landfill organisms on the fate of disposed wastes. A review of the quantitative impact of various organisms on the biological decomposition and removal of solid wastes buried in landfill soil allows some insight into whether organisms other than the fungi, yeasts or bacteria can have a substantial impact on removal or conversion of MSW at a closed or open landfill. The importance of the interactions of these various organism groups on the landfill biological process cannot be overstated, as decomposition products from disposed wastes can introduce toxic or harmful substances into air, soil, groundwater or surface water.

As seen from the previous general and somewhat speculative discussion with regard to potential trophic/environmental relationships among waste site organisms, it is likely that the organisms participating in the early stages of disposed waste decomposition may be linked by a nominally functioning food web, although the environmental and operational conditions at a landfill or dump are such that classic trophic relationships would not develop. However, from the point of view of examining the waste site as a biologically functioning unit, the effect of various organisms present is cumulative, hence the value of assessing the effect of any group of organisms.

In contrast to modern landfills at which municipal (town or village) wastes may be covered on a daily basis with several inches of soil by heavy equipment, open dumps — relatively rare in the U.S. in recent times but more frequently used in less industrialized regions — are a more favorable environment for many of the organisms described. The open dump, as is the case for the composting operation, may also be more representative of a likely waste site ecological system than a modern landfill.

The particular conditions at large modern landfills are obviously selective for opportunistic organisms and would discourage the presence of organisms that cannot compete for food materials within a limited range of feeding opportunity. Refuse removal and decomposition by organisms in the soil subsurface would additionally be impacted by surface and subsurface environmental factors such as local moisture content, thickness and compaction of soil and levels of carbon dioxide, methane or oxygen in the soil pores.

Modern U.S. landfills are operated with bulldozer placement rolling of a 4- to 6-inch layer of fine, permeable, locally available soil over layers of split garbage bags (mainly of low-density polyethylene (LDPE)), municipal solid wastes (MSW) and other discarded materials, piled up to 3 feet thick before soil cover is applied. The soil cover is applied within 24 to 48 hours after the wastes are tipped from collection trucks. Most state and federal regulations require this covering of wastes to reduce vermin and related littering and disease. However, this relatively brief time period between waste placement and burial in soil limits each organism's degree of opportunity. It favors larger, daylight-foraging or night-feeding organisms such as animals, birds and rodents that can act before the wastes are buried or can burrow to or excavate shallow-buried wastes. The advantage this feeding opportunity presents can be further reduced by the predation by other species. For instance, rodents are unlikely to engage in extensive daylight foraging on the open terrain of landfills or dumps because of the risk of predation by raptors and other animals.

The quality of the food is another factor determining which soil or surface animals will first at visit these sites. Nutritional value of a disposed waste would determine whether it is worth the energy expenditure if compared with other food sources within foraging range. One of the most common sights at landfills within a few miles from coasts throughout the world, seagulls are attracted to landfills by food and meat wastes; and they use the landfill as a main or alternative food source, depending on the time of the year. They reportedly prefer feeding at landfills to tidal or cyclic feeding on marine organisms (fish, crabs, shellfish, etc.); landfills may be a more reliable source requiring less travel, even if the nutritional level of MSW is not as high. Their quickness, acute sight and hearing and lack of fear of humans and heavy equipment permits them to congregate and feed by sight during the actual dumping of municipal wastes, continuing to do so until the wastes are covered. They will also excavate wastes if not buried deeply enough.

It is possible that soil insects or worms arriving with landfill soils used for cover could survive the burial process and develop populations during a period of unpredictable length before a waste site location is again covered with wastes and soil. However, most of the soil animals described are aerobic, only survive in an upper soil profile of limited depth and are likely to be more interested in the more nutritious municipal waste types. This suggests that the role and the MSW impact of insects and worms at modern U.S. type landfills may be temporary and limited. This has not been verified; and in general, the interactions between organisms causing the disappearance of solid wastes is an unexplored subject.

The progress of waste decomposition inside a landfill has long been studied. Hydrolysis, which is considered a limiting step in organic MSW decomposition, has been definitively associated with the aerobic fungi and a small (cellulolytic and fermentative) group of bacteria. Acidogenesis, the conversion of hydrolytic products such as glucose and acetate to organic acids, has been associated with bacteria; and the final waste conversion steps, methanogenesis and sulfogenesis, have been associated with methanogenic and sulfur-reducing bacteria.

The major players involved in the early stages of landfill soil decomposition are likely to be the aerobic and facultative bacteria and fungi. The former are usually the largest in numbers in waste/soil profiles, followed by fungi. Fungi apparently have a vital role in waste breakdown because this group contains the universal decomposers of cellulose, the major chemical mass fraction of modern refuse. Fungal enzymatic conversion of solid waste or hydrolysis results in the solubilization of solid particles. The soluble products can be used by both these organisms and bacteria to produce acids (acidogenesis), which can be used by other bacterial groups in metabolism and growth, resulting in the production of gases such as methane, sulfur dioxide and carbon dioxide. The processes described are usually essential elements of the landfilled waste decomposition model.

From the previous discussions, the relative order of impacts of organisms on wastes disposed at a landfill or dump can be described as:

- Assimilation, particulization and inoculation: animals, rodents, birds, insects, worms
- Hydrolysis: fungi
- Acidogenesis: bacteria
- Methanogenesis, sulfogenesis: bacteria

It is useful to consider which of the organisms mentioned might be present and may have an impact.

## DEFINITION OF IMPACT OF ORGANISMS AT DISPOSED WASTE SITE

Modeling of the waste site as a bioreactor presumes development and use of parameters that describe the site first as a biological process as well as a physical system amenable to modeling. While it has been established that biological reaction is the key phenomenon in waste decomposition in disposal sites, and key to treatment of the attendant adverse impacts of waste sites on the human or animal environment, an immediate difficulty for the modeler of this unusual system is the large number of unknowns. Of the wide variety of actual or potential waste site organisms discussed in this chapter, only a few have been studied enough in wastewater and waste treatment systems to ascribe a quantitative effect. A quantitative approach opens additional possibilities for modeling a waste site.

The previous discussion of organisms has shown that their effects on biological processes in soil systems and elsewhere are not always direct or linear. While some organisms ingest and thus convert considerable fractions of solid organic materials placed in soils or waste sites, no less important are physical impacts such as mixing, microorganism inoculation, increasing soil porosity and physical reduction of the amounts of various materials. Conditions at a waste disposal site create different niches for organisms and overlapping opportunities for exploitation of the food supply, but the modeler of the site as a reactor would have an interest in how each organism changes the system as a reactor and whether the change is physical or chemical.

For modeling purposes, the change effected on a waste site and its contents by an organism can simply be described as biological, physical or both. The change is also indirect or direct; but if simple relations between impacts of organisms can be established, this distinction may be less important. For bioreactor modeling, both physical and chemical changes would affect system performance. However, one may want to limit the number of considerations for any particular organism to its main impacts on the biological-physical system.

From an engineering point of view, the goal of modeling a waste site is usually to examine means of treatment or remediation (biological, chemical, physical). As the key phenomena for site treatment are biological and for the type of modeling described in this work, it is convenient to limit waste site considerations to those that impact the biological process. In this context, the effect of one or several organisms becomes easier to describe, although information at present may be insufficient to confirm all hypotheses.

The biological process for the waste site additionally has to be defined as extant only for a certain period of time. Organisms at a waste site can only have appreciable impact for a specified time period. In the case of a landfill (a refuse dump or tip may offer a slightly more reliable food source), the time period for impact is even briefer for most of the organic species described in this chapter. While lignocellulosic wastes in a landfill may last for several human generations and some commercial plastics even longer, the biological process eventually devolves to a point where a landfill is considered relatively insignificant to continued water or air quality impact.

Site treatment limits itself to manipulating biological or chemical processes, to shortening the time period for impact or to permanently reduce the potential for impact. Leachate recirculation, aerobic methods such as low-pressure air injection through landfill soils and aerobic operation for composting (e.g., windrow turning or enclosed aerobic composters) increase the waste decomposition rates, thus shortening the time period and potential costs of waste management. The advantage is that, if air and water effluents from the biological process can also be captured and reduced permanently, site management efficiency can be realized. Means to accomplish this are through landfill gas burn-off or energy, landfill leachate collection and treatment and compost leachate recycle or treatment. Considering that several organism species may have impact only within a fraction of the time considered for a biological process model, the task for the modeler is to define the organism, level of its impact and length of time. Impact level and time period are obviously the most difficult to define with any precision, considering the lack of information. However real-world definitions for organism type, period of impact and impact level could improve how waste sites are managed.

## ORGANISMS REPORTED AT LANDFILLS, DUMPS AND OTHER WASTE SITES: CONSIDERATIONS

The landfill or dump attracts soil and surface-dwelling animals of many species by plant or animal food sources. However, in many countries the presence of some insects, larger animals and birds at landfills is considered a public health threat, hence regulations are set to limit their access. Landfill regulations of the U.S. often specify which animals are to be limited by landfill operation practices, but in general they prescribe immediate soil coverage for food and plant waste. Waste burial limits but does not eliminate the presence of birds, insects and various surface-dwelling animals. The waste site (dump, landfill or refuse tip) also attracts large animals and human scavengers.

In most cases, waste disposal is daily or nearly continuous at landfills or large dumps. This provides a regular and reliable opportunity for animals to feed or otherwise benefit, however briefly, on a per-day basis. Once disposal patterns become familiar, surface animals exploit the opportunity. Various site access and pest or animal control practices may reduce animal populations, which may be factored in for estimates of organism impact at a particular waste site.

The organisms at landfills or dumps would thus include — starting from the smaller sizes but perhaps largest impacts during the time period of bioreactor engineering importance and working upward — soil bacteria, yeasts or actinomycetes, fungi, soil fauna (small soil animals), above-ground insects, birds, rodents, various animals including bears and primates including man. A list of these organisms and casual descriptions of their attractant and potential effect on the biological process is provided in Table 4.2.

## WASTE SITE SCAVENGERS

Landfills or refuse disposal sites have attracted human scavengers throughout history. Today, scavenging is a well-recognized activity, particularly in what are described as less industrialized or poorer areas of the world. In these areas, scavenging or, essentially, informal recycling, may support members of the poorest sector of populations and can be well organized and relatively successful in providing this societal group with very basic necessities. This activity is often considered degrading by society, even though it sometimes pays more than ordinary jobs available. For instance, it has been reported that cooperatives of scavengers in Brazil have enabled their members to earn incomes sometimes twice that of ordinary citizens. The same has been reported for some cities in India and the city of Bangkok (Nair, 1994). According to the World Bank, up to 2% of populations in developing countries survive by scavenging (Medina, 1998).

Human scavengers are generally not attracted to plant or animal food materials that may be present at a waste site but rather seek out and remove materials that can be traded for income or reused — for instance, furniture, equipment or appliances that can be repaired. However, the need for scavenging, despite its benefit to waste recycling, is determined for the most extent by local, regional or social group economics; and even when directed, as in some poorer countries, it is not driven by environmental or waste treatment goals.

Medina (1998) notes that the two distinguished types of scavenging are for self-consumption and for sale and that the items scavenged for self-consumption include food items (scraps, etc.), discarded clothes and abandoned furniture and appliances.

Materials scavenged at waste sites such as dumps or landfills are more typically metals, plastics, glass, cardboard, all of which can be traded for cash and wood. UNEP has reported that the thousands who live at the very large Payatas dump in Bangkok, Thailand, make their living by collecting plastic, cardboard, wood, glass, metals and other items for sale to recycling agents (Svenningsen, 1999). Scavenging for metals in dumps for sale has been reported as a common practice in the ex-Soviet state of Kyrzygstan (Reuters, December 4, 2001), while collection, vending and buying of materials from the municipal dump in the city of Hanoi supported a well-organized scavenger population of around 6000 persons between agricultural seasons. In Mexico City, Calcutta, Phnom Penh, Lima and other world cities, many

**TABLE 4.2**
**Waste Site Attractant and Impact of Biological Organisms**

| Organism | Group Composition | Waste Site Attractant | System Niche and Impact | Impact Level* | Impact Modifier (Control) |
|---|---|---|---|---|---|
| Bacteria | Cellulolytic, acidogenic, acetogenic, methanogenic, methanotrophic | Organic plant and animal wastes | Hydrolysis, acidogenesis, acetogenesis, methanogenesis | 1 | Site moisture content, porosity, temperature, gas composition |
| Fungi | Soil and litter fungi | Plant and animal wastes | Cellulose hydrolysis and decomposition | 1 | Site moisture content, porosity, gas composition |
| Soil Fauna | Mites, beetles, Diptera larvae | Fungi, yeast, bacteria, fungi/bacteria inoculated wastes | Fragmentation of wastes, inoculation with microorganisms, soil porosity | 2 | Site moisture, upper soil porosity, protozoa/fungi/bacteria level |
| Worms | Enchytraeids, earthworms | Fungi, bacteria, decomposing plant litter | Ingestion, turnover of rotting plant litter, porosity modification, inoculation | 2 | Site moisture, temperature, plant/fungi/bacteria/litter level |
| Insects | Flies, beetles, mites, scorpions, centipedes, millipedes | Fruit and vegetable wastes, small soil and surface fauna, rotting wood and plant wastes, insects | Soil fauna, larvae and insect predation, fungivory, plant, food, and animal waste consumption | 3 | Soil moisture, temperature, porosity fauna/fungi/plant waste level |

| | | | | | |
|---|---|---|---|---|---|
| Birds | Eagles, gulls, crows, petrels, kea, skua, sparrows | Plant (fruit, vegetable), animal wastes, food wastes, insects, seeds | Opportunistic removal/ ingestion of materials, fecal bacteria inoculation | 4 | Fruit, meat, vegetable, insect, soil fauna, rodent level, soil cover, dispersant density |
| Rodents | Burrowing animals, rats, field mice, etc. | Fruit and food wastes, seeds, insects | Removal of seed, cereal and food materials, predation (worms, insects, bird ova). | 5 | Fruit, meat, vegetable, insect, soil fauna, rodent level, soil cover depth |
| Large Animals | Bears, foxes, primates, canines | Food wastes, animal wastes, rodents, birds, insects | Food (vegetable, fruit, seed) and meat wastes ingestion/ removal | 5 | Fruit, meat, vegetable, insect, rodent level, soil cover depth |
| Humans | Scavengers | Salvageable materials (metals, cardboard, plastic items, wood, cans, bottles) | Removal of salvageable large waste items, porosity modification | 5 | Soil cover depth, site access controls/regulations, alternative economic opportunity, health education |

* Casual ranking of wastes conversion or removal at waste site.

thousands of scavengers make a living by collecting saleable items from landfills; and in Cairo an estimated 50% of household waste is recycled in this manner by a salvaging population, the Zabaleen, estimated at 27,000; however, a percentage of the organic fraction is collected and processed for animal (pig) feed.

In many cases the organic fraction and ash from garbage disposed at open dumps in less industrialized countries are also targeted by scavengers, for manure, aerobic and anaerobic composting, land application and agricultural uses (Medina, 1998).

One effect of the human scavenger on a disposal site is thus physical removal of various amounts of the waste materials suggested. Nair (1994) reported that the activities of scavengers in Thailand resulted in recycling of 7% to 16% by weight of materials dumped. This statistic is likely to be variable, depending on the region. However this shows that scavenging can affect the makeup of wastes in a site. The bioreactor impact would be to concentrate wastes not scavenged, similar to the result of organized recycling programs. In the case of food and plant wastes, concentration could improve performance of the waste site as a biological reactor by making the remaining mixture more organic and bioreactive.

Another effect of scavenging activities might be to make the upper soil profile of a waste site soil more aerobic. Scavenging involving removal of relatively large solid items could also improve moisture and air distribution by reducing the number of system macropores that may act as preferential air or water flow pathways.

However, the practice of uncontrolled and unprotected scavenging by humans is fraught with individual or public health risk (Carpenter et al., 2000), including exposure to serious infections; diseases; to hazardous, toxic and sometimes lethal substances; physical injury; and immune system attack. This should be discouraged in favor of other alternatives such as organized or cooperative recycling with scavenger participants as trained participants. The quantitative impact of human scavenging on wastes after disposal, albeit minor in most cases compared to that of other organisms, should be considered in countries where it still occurs with regularity and where any of these disposal sites come under consideration for bioreactor treatment approaches.

## BEARS

In North America the presence of black and other bears (e.g., black bear or *Ursus Americanus*, brown bear or *Ursus arctos*, grizzly bear or *Ursus horribilis* and polar bear or *Ursa maritimus*) at landfills is relatively well known and is more likely at rural sites than at landfills in urban environments. Their presence has long been reported in Yellowstone National Park in Wyoming and in Alaska (Sparrowe, 1968; Badyaev, 1998). Bears have been reported to be a relatively common sight at Alaskan landfills and black bears (*Ursus americanus* Pallas) have been spotted rummaging at landfills in New York and other states (Vander Linden et al., 1998). Bears are attracted to food at dumps, including human food, livestock food and bird food (Alaska Department of Fish and Game, 2001). They do not usually fear human presence, thus feed while trucks unload (from safe position) (Waldrop et al., 1971). According to the U.S. Fish and Wildlife Service (Mountain-Prairie Region), grizzlies are attracted by smell to compost piles and dumps, where they target food items such as carrots, bulbs, meat, grease, bones, livestock and poultry feed. Capable of digging four to seven feet deep, they will also dig up buried livestock and may prey on sheep and pigs. These omnivorous animals are likely to be targeting the more nutritious portions of the food waste fraction, though no studies appear to be available of degree of their dependency.

Modifiers for the presence of bears at landfills include reduction of human food content and the use of electrified fences, with varying success.

The effect of bears on waste composition at landfills or dumps is likely to be relatively minor, except in rare cases. In regions where bears have open access, however, the percentage removal of the wastes mentioned is likely to be a significant (but undetermined) fraction of the percentage of food disposed.

The potential for bears and other animals to impact the physical or chemical makeup of waste materials in the site soil profile illustrates the value of future landfill ecology studies.

## OTHER LARGE ANIMALS At WASTE SITES

Other animals reported at waste sites include primates in African or Asian countries, deer, foxes and large rodents. These animals predominantly target the food content. At dumps in Diani, Kenya, baboons are reported to have become habituated to this alternative food source (Born Free Foundation). Primates such as baboons have been reported in feeding dependency on nearby garbage dumps, with significant difference in nutritional success than for wild baboons (Altman et al., 1993)

Open conditions or large animals capable of digging through four to six inches of soil to obtain food favors larger mammals feeding opportunistically at active MSW landfill sites.

The presence and types of large animals at waste sites worldwide is likely to be region- or even site-specific and relative impact may rise in importance if a waste site contains attractive materials such as food and is also accessible to populations of large animals within foraging range.

## SMALL ANIMALS

While there is no argument other than size difference to separate surface animals at waste sites into groups, it is obvious that small animals such as rodents and burrowing species may occupy a different food foraging group than black bears. Members of this group reported include cervals and feral cats (*Felidae*) (World Conservation Union, 1996), dogs and raccoons (*Procyon lotor*), striped skunk (*Mehphitis mephitis*), squirrels and shrews (common or masked (*Sorex cinerus*)) and short-tailed chipmunks (*Blarina brevicorda*) and (Eastern, *Tamias striatus*), field mice (white-footed or *Peromyscus leucopus*), meadow vole (*Microtus pennsylvanicus*) and rats of various species.

One of the most frequently reported rodents is the common rat (*Rattus* spp.), which has long been associated with human food sources and with landfills, dumps and composting sites. Many municipal governments throughout the world have developed disposal standards that directly or indirectly require the control of rats in residential and public waste management. Because rat populations have been reported as significant for the few studies of this topic, a detailed discussion of this rodent is included in this section.

Rat species identified most often include the Norway rat (*Rattus norvegicus*), roof rat (*Rattus rattus*) and mice (*Mus musculatus*). The Norway rat, thought to have been introduced into the American continent by early ship trade with Europe, is terrestrial, ground-foraging and will eat plant foods, grains and fruits, meat, bird and reptile ova and large insects; and while its home range might be 100 meters or more, its foraging range may be several hundred meters (Thorsen et al., 2000). The Norway rat can reach nearly two pounds (900 gm) and use waste dumps as a food source (Brush, 1992); but it has been known to weigh 84 to 588 gm, with males heavier than females, to grow to a foot (25 cm) in length, to eat up to 33% of body weight per day and live at dumps when the opportunity presents (Encyclopedia of Animals). Marsh and Howard (1969) found that the principal rodent species at solid waste dumps in California was the Norway rat (*Rattus norvegicus*), followed by the common roof rat (*Rattus rattus*), ground squirrel (*Citellus* spp.) and mice (*Mus musculatus*). Norway rats living at garbage dumps in Hokkaido, Japan have been associated with human disease parasites (Okamoto et al., 1992).

The study by Courtney and Fenton (1976) found that mice (*Mus musculatus*) were the most aggressive and populous rat species (the next most populous rat species were white-footed mice (*Peromyscus leucopus*)) at a small rural garbage dump, where *Mus musculatus* numbers ranged from 700 to 1000 over an area of garbage covering 0.03 hectares (300 square meters). This garbage dump contained the usual wastes (paper, cardboard, bottles and cans, plastic cups and items, as well as kitchen food wastes, mattresses, appliances and hardware).

It was indicated that mice at this dump lived in predator-safe nests (burrows 10 to 36 cm deep; owls and hawks were the main predators) within a home range of less than 100 meters from the dump and were most active late at night (2100 to 2400 hours). Nests were made of shredded paper and leaves located inside containers (bottles and cans) within the garbage. The authors also noted that the garbage dump was more suited to the omnivorous rat and chipmunk species than to shrews and voles (carnivorous and herbivorous, respectively). Storer et al. (1944) show that squirrels and chipmunks primarily target fruits and seeds of various plants and trees.

Mice have been associated with a wide variety of waste dumps and can take advantage of their smaller size (vs. the Norway rat) to survive, even with predation.

## WASTE REMOVAL IMPACT OF ANIMALS AT DISPOSAL SITES

One approach for linking food source availability at a waste site to likely quantitative impact on wastes is to apply food requirement energetics for specific animals involved. Body mass can be linked to daily food requirements through various metabolic relationships and thus to daily food intake. Inaccuracy creeps in, however, because of limited data on exact foods consumed. For surface animals visiting waste sites, other sources of error also include lack of information on (1) foraging success, (2) average time spent and (3) nutritional quality of foods ingested. For animals with significant numbers and foraging activity, assessment of impact can be useful in constructing materials flows, with a reasonable degree of judgment.

Factors influencing foraging success include mass of food materials present per day and site and food accessibility. Food material percentages in landfilling or composting are guides to mass of food substance present. Disposal methods involving bagging, compaction and coverage with soil may discourage animals other than strong or fast species. For most animals, waste sites are unlikely to be permanent or exclusive food sources as food quantity and nutritional value are unreliable, thus time spent will be a (varying) fraction of total foraging period. Nutritional quality of food materials present at a waste site is also likely to fluctuate, introducing the likelihood of higher ingestion rates to meet the same daily requirements or periodic starvation. For animal ecologists, examination of these relations between food sources and food use may reduce uncertainties for site assessment and management.

Storer et al. (1944) state that the food requirement of an animal is proportional to its body weight, with its relative food intake the ratio of daily consumption to metabolic body size. Standard relative food intake determined for one species can thus be used to determine food intake of another. Norway rat was used for standard food intake calibration. The studies by Nagy (1987, 1999, 2001) developed food requirements and predictive relationships for various wild animals, which are now widely referenced. For the animals mentioned in this section, these studies provide some insight into how food intake and thus material removal can be estimated. The average body masses or weights of the animals discussed are shown in Table 4.3.

According to Nagy (1999), the food intake an animal requires per day, in gm/day dry matter (DM), can be related to its field metabolic rate (FMR), which is a measure of the animal's energy requirements for a free-ranging existence. Various investigators have determined field metabolic rates (FMRs) for animal species. FMR rates have been developed most comprehensively by Nagy (1987, 1989, 2001). Nagy states that the daily food requirements can be determined from the ratio of FMR to gross energy content of the food (type) in kJ/gm dry matter:

$$\text{Daily food requirement} = \text{FMR/gross energy content, food} = \text{(kJ/day)/(kg/gm DM)}$$

Food refuse is indicated to have a gross energy content of about 16 to 19 kJ/gm DM and a digestibility in pigs of 69 to 77% (Dominguez, 1992), suggesting a conversion rate of about 11.0 to 13.1 kJ/gm DM. Sibly and McCleery (1983) also reported the energy value of food waste as 2.5 kcal/gm wet weight. The daily food requirements of the animals listed, in gm dry matter/day, may thus be estimated (conservatively) from the ratio of FMR to digestible food wastes:

$$\begin{aligned} &\text{daily feeding rate (food waste, gm dry matter/day)} \\ &= \frac{\text{FMR (species)}}{\text{digestibility (kJ/gm)}} = \frac{\text{FMR (species)}}{11.0\,\text{(kJ/gm)}} \end{aligned} \tag{4.1}$$

The above shows how rough estimates of food removed by animal groups scavenging at a landfill or composting operation may be developed. Important variables in the relationship shown are obviously the energy content and digestibility of the food

**TABLE 4.3**
**Weights and Food Preferences of Some Surface Animals Reported at Waste Dumps**

| Common Name | Classification | Average Weights Grams | FMR* | Foods | References |
|---|---|---|---|---|---|
| Grizzly bear | *Ursus horriblis* | 147,460–385,660 | 44,043 | Fish, carrion, fruits, predated animals, refuse | |
| Brown bear | *Ursus middendorffi* | 95,000–780,000 | 63,235 | | |
| Black bear | *Ursus americanus* Pallas | 90,700–136,000 | 19,580 | Berries, nuts, grass, plants, carrion, small animals, fish | |
| Baboon | *Papio hamadryas* | 13,000–46,000 | 8,831 | Roots, bulbs, fruits, seeds, insects, mice, birds, worms | |
| Cerval (Serval) | *Felis sylvestris* Lybica | 4500 | 2,238 | Food scraps, insects, spiders, mice, vole, carrion | |
| Feral cat | *Felis catus* | 3500 | 1,863 | Various food items (omnivore) | |
| Dog | *Canis, Lyacon pictus* | 25,170 | 7,865 | Omnivore | Nagy, 2001 |
| Raccoon | *Procyon lotor* | 5400–15,800 | 4,183 | Fish, insects, rodents, eggs, kitchen refuse, carrion | |
| Ground squirrel | *Citellus* | 500 | 450 | Seeds, vegetables, flowers, insects | Storer et al., 1944 |
| Striped skunk | *Mephitis mephitis* | 1400–6600 | 1704 | Bees, grasshoppers, beetles, mice, ground squirrels, voles, eggs, corn, berries, plants, fruits, garbage | |
| Eastern chipmunk | *Tamias striatus* | 96.3 | 143 | Fruits, seeds | Nagy, 2001 |
| White-footed rat | *Peromyscus leucopus* | 19–25 | 41.4 | Seeds, nuts, fruits, insects, snails, worms | Nagy, 2001; Storer et al., 1944 |
| Shrew | *Sorex cinerus, Blarina brevicauda* | 18–30 | 49.0 | Bugs, earthworms, snails, plants, grasshoppers | |
| Meadow vole | *Microtus pennsylvanicus* | 36.9–40 | 115 | Green vegetation, bulbs, grass | Nagy, 2001; Storer et al., 1944 |
| Norway rat | *Rattus norvegicus* | 161 (84–588) | 197 | Omnivorous | Storer et al., 1944 |
| Common mouse | *Mus musculatus* | 17–25 | 44.5 | Grains, fruits, vegetables, cereals, garbage | |

* Field metabolic rate = FMR = kilojoules/day.
Calculated using the Nagy et al. (1999) relation for FMR vs. body mass for mammals [4.82 (body mass in grams)$^{0.73}$].

used by the animals involved. The food preferences listed for some animals cover a range of energy content and digestibility but are likely to be of higher nutritional quality than mixed food wastes at a landfill. Thus, estimates must be conservative.

## BIRDS

Actively operating landfills, dumps and composting operations provide a ready source of nutrition in the form of animal wastes, human food wastes from grain products and animal protein, fruits, seeds and plant matter. As shown early in this section, organic wastes also draw insects, worms and small animals, including rodents. This attracts various bird species to landfills. Birds at landfills are well represented at landfills, as shown by many studies of these sites. Their energy requirements are also high, which translates into relatively high food consumption per unit weight; and thus they are likely to have a more significant waste removal impact than most other surface animals.

From a landfill management point of view, birds are often considered undesirables, as their association with landfills has long been linked potentially to dangerous aircraft collisions. Their feeding activities also increase site littering, soiling of the landfill surface and potential exposure of landfill workers to disease-causing pathogens. As with rodents, many municipal regulations require control of bird populations and specify sufficient separation of waste management operations from residential areas and airports.

Birds have significant scavenging advantage over surface animals in that aerial movement provides ready escape from disturbance and allows easier detection of food placement or location. However, the brevity of waste exposure before soil burial at landfills and the open environment favor a relatively small number of bird species that can feed rapidly, are aggressive and of large size or are generally scavengers and omnivores, chief among which are gulls. Other birds, such as crows, starlings and other species, have nevertheless been reported. Many of the world's landfills and dumps are also less than 100 miles from coasts or open waters that are often normal gull feeding grounds and nesting areas.

The range of bird species reported at dumps and landfills include kea (New Zealand) (Bond and Diamond, 1992), alpine chough (corvids) (Delestrade, 1994) and gulls, crows and starlings (Burger, 1985). Landfills along the coastal U.S. have long been foraged by gulls (*Larus* spp.: herring, ring bill and blockheaded), petrels (skua), sparrows, starlings, pigeons and blackbirds. At the Fresh Kills Landfill in Staten Island, New York, for instance, a common sight at or near the daily dumping area during its years of operation has been the thousands of herring and ring-billed gulls, which can reach over 1000 grams in size. Ring-billed gulls had been indicated to be the dominant bird species at the Fresh Kills landfill (Walt Slavin, NYSDEC Delmar Wildlife Unit, 1988).

Gulls are marine foragers and scavengers and are a protected species under New York State wildlife laws. Previous work has shown that their presence at landfills is worldwide (Kear, 1972). The natural foraging environment for these birds (marine waters) suggests that they would be adapted surface feeders, dependent on sight and animal movement to detect and locate food sources. Common foods for gulls are earthworms, fish, crabs and mussels, which are caught alive via specialized strategies. These same strategies equip gulls to feed on wastes at the surface of dumps or organisms sighted while perched at or flying over landfills; also, gulls would be able to associate feeding opportunity with the presence of other birds and animals and the movement of vehicles.

At landfills where daily cover regulations are followed, wastes remain uncovered for a relatively short time, for instance until the end of the working day. Gulls must thus feed rapidly to exploit foods of high energy or nutritional value, up until the wastes are covered with a layer of soil. As gulls are both rapid and discriminating feeders, length of landfill feeding opportunity may not be a constraint. Evidence of this is that gulls typically do not spend long hours foraging on a daily basis. They do, however, travel to and from nesting and habitat areas on a regular and somewhat predictable (daily) basis. Waste deposition and covering is relatively continuous at municipal landfills. This creates a more or less continuous food supply and thus supports the establishment of a daily flow of gulls. Larger dumps and landfills can obviously supply large gull populations.

Length of feeding opportunity is determined by the rate of application of daily cover. The mass of digestible food wastes removed by gulls is thus likely to be closely related to the amount of time wastes are exposed.

In order to arrive at closely reasoned estimates of the waste removal effect of birds at landfills, it is necessary to relate the metabolic requirements of these species to food types eaten, daily consumption rates and population size vs. daily disposal rates. T. H. Pearson (1968) showed that lesser black-backed and herring gulls eat insects (in the observed cases *Coeleptera carabida* carapace beetles), worms, offal and fish. Other foods made up about 10% of foods for black-backed gulls, for which worms were 3% and insects were 1% of diet. Offal (meat scraps, bread, etc.) from dumps in the Farne Islands made up about 19% of gull diet. Walt Slavin (New York State Department of Environmental Conservation) had noted that gulls at landfills would eat grain foods such as bread if it contained or was smeared with butter.

Sibly and McCleery (1983) showed that where landfills were within foraging range (30 to 40 km), gulls found greater caloric and metabolic advantage and more reliable foraging at dumps and landfills; and although the food types within foraging range included earthworms, starfish, marine fish, mussels, garbage, farm animal put-outs, food in open garbage cans and food scraps on streets, gulls preferred landfill feedings, with the highest feeding rates for male herring gulls. Landfill food waste, 18% of that disposed, was of primary interest. Content of gull stomachs, regurgitates to chicks and changes in weights showed food waste had a higher unit weight energy advantage (2.5 kcal/gm wet weight) than worms, starfish and shellfish. Gulls were found to have an average refuse ingestion waste of 12.5 gm/hour. Male gulls ingested at 15 gm/hour and foraged for about 3 hours per day, while female gulls ingested at 11 gm/hour and averaged 2.8 hours at landfills open 8 hours per day and about 43 hours per week. The gulls fed between the times the trucks were dumped and the waste was covered and had a *net* average caloric intake of 97.4 kcal per bird per day. The average basal metabolism rate (BMR) for the gulls was 3.3 kcal/hour and the birds appeared to operate at 2.5 BMR per day. This would indicate that birds feeding solely on refuse would need to digest 76 to 90 gm of refuse per day. Indigestible matter is rejected as fecal material about 6 hours after ingestion (not necessarily at landfills, although gulls may often spend nonfeeding time at a landfill socializing, resting and loafing, which increases the chances of defecation onsite). According to Nagy (1988), digestibility of food wastes for birds is 50% for cooked vegetables, 70% for bread, 70 to 80% for fruit garbage and 80 to 90% for meat scraps. With an average food waste digestibility of 75%, this indicates that gulls would average 71 gm/day of refuse to meet minimum daily requirements. Because this excludes energy for foraging, it is more likely that the daily intake was more represented by the BMR or the birds would need to ingest an average of 111 gm/day of refuse if they fed only at landfills. This means that a gull population of 100 males and 100 females (pairs) could theoretically ingest/remove 269.3 pounds of food wastes per week at a landfill with an operating schedule of 5.5 days (44 hours) per week. Gull populations at landfills of size are usually much larger than this number. The gull population at the Fresh Kills Landfill had been estimated at 10,000 (NYSDEC Delmar Wildlife unit, 1988).

Nagy (1987) also addressed the feeding requirements of birds and other species and showed that dry weight food ingestion could be related with simple mathematical models to body weight of the species involved. Using the Nagy relationships and the data on feeding from Sibly and McCleery (1983), it can be shown that herring male and female gulls of average weights 1053 and 858 gm, respectively, would need to consume 66 and 57 gm dry weight of *usable* food waste refuse per day; for ring-beak gulls (1200 gm average) this would be 73 gm; for petrels (1400 gm) this would be 81 gm; for kittiwakes (386 gm, rare at landfills) this would be 33 gm; for sparrows (17 to 19 gm), theoretical dry weight ingested would be 4 to 5 gm and for starlings (74–85 gm) food ingestion would be 16 gm.

Sibly and McCleery (1983) also estimated the numbers of gulls at eight landfills of operational times of 34 to 43 hours/week and showed that gull numbers were closely related to the tonnage of wastes disposed per week. The data of these authors can be used to develop a relationship between wastes disposed and gull number. Regression with gull numbers and tons/hour at the eight landfills studied yields the relation:

$$G = 3.57 \times (\text{TFWW})^{1.392}$$

for a correlation of 0.98, where G = gull number and TFWW = tons of food waste per week (44 hour work week).

The above relation indicates that even for U.K. municipal waste of 18% food content (U.S. landfilled waste is usually 8 to 12%), only a small percentage of the food waste would be available to birds such as gulls. Using this relation and the case of an average-sized U.S. landfill with an input of 400 tons/day, of 11% food refuse, 242 tons of food-related refuse would be received in a 44-hour week, which could support a gull population of 7430 *herring gulls*. Based on the consumption rates of the latter, 110 gm/day, they would ingest/remove 4.95 tons of food wastes in a 44-hour week if they fed only at the landfill. This amounts to a removal rate of 2.1%. The amount removed can be further reduced by bird control methods and more rapid placement of daily cover; thus, a more conservative removal range may be 0.2 to 3% removal for food wastes. For birds other than gulls, the volume of available food waste also depends on length of waste exposure. With the typical activity level favoring large, aggressive bird species, the role of smaller birds is likely to be more minor than for gulls. The former, however, may be attracted by the exposed insect and worm species.

The implications of the above to waste site analysis are that the role of birds in the removal of food wastes may be minor but are of statistical and operational importance. Long-term complaints from residents near Fresh Kills Landfills during its operation have included that their yards and households were *bombed* by gulls in their flight/foraging patterns — which, though a nuisance, also suggests that landfill surface inoculation by gull fecal bacteria may be more than expected. Possible landfill management options include the exclusion of food wastes altogether, not puncturing food waste bags and rapid covering of food wastes, with potential advantage to nearby citizens and to aircraft. The implications for landfill initial decomposition/removal rates, however, are that the presence of gull species at landfills is likely to be of degradation benefit.

## WASTE REMOVAL BY INSECTS AND SOIL MESOFAUNA

The role of insect populations at sanitary landfills and waste dumps is a relatively neglected topic. The influence of insects on the colonization of shallow buried wastes by smaller organisms is likely to be significant, especially in undisturbed conditions. Many insects can feed directly on waste or litter microflora and microfauna and relative influence depends on the type of insect, waste, accessibility and insect breeding habits. In one of the few studies of soil and waste fauna at landfills, Chan et al. (1997) reported the following species for two large Hong Kong landfills with (temporary) soil cover of compacted granite soil of 50 cm depth over municipal wastes.

**TABLE 4.4**
**Landfill Cover Soil Fauna at Two Hong Kong Landfills**

| Species | Name | Numbers/Cubic Meter |
|---|---|---|
| *Collembola* | Springtails, 0.16–6 mm | 27,000–63,400 |
| *Psocoptera* | NA | 0–3680 |
| *Thysanoptera* | NA | 0–923 |
| *Hemiptera* | Milkweed and stink bugs | 0–437 |
| *Lepidoptera* | Butterfly, moth larvae | 0–1680 |
| *Diptera* | Beetles, fleas | 1000–6140 |
| *Hymenoptera* | Wasps, 11–44 mm | 8–37/m$^2$ |
| *Coloeptera* | Wood eaters, 4–30 mm | 1960–4270 |
| *Nematoda* | Roundworms, 0.5–2 mm | 0–7650 |
| *Diplopoda* | Millipedes, 1.3–40 mm | 0–3940 |
| *Oligochaeta* | Earthworms | 0–591 |
| *Isopoda* | Sowbugs, pill bugs, 1.3–20 mm | 16,100–32,900 |
| *Acarina* | Soil mites, 0.16–3 mm | 10,300–20,400 |
| *Aranae* | Spiders, 0.5–7 mm | 0–6 |

The most populous species are clearly *Collembola, Isopoda, Acarina, Diptera* and *Coleoptera.* Also carbon dioxide and methane levels in landfill soil at these landfills were reported as high, with the higher levels associated with lower numbers of *Collembola.*

Gray, Sherman and Biddlestone (1974) characterized the microfauna of (aboveground) composting refuse as ants, springtails (*Collembola* spp.), millipedes, worms and nematodes.

Wastes buried and compacted under soil are unlikely to be accessible to many of the small soil animals discussed. Degree of accessibility depends on depth in the soil profile and degree of compaction. Fauna that can burrow through soil include ants, earthworms, millipedes, springtails and larvae of insects. The duration of refuse impact depends on distance of the waste from ground surface and length of time before the soil layer is recovered with waste and soil. It would be a futile effort to attempt to analyze these effects at this point, considering the lack of data and limited investigation; nevertheless, a general discussion provides some insight into the possible quantitative role of insects and soil fauna in the removal/decomposition of refuse buried in waste sites.

The feeding habits of soil mesofauna range from predatory to generalistic. Types of food preferred and metabolic consumption rates can provide some insight into quantitative effects on MSW buried in the soil.

## IMPACT OF WORMS AND NEMATODES

Worms, including the nematodes, the enchytraeid worms and the lumbricids (including the earthworms) are ubiquitous in most soils with organic content. The presence of earthworms can significantly increase the rate of decomposition of residues near the surface or in soil through burrowing, feeding and casting (Mackay and Kladivko, 1985) and earthworms can increase the water retention capacity of soils (Stockdill, 1982), soil grain stability, pore size and infiltration rate. Earthworms are active in fragmentation of feeding materials, which can hasten decomposition by smaller organisms. The fragmentation occurs as a result of grazing on bacteria, fungi, plant roots and protozoans (Coleman and St. John, 1987).

Actual numbers, biomass and contribution to wastes conversion or removal depend on soil characteristics and how those characteristics were modified by environmental influences. Representative counts such as 2.5, 40 and 220 earthworms/square meter have been reported in acid, sandy soil; acid brown fertile deciduous forest soil; and in moderately acid, fertile soil with good crumb structure, respectively (Staaf, 1987). Development of soil crumb structure can take a long time, even when soils are undisturbed, due to the difficulties of soil organisms colonizing the site (Weidemann et al. 1980). Dunger (1968) and Schriefer (1976) estimated earthworm migration from surroundings to soil at wastes dump sites at 10 and 4 meters per year, respectively; and Schreifer found colonization only by species with horizontal, near-surface burrows. The results of Staaf (1987) generally indicate the unfavorability of low pH soils to worm species.

One of the few discussions of small soil animals at landfills has been by Weidemann, Koehler and Schriefer (1980). This study, which focused on needs for establishment of viable biotic communities at abandoned dump sites, stated that adverse conditions for nutrient recycling within the biotic community and for biomass rate regulation, include shape of the dump, degree of soil compaction, the widely fluctuating soil temperatures present until vegetative cover has been established and level of methane production.

Methane production can be accompanied by anaerobic or near-anaerobic conditions near the upper part of the waste-soil profile. These anaerobic conditions can influence the availability of important nutrient sources such as fungi, bacteria and protozoans to worms, as anaerobic conditions are likely to limit the biomass quantity of these groups. While many bacteria are capable of anaerobic growth, fungi — which are obligate aerobes — are relatively sensitive to reduced oxygen levels (Griffin, 1972), which would limit the presence of fungivores. The presence of large numbers of anaerobes does not necessarily create favorable conditions for the presence of bacteriovores because of its coupling with reduced oxygen levels.

The co-production and relatively elevated level of carbon dioxide at inactive landfills can also affect the pH of the liquid phase with which soil animals such as worms must come into contact. Carbon dioxide can be sorbed from landfill gas into soil water as the gas migrates through the soil. Because this strips carbon dioxide from the gas, it increases its relative methane content and increases the acidity of the soil water, as $CO_2$ sorption results in bicarbonate ion formation.

The toxicity of carbon dioxide to soil animals and flora depends on its ratio to oxygen present. Russell (1971) reports that for fungi, partial pressures of 0.21 atm of oxygen and above 0.1 atm of $CO_2$ were toxic. According to Dickinson (1974) most soil animals avoid (soil) regions of low oxygen concentration. The result of these factors may be the creation of *gas deserts* (Neumann, 1976; Schreifer 1981), which could possibly be rectified by *gas drainage* (Pierau, 1975; Tabsaran, 1976) (reported by Weidemann, Koehler and Schriefer, 1980). The results of the Hong Kong landfill studies by Chan et al. (1997) also suggest the negative effect of landfill decomposition soil gases such as $CO_2$ and methane on the presence of earthworms. A combination of unfavorable pH and low oxygen concentration in the liquid phase of landfilled wastes and soils may be one of the most serious limitations for colonization and removal of wastes in the landfill by soil fauna. Conversely, where these conditions are not severe or mitigated by aeration for sufficiently long periods and with the presence of sufficient moisture, the landfilled wastes decomposition effect of worms and soil animals could be increased.

The raised shape of landfills or dumps or compost piles also causes microclimatic variations at the surface as well as more exposed conditions, which might affect the capacity of small soil-dwelling organisms to survive. Some effects of this could be more rapid moisture and temperature fluctuations near the surface. Establishment of complete vegetation cover on the surfaces of inactive landfills would, however, dampen biotically adverse temperature fluctuations. Unfortunately this would occur late after waste coverage.

In many cases local soils used for daily cover are more sand than clay. Such soils may not retain sufficient water to support significant populations of worms and other soil animals, especially under exposed sun and wind conditions.

Weidemann et al. (1980) have referred to refuse dumps as isolated islands within urban or industrial areas that retard site colonization by slow-running hygrophilic and epigeic animals such as soil worms. Animals such as the drought-resistant *Collembola* can colonize landfill soils more rapidly, as they are easily transported by wind. This probably indicates the greater likelihood of landfill upper waste-soils colonization by *Collembola* rather than by worms during the early stages of landfills, before covers and subsequent layers are placed. Daily covering with soil adds biota to the buried waste profile, as the soil placed can contain numerous species. Colonization as a result is, however, likely to be limited because soil mixing, placement and compaction expose these species to extreme environmental conditions. Most earthworms introduced with daily landfilled cover and associated with plant roots are likely to be killed by mixing and compaction (Weidemann et al., 1980).

Earthworms are the only soil animals that can form new pore space (Weidemann et al., 1980), which indicates that their presence in the upper waste profile could hasten decomposition. It has been reported, however, that in terrestrial soils with a crumb structure, with space to move in water films on the crumbs and with a continuous water film present, relevant worm species should have little difficulty in movement as the vermiform shape is well suited to passage through narrow openings. Thus, plant material can readily become colonized by smaller worm species and by larger species once the soil becomes more porous.

Soil compaction as practiced with daily MSW covering and final cover placement increases runoff and evaporation while preventing normal moisture infiltration. Compaction can also result in lower landfill layers becoming waterlogged. It is possible that tightly packed soils could prevent the migration of larger worms to buried food sources. Capping as practiced with a final compacted clay or clay geotextile cover is likely to discourage the vertical migration of worms beyond the topsoil layer because the fine, stiff material presents resistance to burrowing. The influence of worms in such landfills is thus likely to be insignificant during the period of inactivity or after a final cover has been placed. However, the presence and utilization of earthworms in composting is well known in vermicomposting; therefore, the compost site might be a better microcosm to study the quantitative impact of earthworms.

Considering the likelihood of their presence at some stage of landfill existence, worms could still contribute a sizeable influence on landfill waste removal dynamics if suitable conditions are present. Such conditions, as the above discussion suggests, would include sufficient oxygen and moisture, sources of nutrition and normal temperature ranges.

It has been noted that the abundance and presence of soil-dwelling animals in horizons of aerated soils is related to the presence of nutritional sources within that profile. Common nutritional sources include bacteria, fungi, protozoans, roots and plant matter in late stages of decomposition. While other environmental factors are likely to be at least as important to the presence or absence of worms at landfills, it is likely that nutritional sources would not be limited. A nutrient limitation, however, could be the influence of overall high C/N ratio, considering the dominance of MSW by low-nitrogen paper wastes. The strong presence of fungi and bacteria in the more aerated soil fractions, especially in uncapped landfills with low gas production after lengthy activity, could abet the presence of worm species. Moore, Walter and Hunt (1988) report that except for fungivores, nematodes potentially derive most (99.5%) of their energy from bacteria; and their production ratio (mg N nematode/mg N bacteria), excluding root feeders, was on the order of 2.2/100.

Important environmental influences likely to limit the numbers of worms at or in landfills such that their relative contribution to MSW conversion is small include resistance to worm movement through compacted and poorly aerobic soil profiles to reach nutritional sources, the need for relatively high soil humidity and oxygen levels, adverse carbon dioxide levels and unsuitable soil temperature.

Nematodes are essentially aquatic, requiring a humidity level such that a liquid water film is present to maintain their activities and movements through soils. Rewetting soil is known to increase the numbers of nematodes present, which can be due to reactivation of resting worms, increased nutrition due to increased growth of bacteria and fungi, and the hatching of eggs over a sufficient period of time. The effect of moisture changes on nematode numbers is said to be related to increase in soil oxygen and increased bacterial activity. Soil nematodes can survive dry conditions by coiling up and entering a state of anhydrobiosis (Schnurer et al., 1986) and activating rapidly after wetting (Whitford et al., 1981); but generation times are relatively long for eggs deposited in soil. Mackay and Klavdivko (1985) report that earthworms have their greatest activity periods in spring and late autumn, which indicates the importance of sufficient moisture and favorable temperatures to the existence and contributory activity of worms in landfills. According to Griffin (1972) the motility of nematodes and enchytraeid worms is directly affected by water content. This would limit the favorable humidity range for most worms to landfill areas sufficiently wet. The limiting water content for the presence and activity of worms is unavailable, but Stockdill (1982) indicated no earthworm activity below -1500 kPa soil tension. For soils this is relatively moist, which would indicate that, even under aerobic conditions, the overall presence and effect of nematodes on landfilled wastes are likely to be small.

The presence and effect of worms are likely to be more important for other types of waste sites such as composting sites, biofilters, etc. However, heat buildup and moisture loss in compost sites could discourage development of large worm populations.

## SPRINGTAILS (*COLLEMBOLA*)

The adaptability, versatility and soil movement capacity of *Collembola* make it likely that this species would be well represented, though of limited waste removal impact, in waste site soils (compost piles are again an exception). The springtails (taxon *Collembola*) are well-known mandibulate arthropods of average live weight size range 3 to 718 mg (full range 30 to 3160 mg), which preferentially feed on decaying moist leaf and grass materials (and microbes in fecal pellets of other organisms). Litter feeding and preference is often species-specific and dependent on position in the soil or litter profile. Materials on which springtails feed include decaying roots, twigs, leaves and branches, moss, lichen, algae, cellophane, filter paper and nematodes. The larger springtails such as the *Tomocereus* spp. tend to feed deeper in soil and may prefer fungi, with decaying leaf litter as an alternative (Anderson and Healey, 1972). Anderson and Healey (1972) also showed that smaller species such as *Orchesella flavescens* fed almost exclusively on unmodified leaf litter, while larger species such as *Tomocereus* spp. had gut content of about 40% fungal material.

The importance of fungi as feed for larger springtails suggests that, for shallow buried wastes, a solid waste soil profile rich in hyphal and conidiophoric fungi and leaves could support a sizeable population of larger springtails. The relative importance of this association and the potential for springtails to be present in MSW composts, vermicomposting operations, in decaying litter, in soil biofilters and in stable soil situations where litter is buried or near the surface suggests the usefulness of further review of the potential impact of this group.

Naglitsch (1966) showed that springtails were associated with the humification state of grass and straw decomposition (Naglitsch, 1966); and Curry (1979) showed that springtails multiplied rapidly in fine mesh nylon bags of litter at the surface, but more gradually in litter bags buried 3 to 4 inches below the soil surface. In the latter case the *Collembola* built up large populations in a microorganism-dominated environment, after about 9.5 months (Curry, 1969). Werner and Dindal (1987) have noted that *Collembola* migrate into litter in search of fungi, bacteria and yeast and that surface *Collembola* tend to be larger and more active, while soil dwellers tend to be smaller, less productive and adapted to a constant supply of poor food. The food quality of refuse would not necessarily be poor, thus this contention may not hold for some *Collembola* species at a landfill. The foregoing conditions suggests that the ideal situation for *Collembola* in waste below the surface of a landfill would be aerobic, moist, yard-waste-rich and have relatively undisturbed soil conditions, such as at an incompact or lightly compacted soil-waste profile at an inactive section of a landfill without final or intermediate cover. Operational factors such as soil compaction and daily cover depth are likely to limit these conditions such that, if they exist, they are likely to be patchy and relatively near the waste site surface.

The role of *Collembola* in the sizing, shredding and colonization of buried MSW is thus likely to be limited, even though these soil fauna are one of the most common groups in all types of soil. This is because packed soils of any depth, with rich decomposition gas content, are not likely to be hospitable environments for *Collembola*. Consideration of quantitative impact is useful because of the ubiquity of this species.

*Collembola* are known for cutting and skeletonizing leaf and litter wastes, increasing the decomposition rate as surface is increased and colonizing and disseminating microorganisms through the ingestion and egestion of fungal spores and bacteria (Moore and Walter, 1988). Consumption rates for *Collembola* on wastes relate to body size, metabolic value of wastes and rate of respiration. Some *Collembola* consumption rates reported are 0.045 mg/gm body mass for feeding on forest litter (Kowal and Crossley, 1971); 0.049 mm/day on various tree leaves for *Folsomia fimetaria* (Dunger, 1956); 0.07 mm/day on leaves for *Onychiurus armatus* (Witkamp, 1960); 20% of dry body weight/day at 40 to 70% assimilation (Petersen and Luxton, 1982); 7% of dry body weight on fungi and 2.5% as predators (Werner and Dindal, 1987); and 0.002 to 6.9 mg/day/gm body mass (Slansky and Scriber, 1985). The dry weights of these animals are typically 45% of the live weight.

Work done by various authors on consumption, respiration, production and caloric value show that ingestion can be related to total *Collembola* biomass and caloric value of the material. Straightforward estimates using the relations developed by Harding and Stuttard (1968), Reichle (1971) and McNeil and Lawton (1970) allow one to arrive at the estimation for *Collembola* of ingestion of 0.007 to 1.62 calories/individual, 0.6 to 31.4 micrograms/individual/day for *Collembola* feeding on refuse at the caloric values listed and 0.4 to 21.4 micrograms/individual/day for Collembola feeding on leaves. Obviously leaf matter might be preferable to refuse, but this depends on the comparative degree of waste or leaf colonization by soil fungi or bacteria. Older, shallow buried digestible wastes are likely to be better colonized than fresh MSW, so soil *Collembola* population growth might be favored for wastes buried for some time.

In terms of quantitative landfill site impact, though minimal, one approach is to relate these estimated consumption rates to populations likely to be present at some stage of waste burial. Dunger (1968) and Lebrun (1971) have listed the range of *Collembola* at 30 to 3160 mg live weight/m$^3$ in terrestrial habitats; and Curry (1969) has reported numbers of *Collembola* in buried coarse- and medium-mesh bags with grass and plant litter at 1500 to 2500 average (to 30,000 in fine bags). A rough estimate with the above and the mass of litter yields an average of 100 *Collembola*/gm of buried litter or 1000/gm for smaller *Collembola* species. With the various adverse environmental factors likely to be present at landfills, a conservative estimate might be 200 mg live weight per square m. This suggests a lower soil range of about 6000 individuals/square m for smaller *Collembola* species (e.g., *Isotoma, Onchyiurus, Procampatus* and *Folsomia*).

Ingestion and assimilation rates of 5 micrograms/individual and 40%, respectively, and individual average weights of 30 micrograms suggest a potential refuse removal rate of 13.3 mg/m/day; or that a landfill section of size 10,000 m (2.47 acres) could have a refuse removal rate of 133 gm/day (53.8 gm/acre/day) due to *Collembola*. This amount would be large for compost pile operations due to more favorable conditions and greater organic waste content. At a possibly conservative two months per year when this could happen (previous work discussed stated that several months [9.5] would normally pass before *Collembola* in deeper soil layers become acclimatized and establish populations of size), it is suggested that removals/year due to *Collembola* might be of the order of 3 to 4 kg/year.

The accuracy of this estimate cannot be determined at this time. In terms of relative impact of *Collembola* and other species, including worms, it can be easily seen that the population sizes and thus the order of accuracy of this type of discussion would depend largely on the operational history of the landfill. A site with thinner and less clayey soil covers and without high degrees of soil compaction could experience much higher rates of refuse removal, possibly of one order or more of greater magnitude.

## WASTE SITE MICROORGANISMS: FUNGI, YEAST AND BACTERIA

The various microorganisms found associated with landfills, decomposition of solid wastes and refuse composting are tabulated in Table 4.5.

## SOIL FUNGI

Though the role of fungi has long been considered fundamental to aerobic decomposition of vegetable and animal matter above and at ground surface, for example in composting of organic wastes, the landfill presents a specialized environment. Only the upper soil layers are sufficiently aerobic to support significant fungal mass. The aerobic landfill zone can reach three meters depth (Palmisano et al., 1993) and waste layers are likely to always be included in this depth profile.

The above suggests that soil fungi — which can grow under relatively harsh conditions (drier than bacterial and acidic conditions) — could play a key role in decomposition via waste hydrolysis before the organic waste environment becomes anaerobic through depletion of soil oxygen, burial by additional layers of landfill, during drier landfill conditions and under imposed landfill aeration.

The critical relation of soil dissolved oxygen content to mass of fungi likely to be present in landfills suggests that fungal mass and waste surfaces concentration in MSW would be highly dependent on availability of oxygen in the soil/waste matrix. Air and oxygen levels depend on accessible porosity (to air) of the landfill. Accessible porosity depends on the moisture content of the soil/waste matrix as well as its packing and density characteristics. The impact of fungi on landfilled solid wastes must thus be estimated using reasonable assumptions about the physical and dynamic characteristics of landfill porosity.

In the near-surface profile at uncapped landfills, fungal growth can be aided by plant roots penetration (inactive landfill area), burrowing by soil fauna and moderate to light soil compaction.

In a landfill still generating abundant carbon dioxide, high levels of this gas compared to the $CO_2 - O_2$ optima for fungi could be inhibiting to the growth of fungi in the moister, deeper aerobic soil zone, even if sufficient moisture is present, because the greater solubility of carbon dioxide than oxygen in the water film (Garrett, 1936) around hyphae would increase the stress on the fungi. Dickinson (1974) notes that most soil animals avoid regions of low oxygen concentration but that $CO_2$ can stimulate fungi growth and reproduction below. However, if oxygen is present at 0.21 atmospheres, $CO_2$ could be toxic at partial pressures above 0.10 atmospheres. Dickinson reports that such $CO_2$ concentrations are uncommon in most soils. It is likely that landfills provide the exception to this, as $CO_2$ levels are likely to be elevated compared to most other soils; thus, inhibition of fungal activity could result for most of the landfill.

Landfill liquid phase pH could also be a limiting factor if low in value and if an abundance of carbon dioxide were present; as pH decreases, the amount of dissolved $CO_2$ in the water film would increase, decreasing the availability of oxygen. Soil fungi are also unlikely to grow very near the surface of landfill soils because of humidity constraints, such as relative humidity levels in soil below 95%, in equilibrium with

**TABLE 4.5**
**Landfill and Solid Waste Microorganisms**

| Organisms and Conditions | Species or Genus | Soil Count | Old MSW No. ($\times 10^{-6}$) | New MSW No. ($\times 10^{-6}$) | Reference |
|---|---|---|---|---|---|
| Macrofauna, aerobic compost | Genus: Worms | NA | NA | NA | Alexander, 1961 |
| Macrofauna, aerobic compost | Genus: Millipedes | NA | NA | NA | Alexander, 1961 |
| Macrofauna, aerobic compost | Genus: Springtails | NA | NA | NA | Alexander, 1961 |
| Macrofauna, aerobic compost | Genus: Nematodes | NA | NA | NA | Alexander, 1961 |
| Macrofauna, aerobic compost | Genus: Ants | NA | NA | NA | Alexander, 1961 |
| Macroflora, aerobic compost | Genus: Fungi | NA | NA | NA | Alexander, 1961 |
| Microfauna, aerobic compost | Genus: Protozoa | NA | NA | NA | Alexander, 1961 |
| Microflora, aerobic compost | Genus: Bacteria | NA | NA | 100–1000 | Alexander, 1961 |
| Microflora, aerobic compost | Actinomycetes | NA | NA | 0.1–100 | Alexander, 1961 |
| Microflora, aerobic compost | Fungi | NA | NA | 0.01–1 | Alexander, 1961 |
| Microflora, aerobic compost | Algae | NA | NA | 0.01 | Alexander, 1961 |
| Microflora, aerobic compost | Viruses | NA | NA | NA | Alexander, 1961 |
| Predominant fungi: aerated refuse | Trichoderma | NA | NA | 8–40 | Mahloch, 1970 |
| Aerated refuse | Geotrichum | NA | NA | 30 | Mahloch, 1970 |
| Aerated refuse | Rhizopus | | NA | 7 | Mahloch, 1970 |
| Aerated refuse | Penicillium | | NA | 11–18 | Mahloch, 1970 |
| Aerated refuse | Aspergillus | | NA | 10 | Mahloch, 1970 |
| Aerated refuse | Cladosporium | | NA | 4–20 | Mahloch, 1970 |
| Fungi, composts | Monilia | | NA | | Mahloch, 1970 |
| Fungi, landfills | Fusarium | | NA | | Eliassen, 1942 |
| Fungi, old compost | Scopulariopsis | | 54.3 | 47.8 | Ahrens et al., 1965 |
| Fungi, old compost | Mucor | | 11% | | Mahloch, 1970 |
| Fungi, old compost | Penicillium | | 40% | | Mahloch, 1970 |
| Fungi, old compost | Geotrichum | | 33% | | Mahloch, 1970 |

*(Continued)*

**TABLE 4.5**
**Continued**

| Organisms and Conditions | Species or Genus | Soil Count | Old MSW No. (x $10^{-6}$) | New MSW No. (x $10^{-6}$) | Reference |
|---|---|---|---|---|---|
| Fungi, old compost | Sporotrichum | | 4.5% | | Mahloch, 1970 |
| Fungi, old compost | Trichoderma | | 4.5% | | Mahloch, 1970 |
| Fungi, old compost | Aspergillus | | 6% | | Mahloch, 1970 |
| Yeasts, aerobic compost | Streptomyces | | | | Ciencaile, 1966 |
| Yeasts, aerobic compost | Micromonospora | | | | Glathe, 1959 |
| Yeasts, aerobic compost | Micromonospora | | | | Goleuke, 1954 |
| Bacteria | | | | 17.2–31.4 | Burchinal, 1966 |
| Bacteria | | $10^{6} - 10^{9}$ | | | Burges and Raw, 1967 |
| Bacteria, anaerobic, landfill | E. coli | | 0.33 | | Eliassen, 1942 |
| Bacteria, anaerobic, landfill | Coliforms | | 0.74 | | Eliassen, 1942 |
| Bacteria, anaerobic, landfill | Lactose fermenters | | 1.46 | | Eliassen, 1942 |
| Bacteria, landfill | | | 12,060 | 27,602 | Ahrens et al., 1965 |
| Bacteria, aerobic, composts | Micrococcus | | | | Cienciala, 1966 |
| Bacteria, aerobic, composts | Sarcina | | | | Cienciala, 1966 |
| Bacteria, aerobic, composts | Staphylococcus | | | | Cienciala, 1966 |
| Bacteria, aerobic, composts | Proteus | | | | Cienciala, 1966 |
| Bacteria, aerobic, composts | Pseudomonas | | | | Cienciala, 1966 |
| Bacteria, aerobic, composts | Bacillus | | | | Cienciala, 1966 |
| Bacteria, aerobic, composts | Corynebacterium | | | | Cienciala, 1966 |
| Bacteria, aerobic, composts | Bacillus | | | | Glathe, 1959 |
| Bacteria, aerobic | G– nonsporulants | | | | Burchinal & Wilson, 1966 |
| Bacteria, aerobic | G+ nonsporulants | | | | Burchinal & Wilson, 1966 |
| Bacteria, aerobic | Lactobacillus | | | | Burchinal & Wilson, 1966 |
| Bacteria, aerobic | Propionobacterium | | | | Burchinal & Wilson, 1966 |

(drier) air (Griffin, 1963). Work by Garrett (1938) has shown that forced aeration of soil stimulates fungal growth; thus it is possible that, in inactive landfill sections, significant changes in barometric pressure as well as rapid flooding and draining might stimulate soil fungi. All of this suggests a reduced effective surface depth profile for natural fungi at landfills; i.e., a possible upper aerobic, pH-neutral zone.

Fungi abundant in soil include well-known waste decomposers such as *Aspergillus* sp. and *Penicillium* sp., with the former more common in warmer soils and the latter more frequent in temperate, wetter soils (Pugh, 1974). Both groups are quite important because of potential adaptability to climate-related temperature-moisture changes possible in upper landfill soils, including ability to withstand drier conditions than most organisms and adaptability to drying-wetting cycles. For example, *Aspergillus* has been reported to show xerophilicity at temperatures greater than 30°C, mesophilicity between 20 to 30°C and hygrophilicity below 15°C (Bonner, 1948).

The above can be interpreted as meaning that landfill *Aspergillus* species can be active in dry, relatively hot conditions, thrive at normal temperatures and moisture content and be sensitive to water content of the landfill layer in colder conditions. This would ensure activity of lower consumer species such as bacteria because the fungi have an important primary role as waste hydrolyzers.

In the absence of landfill soil fungi studies, the relative importance of *Aspergillus* sp. cannot be confirmed. Other waste-decomposing species potentially present include mesophiles such as *Fusarium, Rhizopus, Phycomycetes, Trichoderma* species and hydrophiles such as *Verticillium* (Schmedieknecht and Pilsk, 1960). Pugh (1974) reports soil fungi species as also including *Cladosporia, Alternaria* and *Mucor*. Maloch (1970) reported the predominant fungi present during aerobic decomposition of solid wastes as *Trichoderma, Geotrichum, Rhizopus, Penicillium, Aspergillus* and *Cladosporium*, with numbers in rotting compost and MSW ranging from about 100,000 to 100 million per dry gram of substrate (waste paper or grass and yard wastes).

Russell (1923) reported that fungi numbers dropped very rapidly with depth in soils, with numbers in (undisturbed) soils relatively constant in the top 4 to 6 inches, rapidly declining between 5 to 9 inches into the soil profile and dropping to only a few, or absence, at 20 to 30 inches — with fertilized, acid conditions favoring higher numbers and minerals and lime favoring the lowest numbers. The lower range of numbers was about 26,000/gram of soil to about 100,000/gram of soil. It is clear that landfill upper soil conditions — except for thin, incompact or loose soil conditions, including limited soil cover and mixed soil conditions of the early life of landfills, and the potentially adverse landfill soil conditions mentioned above — should not favor great numbers or mass of fungi; thus the lower numbers might be of greater interest to calculation of fungal impacts on landfill decomposition. This may be offset to some degree by the nutrient value of rotting wastes in the upper profiles and the increased porosity provided. A cautionary approach might be conservatism in the choice of soil fungal mass numbers to represent landfill soil fungi. It has been shown that in leaf litter treated with pesticides to exclude the effect of soil mesofauna, the litter fungi present might have been responsible for about 35 to 40% loss of dry weight after 48 weeks (about 0.1% dry weight loss per day (McCauley, 1975)). As organic MSW fractions such as cellulosic material are likely to be more nutritionally refractive than leaves, weight loss due to fungi might be slightly less than this.

A simple calculation using conservative figures — such as a 1 meter thick soil/waste layer of 25% soil covering a hectare (100 × 100 m), 75% MSW, density 1500 kg/m, 90,000 fungi (about 0.06 mg)/gram (Russell, 1923) of dry organic waste about 60% degradable by weight and waste and soil moisture contents of 50% and 21%, respectively — a degradable MSW mass of about 2.5 million kg of dry organic MSW/hectare (density 653 kg/cu m compacted) might have a fungal mass of at least 116 kg/hectare (0.116 $kg/m^2$) for a one meter soil/waste depth. The accuracy of this estimate cannot be determined.

### LANDFILL BACTERIA

The critical qualitative and quantitative role of landfill bacteria (aerobic, facultative and anaerobic) in any discussion and the complexity of the topic merit a more thorough study than a discussion of this kind. Greater than 50% of all daily or annual landfilled waste decomposition rates are likely to be directly associated with soil bacteria, which represent hydrolytic, acidogenic and methanogenic aspects of waste decomposition. Palmisano et al. (1993) have reported the number of total bacteria associated with 3 meter deep samples of MSW buried in three modern U.S. sanitary landfills (Fresh Kills Landfill, Staten Island, New York; Los Reales Landfill, Tucson, Arizona; and Naples Landfill, Naples, Florida) at $10^{10}$ bacteria per gram dry weight of refuse, with aerobic bacteria ($10^4$ to $10^7$ per gram MSW dry weight) being as well represented as anaerobic bacteria ($10^5$ to $10^8$ per gram dry weight MSW); and hydrolytic bacteria were only 0 to 15% of the total bacterial count. This study also discussed work by Donnelly and Scarpino (1984) and Pahren (1987), in which the incubation of MSW in a model landfill for 20 months resulted in the presence of $10^5$ to $10^6$ fungi, $10^6$ actinomycetes and $10^6$ to $10^8$ aerobic proteolytic bacteria per gram dry weight of MSW. These counts are valuable to any qualitative and quantitative estimate of landfilled waste ecology and decomposition.

## SUMMARY

The examination of organisms reported or possible at waste sites suggests important and primary influences on the rotting and conversion of waste material. The roles of these organisms are highly interdependent. Physical and environmental factors are important constraints to feeding and growth success of nearly all organisms, although the waste site should otherwise provide a rich source of nutrition for many of the organisms discussed. For surface-dwelling animals, dominant species are likely to be omnivores, opportunistic, aggressive and fast-moving or -feeding. Birds and rodents are likely to be the dominant groups of large animals, while insects and flies are likely to dominate among smaller aboveground organisms. For belowground animals, soil moisture content, heat and soil can determine numbers and level of impact. Most belowground fauna are favored by moist, aerobic and porous conditions as well as the presence of fragmented and decaying plant matter that can support extensive colonization by bacteria, fungi and protozoa. Soil fauna groups that are highly adaptable to disturbed conditions such as mites, nematodes and *Collembola* are likely to dominate, with an uneven distribution pattern reflective of organic richness in particular locations at the waste site.

To derive quantitative estimates of the effects of any organism species present at a waste site, key factors to be defined include (1) types of materials targeted by the organism and average nutritional value, (2) feeding opportunity or niche and its duration, (3) daily food requirements and (4) degree to which feeding on or removal or conversion of wastes is limited by site location, waste distribution, climate and the impact of other organisms. Quantitative impacts of large animals and scavengers are also important where waste sites or dumps are within foraging and are accessible to large animals and where human scavenging is allowed. These quantitative impacts can be estimated using simple approaches and permit consideration of whether particular site conditions or operation standards increase or limit the impact of certain types of animals. While the waste removal impact of many organisms may be minor, in terms of individual species, the discussion has shown how the mixture of wastes disposed creates the involvement of a host of organisms, the cumulative impact of which is of relatively high importance to the waste site modeler. The anecdotal or limited supporting data on interdependency and interactions between organisms at waste sites, however, limits accuracy of estimation of quantitative impact, with organisms such as fungi and bacteria as the exception. Future ecological studies at waste sites should illuminate the trophic roles of individual organisms and how they affect waste decomposition and thus the wastes management process.

# 5 Moisture and Heat Flows

## MOISTURE AS A CONTROL OF PROCESSES IN THE WASTE SITE

A key influence over decomposition processes in waste sites (landfills, dumps and compost operations) is water content of the soil-waste environment. Availability of water to microorganisms controls growth, the conversion and transport of substrates, and heat generation and transport. The limiting effect of water content has long been noted in landfilling and composting studies. In the latter, water is often supplied to ensure a moist environment for optimum aerobic decomposition. If a focus of modeling is simulation of waste sites as reactors, the impact of water content on biological processes cannot be ignored as a topic.

*Moisture availability*, as important as *moisture content*, is often a major consideration for microbiological processes. Drying out of waste material is known to slow rates of decomposition. In many excavated landfills, paper and food wastes have been found essentially intact and dry or relatively unexposed, even after decades of burial. This is particularly the case for waste materials sandwiched between materials under soil pressure or enclosed in bags. As dehydration affects soil microorganism activity, it will also affect the waste site system capacity for conversion chemicals contained in wastes and regarded as air or water pollutants that can be released via waste site gas or leachate. The significant influence of moisture on microorganism activity on wastes buried in soil makes a discussion of moisture availability necessary.

Microorganisms in waste sites are essentially attached to pore/outer surfaces of soils and solid waste materials, as colonies or in biofilms. Because of the unevenness of metabolic value of wastes or soil organics and of water uptake and distribution (varying with waste type and local porosity), microorganism concentrations are likely to vary with location and biofilms may be discontinuous. This patchy condition will reflect a wide range of water availabilities, even within a small segment of a waste site. While this complicates predictions, water availability to soil organisms is a property of the materials in the microorganism environment and governed by water supply. Thus the availability of water can be considered in terms of sorption behavior and moisture supply.

Water availability to microorganism colonies in soil depends on the energy gradient between the potential energy of water at a point in an environmental system and the potential energy of free water (potential energy $= 0$ ). In unsaturated soil and organic materials, water is at a lower potential energy than free water, or it takes more energy to extract. Thus these would have a negative ($< 0$) potential energy (Papendick and Campbell, 1981). The potential for water held in soil at field capacity (water content slightly less than level for gravity drainage) is near zero, e.g., $-8.0$ cm and is typically referred to as the *air entry potential* or *bubbling pressure*.

The air entry potentials for sand, clay and loam soils were −0.007, −0.02 and −0.17 bars (−7.14, −20.4 and −173.3 cm water, respectively) (Campbell, 1974). By contrast, the water potential of a nearly dry soil can be −60 to −100 m, at which potential a large cell energy expenditure would be required for diffusion of solutes across the cell wall.

For water adsorbed by a soil system, water potential $\psi$ is the sum of soil osmotic potential $\psi_\pi$, matric potential $\psi_m$ (sum of the solid material absorption and capillary effects), the gravitational potential $\psi_g$, pressure potential $\psi_p$ and overburden (effect of weight of material above on water movement) potential $\psi_\Omega$:

$$\psi = \psi_\pi + \psi_m + \psi_g - \psi_p + \psi_\Omega = \text{water potential} \tag{5.1}$$

Matric and osmotic potentials are considered the major influences on water potential in plant and soil systems, influencing water flow and availability; and of these, the matric potential is the largest component in unsaturated soils, crop residues and organic materials, while the osmotic is only significant in soils that are saline or amended with organic wastes or fertilizer (Papendick and Campbell, 1981). For soil systems, the water potential is typically taken as the same as the matric potential, as the contribution from the osmotic potential is often neglected.

For water potential of organisms, however, total water potential is the sum of matric $\psi_m$, osmotic $\psi_\pi$ and pressure potential $\psi_p$. The pressure potential is the difference between fluid turgor pressure inside the cell and the liquid phase outside the living cell. The matric potential is the largest, and, in solid materials, such as soils and wastes, is a function of and is controlled by the water content of a material. As water content of a particulate system such as a soil-waste site also determines transport (flow and diffusion) of nutrients to organisms, a reduction in water content affects nutrient availability. At the surfaces of organic wastes, dehydration processes would reduce the water flow and the transport of nutrients into the organism cell. Reduced water flow would increase the concentration of the cell's internal solution, which decreases its osmotic and pressure potential and thus its internal water potential. Organism cells typically attempt regulation for equilibrium with the surrounding environment, but this is likely to involve the loss of cell water to the environment under drying conditions. Cell wall distortion is one response to decreased cell pressure potential, which some organisms such as yeasts with thicker cell walls can better resist than bacteria. Increased internal solute concentration can draw in salts (to a neutral pH internal environment), which can have inhibitory effects on cell enzymatic activity. Sensitivity to drying effects varies with organism groups; for instance, Harris (1980) notes that bacteria are more sensitive to decreasing water content than filamentous organisms and yeasts (with a thicker wall). One reason is that bacteria need a water film (on the surface of materials) for locomotion, while filamentous microorganisms do not (they can bridge air spaces). Campbell and Papendick (1981) report that the motile bacteria, nematodes and Phycomycetes are limited in movement by the size of material surface pores filled with water, with Phytophora fungi requiring water-filled pores of minimum size 40 to 60 $\mu$m for dispersion; eelworm larva (*Heterodera scachtii*) requiring pore sizes of 30 to 60 $\mu$m, corresponding to a water potential of

$-0.25$ to $0.04$ bars ($-255$ to $-20.8$ cm water); and bacteria (*Pseudomonas aeruginosa*) requiring pore sizes of 1 to 5 $\mu$m.

Harris states that the effect of a solid material's matric potential on water film thickness is that drying reduces film thickness — which, apart from reducing the surface area over which bacteria might move, also reduces the diffusion of solutes and removal of metabolic products. With water potential of a material in the dry zone, availability of moisture to organisms as well as organism movement are limited by the strength of surface adsorption forces. The lower the value of the water potential, for instance, $-200$ cm as compared to $-5$ cm, the harder it is for living organisms in soil to extract water or to be surrounded by a film of water as is necessary for decomposition bacteria at the surface of soils or wastes. The need for a water film for movement additionally affects metabolic activities of motile bacteria because, apart from restricting movement to more favorable locations, a thin or decreased water film is coincident with stronger water sorption forces that can trap the organisms and its metabolic products. Campbell and Papendick (1981) report that sorption forces at the air-water (film) interface "can hold the organism down like a rubber membrane."

Thus, actual value of water potential for a type of waste or soil affects microorganism metabolic processes including growth. Water potential can be translated into water film thickness and water content. The work of a modeler of waste decomposition as affected by water content could therefore involve relating water potential to the kinetics of microorganism growth and substrate consumption (or decomposition).

## WATER FILM THICKNESS ON SOLID MATERIALS UNDER SORPTION REGIME

Freely draining material in a waste-soil matrix is subject to removal of free moisture through gravitational downflow (as well as moisture evaporation from the surface). Surfaces remaining are covered with a moisture film, the thickness of which correlates with humidity of gas in the unsaturated pore system. Water film thickness ranges from a minimum correlated with near-dry material conditions to a maximum correlated with the amount of water that can be held in air against gravity, usually directly correlated to surface tension of the liquid involved and coincident with a matric potential at or near zero.

The theoretical average thickness of this moisture film on the surface of a porous material can be described by physical-thermodynamic relationships that have been developed by various investigators to define and estimate the volume of adsorbate (liquid) in a solid material under dry to saturated conditions. As the adsorbate has to cover the surface as single to multiple layers, between dryness and saturation, the amount of adsorbate covering the solid substance in a single layer, one molecule thick, can be related to the surface area by the simple proposition that the volume is a multiple of the surface area and the thickness of the molecule.

Moisture films thus reach a maximum thickness at or near maximum surface tension of the liquid. Bridges of water drops between soil particles, or drops of water held against gravity at the underside of soil particles, achieve a maximum volume or thickness determined by surface tension. The thickness of the water film is thus a function of surface tension at the air water surface for the particular material. Because the air–water interface of the film must be in equilibrium with air or soil gas humidity, thickness also relates to relative vapor pressure of air to water. For water as the liquid adsorbate, equilibrium vapor pressure is reduced below that for saturation by negative pressure from surface tension. For instance, relative vapor pressure at the air–water interface can be 1.0 at saturation, whereas matric suction is negative for a water film held against gravity under surface tension, meaning that its relative vapor pressure is less than 1.0. The pressure reduction under surface tension relates to the curvature of the liquid surface. This pressure-curvature relation, which has often been modified and used in studies of the retention of liquids in porous media (Fisher and Israelachvili, 1979), was first defined by Lord Kelvin (Thompson, 1871), as:

$$r_k = \frac{2\sigma V_m}{(R_g T)\ln(\frac{p_0}{p})} \tag{5.2}$$

where $r_k$ = Kelvin radius = average diameter of open inner pore when pore walls are coated with liquid, $\sigma$ = surface tension of sorbate (for water, $\sigma$ = 72.6 mJ/mole; Naono and Hakuman, 1993); $V_m$ = molal volume of sorbate (for water, $V_m = 18.07 \times 10^{-6}\text{m}^3$/mole; Naono and Hakuman, 1993); $R_g$ = universal gas constant = 8.314 Joules/(mole °$K$); $T$ = temperature in degrees Kelvin; $P_0$ = liquid saturation vapor pressure (equivalent to relative humidity = 1.0 = $a_w$); and $P$ = liquid vapor pressure.

The term $r_k$ only refers to the (average) radius of the material pore not filled with water, so it can be related to the radius $R$ of the pore when dry and the average thickness $t$ of moisture film on the pore surface. The contributions of Coenlingh (1938) and Wheeler (1945) were to provide the more correct interpretations of the Kelvin relationship (Ngoddy and Barker-Arkema, 1970). They stated that, when the (dry) average pore radius was $R$, the thickness $t$ of condensed liquid of cylindrical curvature (along the pore) rather than hemispherical (end) curvature on the pore surface was given by:

$$R = (r_k + t) = (r + t) \tag{5.3}$$

where $r$ defines the inner open radius of the pore when its inner surface is covered with a film of thickness $t$; $r = r_C$ inner pore radius under condensation, for (moisture) adsorption on the surface and $r = r_k$ for moisture desorption ($t = 0$). This modifies the Kelvin (1871) relationship to:

$$r_c = \frac{2\sigma V_m}{(R_g T)\ln(\frac{p_0}{p})} = R - t \tag{5.4}$$

## METHOD I FOR LIQUID FILM THICKNESS DETERMINATION

Halsey (1948), working from the BET sorption model previously described, derived a semi-empirical model for the liquid thickness $t$ and based on the Wheeler (1945) model above, as:

$$t = \tau \left( \frac{Q_{st}}{(R_g T)\ln(\frac{p_0}{p})} \right)^{\Omega} \tag{5.5}$$

where $t$ is in angstroms (Å, $= 1 \times 10^{-10}$ m), $Q_{st}$ = isosteric heat of adsorption of the liquid and $\tau$ and $\Omega$ are empirical constants. Using sorption isotherms of several biological materials, Halsey (1948) estimated the values of $Q_{st}$ and $\Omega$ to be 1.75 and (1/2), respectively, over the relative pressure ($p_0/p = a_w$) range of 0.2 to 0.98. The value of $\tau$ was taken to be roughly equal to the diameter of a molecule of the liquid. In the case of water and the spherical molecule as recommended by Stamm (1964), the water molecule diameter has been estimated to be $3.673 \times 10^{-10}$m (Caurie, 1981) or $\tau = 3.673$ Å. Thus the Halsey (1948) relation for liquid thickness in moisture sorption can be restated as:

$$t = 3.673 \left( \frac{1.75}{(R_g T)\ln(\frac{p_0}{p})} \right)^{\frac{1}{2}} = 4.859 \left( \frac{1}{(R_g T)\ln(\frac{p_0}{p})} \right)^{\frac{1}{2}} \tag{5.6}$$

where $t$ is in angstroms ($10^{-10}$m). Halsey's more general relationship related relative coverage by liquid (to amount adsorbed on the monolayer) to relative pressure; i.e., water volume adsorbed/water in monolayer vs. $P/P_0$:

$$V/V_m = [\alpha/(P/P_0)]^{(1/r)}$$

where $a$ and $r$ are constants. By varying $r$ between 1.0 and 15.0 for adsorption on various materials, Halsey found that for mercury adsorption, the best fit was ($r = 3$). This has been approximated for the layer thickness estimation in angstroms as:

$$t = 3.54 \left( \frac{5}{\ln(\frac{p_0}{p})} \right)^{\frac{1}{3}} \tag{5.7}$$

This relationship for $t$ must be modified for moisture sorption as the value of $\tau$ ($\tau = 3.54$) applies to nitrogen; also, the value 5.0 relates to the ratio $Q_{st}/R_g T$, where $Q_{st}$ and $R_g$ values are specific to nitrogen vapor. Applying the previously discussed water molecule diameter value, the isosteric heat of adsorption for water, the above expression becomes:

$$t = a \left( \frac{Q_{st,water}}{(R_g T)\ln(\frac{p_0}{p})} \right)^{\frac{l}{r}} = 3.673 \left( \frac{Q_{st,water}}{(R_g T)\ln(\frac{p_0}{p})} \right)^{\frac{1}{3}} \tag{5.8}$$

The value of $Q_{st}$ can vary between 20 and 0.0 kJ/mol for water sorption, varying according to water content from dryness to saturation (Lim et al., 1999). The thickness of the film can thus be calculated.

## Correction of Errors in Calculation of t by Method I

Halsey's (1948) relationship for moisture film thickness has a source of error in the constant $r$ (Conner et al., 1994). Other investigators found $r$ to be closer to $r = 2.7$ and a better match of the standard sorption isotherms developed by Gregg and Sing (1982) for which $r = 2.5$ and for BET isotherms where $r < 1.5$ (Conner et al., 1994). The Halsey (1948) relationship was shown to give unrealistic values when estimated values of the constants $a$ and $r$ were used. Conner et al. (1994) suggest that, because the values of ($V/V_m$) and ($P/P_0 = a_w$) are normally known form sorption data, the values of $a$ and $r$ can be easily and accurately calculated for use in the Halsey equation.

For instance, in the Halsey general relationship $V/V_m = [a/(P/P_0)]^{(1/r)}$ where $V$ = volume of liquid adsorbed, $V_m$ = volume of liquid adsorbed in the monolayer $= w_m \times$ (1gm/cc), $P/P_0 = a_w$; regression of ln $(V/V_m)$ against ln $(a_w)$ gives that intercept = ln $(a)$ and slope = $(1/r)$, for example:

$$\frac{V}{V_m} = a\left(\frac{1}{(p_0/p)}\right)^{\frac{1}{r}}$$

$$\ln\left(\frac{V}{V_m}\right) = \ln(a) + \left(\frac{1}{r}\right)\ln\left(\frac{1}{(p_0/p)}\right) = \ln(a) + \left(\frac{1}{r}\right)\ln(a_w) \qquad (5.9)$$

These values are easily obtained from regression programs. In Table 2.2 in Chapter 2, the relation between water sorption and water activity of several solid materials was provided in terms of the estimated constants $A$, $B$ and $C$, the water activity $a_w$, the thermodynamic constants $c$ and $K$ and the monolayer capacity $w_m$ in g/100 g dry material, where:

$$\frac{a_w}{w} = A + Ba_w + Ca_w^2$$
$$w_m = \frac{1}{ACK} \qquad (5.10)$$

which is the same as:

$$\frac{w}{w_m} = \frac{(ACK)a_w}{A + Ba_w + Ca_w^2} \qquad (5.11)$$

Because all of the constants have been provided by Table 2.8 and $w/w_m$ has the same meaning as $V/V_m$ for water and the value of $a_w$ can be varied between 0.0 and 1.00, the expression for finding $a$ and $r$ for thickness of moisture film can be stated as:

$$\ln\left(\frac{(ACK)a_w}{A + Ba_w + Ca_w^2}\right) = \ln(a) + \left(\frac{1}{r}\right)\ln(a_w)$$
$$t = a\left(\frac{Q_{st,water}}{(R_gT)\ln(\frac{p_0}{p})}\right) = \text{water film thickness} \qquad (5.12)$$

Conner et al. (1994) also recommended increasing the accuracy of the expression for $t$ developed from the Halsey relationship by (a) analysis of pore shapes by developing ratios of desorption vs. adsorption for values of $a_w$ above 0.5 and (b) use of the relationship for $t$, modified as above, for $a_w$ below 0.5.

No examples of hypothetical thickness of a water film on a solid material are provided here, as this estimate may or may not be directly linked to biodegradation rates in the system. Water activity of the material and implicitly water content or water potential is likely to be a more direct determinant of biodegradation rate; and water content of a particular material can be determined by other means described in various chapters.

Also, for the case of the adsorption data for the various potential waste materials, it is believed that, considering the scale of the soil-waste system and the expected gas or air laminar flow rates involved, adsorption rates and desorption rates along the moisture characteristics curves are roughly equal. Thus, values of the constants $a$ and $r$ estimated as above may be roughly applicable, with better approximations of actual water film thickness $t$ under drier conditions ($a_w < 0.5$) and more *averaged* approximations of $t$ in the wetter range of $a_w$ (0.5 and above).

## METHOD II FOR MOISTURE FILM THICKNESS

Like Conner *et al.* (1994), Naono and Hakuman (1993) studied the difficulties of accurately determining the porous texture (adsorption areas and liquid thickness) of a material based on adsorption data and isotherms when a hysteresis loop (curve mismatch) could be expected to exist between adsorption and desorption curves (especially for water adsorption isotherms). Naono and Hakuman (1993) proposed and used the relations:

$$\begin{aligned} \text{adsorbed water/unit area} &= 26.8 \times (V_{gas}/S_{\mathrm{BET}}) = \mathrm{gms}/(10^{-18}\mathrm{m}^2) \\ \text{thickness of sorbed water} &= t = (V_{liq}/S_{\mathrm{BET}}) \times 10^{-3}, \ \text{in nm}(10^{-9}\mathrm{m}) \end{aligned}$$

where $V_{gas}$ = adsorbed volume of water vapor, in $\mathrm{cm}^3$ at STP; $V_{liq}$ = adsorbed volume of water in liquid form, in $\mathrm{cm}^3/\mathrm{gm}$; and $S_{\mathrm{BET}}$ = BET sorption model calculated specific surface area (SSA) as previously described.

The above approach is much simpler than Method I and its corrections, but it does not account for the effect of temperature variation on thickness of the water film. It is believed that for a thin film, gas temperature (and humidity) could have a relatively strong effect on water loss and subsequent thinning of the water film. Use of the expression by Naono and Hakuman (1993) must be with the assumption that the waste site or a particular material undergoing moisture sorption is within the ambient temperature range.

## WATER POTENTIAL VS. WATER ACTIVITY OF SOILS AND SOLID POROUS MATERIALS

The foregoing discussion of water film thickness reintroduced the concept of water activity. Water activity and water potential are both descriptors of moisture uptake by materials likely to be present in a waste site.

The *water activity* of water-imbibing (hygroscopic) solid materials can be directly related to their *water potential* by considering the forces holding water to the material. The water potential is a sum of these forces and is a direct quantification of the force or pressure needed to extract moisture from the material. This concept is important to considerations of microorganism access to moisture, evaporation and sorption and diffusion.

The forces summing the *water potential* $\psi$ of a soil or solid material are (Papendick and Campbell, 1981): the *osmotic potential* $\psi_\pi$ (negative, affected by the acidity or alkalinity of the moisture); *the matric potential* $\psi_\pi$ (negative, typically the dominant potential, comprising the adsorption and capillary forces intrinsic uptake by the solid material and typically the dominant potential); the *pressure potential* $\psi_p$ (positive or negative, representing the gas [atmospheric vs. pore] pressure difference, or hydraulic head pressure); *gravitational potential* $\psi_g$ (related to relative elevation from the reference point); and *overburden potential* $\psi_\Omega$ (a measure of the effect of weight above on water present in the solid material). The water potential is thus (Papendick and Campbell, 1981):

$$\psi = \psi_m + \psi_\pi + \psi_g + \psi_\Omega - \psi_p \tag{5.13}$$

For waste sites, the most important of these are likely to be the matric and osmotic potentials, particularly with regard to biological activity.

The matric potential $\psi_m$ of soil has long been studied by soil scientists to predict how much water soils may retain or drain and how water potential (suction) or soil wetting varies with soil depth. The value of $\psi_m$ ranges from 0.0 at saturation to a high negative value for very dry material. Saturated sand can have a matric potential of −8 cm when near saturation (at field capacity) but can have a matric potential or suction of over 10,000 cm at dryness. In waste sites such as landfills, composting operations and contaminated soils containing chemicals of biological treatment interest, the matric potential determines the capacity to retain moisture inflow and thus affects the optimum moisture for treatment processes.

The osmotic potential $\psi_\pi$ is affected by which substances are dissolved in the water in a system. The presence of dissolved salts or acids, especially if in significant concentration, will lower the vapor pressure of water present and thus reduce the water potential. Microorganism or plant or animal cell walls are not permeable to all solutes present and will admit water more readily than other substances, thus, an osmotic potential can exist between water outside the living organism and the cell interior. As would be expected, this osmotic potential is only important when the solute concentration is relatively high, for instance in a saline or hypersaline environment. However, Papendick and Campbell (1981) point out that osmotic potential can also be significant in soils where organic wastes, fertilizer or crop residues have been added. While the landfill interior could be a good example of this, little information is available on how organic acid concentration of moisture sorbed by landfilled materials should affect the osmotic potential of organisms (bacteria, fungi, yeast) present on the surface of materials. In most cases moisture sorption in a waste site would involve comparatively dilute solutions. Or and Wraith (1999) state that for dilute solutions, the osmotic potential in kPa is proportional to the concentration of solute and the environment temperature, as:

$$\psi_{\pi} = -RTC_{solute} \tag{5.14}$$

where $R$ = universal gas constant = $8.314 \times 10^{-3}$kPa.m$^3$/(mol.°$K$), T = °$K$ and $C_{solute}$ = solute concentration in moles/m$^3$ of solvent volume. As finding the electrical conductivity of soil solutions is relatively straightforward, another useful expression for osmotic potential, in kPa, of dilute liquid sorbed into solid materials in terms of conductivity is (Or and Wraith, 1999):

$$\psi_{\pi} = -36EC_{solute} \tag{5.15}$$

where $EC_{solute}$ is the electrical conductivity of the liquid sorbed at saturation.

The pressure potential describes the difference in gas pressure or hydrostatic pressure (hydraulic head) between a reference point and the free water surface. Hydrostatic pressure ($P = \rho gh$) is known to vary according to vertical distance $h$ above a reference and liquid density $\rho$, but it only has meaning for standing water. In waste sites this situation is highly unlikely, unless for the case of groundwater mounding in a landfill or leachate ponding in a composting pile. In any case waste sites are likely to be both unsaturated and highly permeable, meaning that hydrostatic pressure would essentially be zero. Differences between gas pressure (and temperature) inside a waste site and atmospheric pressure are known to exist and are due to gas buildup, system gas permeability and temperature differences. Gas pressure differences between waste site interior and the outside atmosphere has been associated with passive outflow rates of gas from landfills. It is well known that the rate at which solid waste landfills pump out gas is affected by rapid change in barometric pressures outside the landfill. Because gas outflow rate and temperature would affect the rate of moisture evaporation along preferential paths of gas migration (connected pockets and macropores), pressure differences (potential) between the inside and outside of waste sites could affect water loss or gain rates at waste material surfaces open to paths of gas movement and thus local solute concentrations. However, the reference point for gas pressure is usually the atmosphere outside a waste site; and as the gas flow direction is generally toward the outside atmosphere, the pressure potential can be considered positive, compared to negative (suction) values for matric and osmotic potential.

In relatively deep waste sites, with heavy soil-waste overburdens, the maximum water content held by a material would be affected by the weight above it, as the overburden acts to squeeze the material and thus reduce its water uptake capacity. However, at the microscopic level, very small organisms such as bacteria survive in thin films of water on pores or on open surfaces of solid materials, provided nutrients and water are otherwise available. Overburden pressure or overburden potential $\psi_{\Omega}$ might affect the fraction of material surface available to decomposition microorganisms, as compression of a material would make less of its pore surface available; but it is unlikely to affect the water potential of thin moisture films.

Gravitational potential is only likely to affect the growth of large plants over a waste site because roots must draw water from the interior of the site. Microorganisms are unlikely to be affected by gravitational potential, as there would be essentially no difference in elevation between the organism and its environment.

The above suggests that for biological treatment considerations, the main components of total water potential for a material in a waste site would be the matric, osmotic and pressure potential:

$$\psi = \psi_m + \psi_\pi + \psi_p \tag{5.16}$$

Additionally, Papendick and Campbell (1981) take the view that, for living cells including those of microorganisms, the water potential is the sum of the osmotic and pressure potential; but cell pressure potential is due to turgor rather than solute present in the cell and the thinness of the cell wall excludes the likelihood of significant pressure potential between it and the surrounding environment. This has an important implication for microorganisms of waste site treatment interest. If these factors result in little or no difference between the potential inside the living cells and their outside environment, a microorganism on a material must be at the same water potential as at its location on the surface of the material. Thus the water potential of the environment immediately around the organism determines its water potential.

As the water potential determines the availability of water and thus nutrients from the waste environment around a microorganism, relating the water potential to water activity of a waste material is a convenient approach for relating solid materials water content to degradability by organisms present.

At equilibrium, solid materials inside a waste site are likely to have water potentials equal to the potential of water vapor of the gas surrounding them (landfill gas or air). Water vapor potential of the gas or air can be related to its relative humidity (RH) by the Kelvin equation:

$$RH = \frac{p}{p_0} = a_w = \exp\left(\frac{M_w \psi_w}{\rho_w RT}\right)$$

$$\psi_w = \frac{\rho_w RT}{M_w} \ln(a_w) \tag{5.17}$$

where $p$ = water vapor pressure, $p_0$ = saturated water vapor pressure, $\rho_w$ = density of water = 1000 kg/m$^3$ at 20°C, $R$ = universal gas constant = 0.008314 kPa.m$^3$/(mol.°K), T = °K, $M_w$ = molecular weight of water = 0.018 kg/mole.

As the above relations are stated in terms of kilopascals, but much of soil matric suction information is reported in bars, atmospheres (Atm), cm of water or mm of Hg (mercury), conversions should be made as needed. It should be noted that 1 Atm = 101.325 kPa = 760 mm Hg = 33.9 ft of $H_2O$ = 1.01325 bars.

It can also be reasoned that moisture at the surface of any solid material, or the liquid-gas interface should have essentially the same water potential and thus the same water activity as its surrounding environment, including that of the pore gas in the waste site. As shown earlier, because material moisture content is well described by its relation to its water activity, the water content of a material can be related to its water activity and thus its water potential. For the GAB moisture sorption relation, as previously noted:

$$a_w/w = A + Ba_w + Ca_w^2$$
$$A = 1/(w_m ck),$$
$$B = (1/w_m)[(c-2)/c]$$
$$C = (k/w_m)[(1-c)/c].$$

The values of $A, B$ and $C$, estimated and tabulated in Table 2.8, along with values of $w_m, c$ and $k$ as determined from moisture sorption isotherms and if the relative humidity at its location is known, the specific water potential at a waste material's surface, $\psi_{w,i}$ can be calculated. Conversely, if the water content of a material is approximately known, its specific water activity and water potential can be so determined. If the constants provided are used:

$$\psi_w = \frac{\rho_w RT}{M_w}\ln(a_w) = \frac{\rho_w RT}{M_w}\ln(RH)$$

$$RH = \exp\left(\frac{M_w}{\rho_w RT}\psi_w\right)$$

$$\frac{a_w}{w} = A + Ba_w + Ca_w^2$$

$$\frac{w}{w_m} = \frac{cka_w}{(1-ka_w)(1-ka_w+(Ck)a_w)}$$

$$w = \frac{(w_m ck)a_w}{(1-ka_w)(1-ka_w+Cka_w)} = \frac{a_w}{A+Ba_w+Ca_w^2}$$

$$= \frac{RH}{A+B(RH)+C(RH)^2}$$

$$w = \frac{\exp(\frac{M_w}{\rho_w RT}\psi_w)}{A + B\exp(\frac{M_w}{\rho_w RT}\psi_w) + C\exp(\frac{2M_w}{\rho_w RT}\psi_w)} \tag{5.18}$$

The above shows that if the relative humidity of the waste-soil environment is known, the matric potential $\psi$ can be found and subsequently the water content of the material, using the constants provided in Table 2.8.

## THE ISSUE OF MIXED WATER SATURATION OR VARIED WATER POTENTIAL IN WASTES

The above relationships also suggest that various materials in a mixed waste site could show quite different degrees of wetting and water potential because moisture content could vary according to water activity and is material-specific. This poses the problem that, though materials may be adjacent, for the same surface air-water interface, the matric potential of each material would be the same; but each water activity (and moisture content) would be different, which poses the problem of moisture content

**TABLE 5.1**
**Water Contents of Waste Materials at the Same Humidity ($a_w = 0.89$)**

| Material | $w_m$ | c | k | A | B | C | Water Content Gm/100 gm | Saturation Gm/100 gm |
|---|---|---|---|---|---|---|---|---|
| Cauliflower | 7.0 | 25.2 | 1.03 | 0.0055 | 0.132 | –0.141 | 83.73 | 90 |
| Newsprint | 3.6 | 10.9 | 0.69 | 0.037 | 0.227 | –0.174 | 8.82 | 11.3 |
| Cardboard | 4.8 | 20.1 | 0.73 | 0.014 | 0.188 | –0.145 | 13.35 | 17.0 |
| Leaves (Oak) | 9.0 | 31.2 | 0.81 | 0.0044 | 0.104 | –0.087 | 31.85 | 45.0 |
| Cotton | 5.2 | 8.69 | 0.718 | 0.031 | 0.148 | –0.122 | 13.53 | 17.6 |
| Leather | 12.9 | 9.5 | 0.65 | 0.013 | 0.061 | –0.045 | 28.43 | 34.5 |

causing different kinetic rates for different waste types in a site. Table 5.1 shows that at the same relative humidity, the water content (at the surface) of various materials will vary considerably, suggesting variation of the availability of water at the surface of buried materials with different sorption properties. Note that thickness of the water film varies according to sorbed water content.

## MAXIMUM MOISTURE SORPTION BY A MATERIAL

It is useful to have an idea of the maximum water content of a material for various reasons, including porosity and water retention capacity under normal environmental conditions. Caurie (1970) states that moisture condensation begins when surface adsorption forces no longer apply; generally above 90% humidity ($a_w = 0.9$), Raoult's Law applies. Raoult's Law for dilute solutions (adapted to the capillary condensation region of the adsorption isotherm for a solid material) is

$$\frac{1-a_w}{w_w} = \frac{M_{w,1}}{M_{w,2}}\left(\frac{W_2}{W_1}\right)$$

$$w = \left(\frac{W_1}{W_2}\right)$$

$$\frac{1}{w} = \frac{M_{w,2}}{M_{w,1}}\left(\frac{1-a_w}{a_w}\right) \tag{5.19}$$

where $a_w$ = water activity $M_{w,1}$ = molecular weight of solvent (water), $M_{w,2}$ = molecular weight of solute (solid material), $W_1$ = weight of solvent (water) in g/g solids and $W_2$ = weight of solute (solid material).

Seve et al. (2000) state that the total water content in a material (sediment) is the sum of the BET adsorption water uptake and condensation water. The limiting humidity level of the BET isotherm to be applicable is at a humidity of 75%. This corresponds roughly with the limit for bound water for the GAB isotherm (from the argument that even xerophilic organisms such as fungi are not able to grow on materials at below 75% humidity, or $a_w = 0.75$). Applying a value of $a_w = 0.75$ to one of the materials listed in Table 5.1 (oak leaves) indicates that water content would be 22.45 g/100 g and at $a_w = 0.89$, the condensed water would be 31.85–22.45 = 9.4 gm/100 gm.

This suggests a relatively small coverage of the material or a relatively thin film, thus below optimum for bacterial growth.

Raoult's Law applied to the condensation region of the adsorption isotherm is a special and limiting case of the BET relationship (Caurie, 1970). Caurie states that at low humidities and low adsorption, all adsorption models should reduce to Henry's Law, for which water vapor adsorbed is directly proportional to relative pressure (or $a_w$), in the form:

$$w = k_2 a_w \tag{5.20}$$

which simplifies to $w = k_3 a_w^S$, where $S$ is an exponent representing sorption at low humidity and $k_3 = cw$, where $c$ is some constant. Caurie (1970) further states that the Henry's Law expression for adsorption under very dry conditions is of the same form as the Raoult's Law expression for adsorption in the high humidity range and they are special cases of the general form of the full humidity range expression. The modified BET model thus derived by Caurie (1970) for multilayer adsorption or the full range was thus:

$$\frac{1}{w} = \frac{1}{cw_m}\left(\frac{1 - a_w}{a_w}\right)^{\frac{2c}{w_m}} \tag{5.21}$$

It is useful to have an estimate of the saturated moisture content of a material. Caurie (1970) notes that near saturation, the form above approximates a series expression:

$$\ln(w) = \ln(cw_m) + \left(\frac{2c}{w_m}\right)\left(a_w + \frac{a_w^2}{2} + \frac{a_w^3}{3} + \cdots \frac{a_w^n}{n}\right) \tag{5.22}$$

The above expression requires substitution of a term for the series on the right-hand side and determination of the Caurie (1970) adsorption model constants $w_m$ and $c$. Note that c in the Caurie model does not have the same value as the constant $c$ in the GAB adsorption model. Testing the series model above, compared to the GAB model, indicates that this Caurie model does not give satisfactory results near $a_w$ 1.0, which suggests that the series relation proposed may not be correct for the isotherm in the near saturation range.

Comparing the GAB model described in this section, when $a_w \approx 1.0$, or saturation, including condensation range or liquid on the surface of the material, the GAB model is:

$$\begin{aligned} \frac{w}{w_m} &= \frac{cka_w}{(1 - ka_w)(1 - ka_w + Cka_w)} \\ w_{max} &= \frac{w_m ck}{(1 - k)(1 - k + ck)} \end{aligned} \tag{5.23}$$

While the above relation is relatively accurate in estimating $w_{max}$, the maximum adsorbed water content, when compared to graphical plots of $w$ vs. $a_w$, the degree of accuracy is unknown because very few water sorption isotherm investigations approach material saturation. Various investigators have noted that for adsorption models, the water content curve becomes infinite near saturation humidity and that water can be held in pores above the sorption saturation point. However Caurie (1970) states that water uptake cannot be infinite and must approach a limit at or near saturation. Notably, most of the waste materials water sorption information examined for Table 2.8 shows maximum water contents well above those reported in literature for wastes as received. Thus saturated water content may be taken as at $a_w \approx 0.9999$.

## EFFECT OF WASTE MOISTURE CONTENT ON SOIL ORGANISMS

One means to assess whether water availability has any effect on overall degradation rates is to determine whether the conditions present could retard or enhance biodegradation. For most organisms present in soil, favorable conditions include optimum water content and favorable water suction (or potential). Unfavorable water availability would limit transport of nutrients, of organisms and the movement of organisms over the waste surfaces.

For waste sites such as landfills, the typical bulk water content at which drainage might occur is about 0.29 volume/volume, or field capacity. Landfilled refuse also has an average field (moisture retention) capacity of about 385.16 kg/cu m and a dry density of about 456.4 kg/cu m (averages of 19 samples, Holmes, 1983). Also, according to Zeiss and Major (1992), the dry or irreducible moisture content of refuse is about 0.0128 vol. water/vol. refuse. In between such extremes, the thickness of a water film on the material surface may be too small to support microbial activity, or the moisture suction may be too high.

To address how moisture content variation between these extremes might affect organism or microbial activities, the moisture content has to directly describe some organism effect. Typically, effects of moisture content on soil organisms are defined in terms of moisture or water availability (water potential or moisture suction), rather than water content (Harris, 1980).

The actual moisture content of porous materials can be described in terms of the thickness of the water film covering available surface, thus dry and wet extremes of moisture and conditions in between can be represented in terms of the thickness of water films. At the dry extreme, the water content represents a very high moisture suction or low relative humidity (= low water activity $a_w$ : $0.0 < a_w < 1.0$) and a film thickness approaching that of bound water. At the wet extreme the water content represents a low moisture suction or high humidity (= $a_w$ near 1.0) and a film thickness approaching that of surface tension maximum. Within this range of film thickness there is film thickness at which biodegradation is limited because of the high suction required for water availability. One discussion of effect of limiting water film thickness by Harris (1980) reports values found by various investigators as follows.

It can be seen from the values of Tables 5.2 and 5.3 that relatively high suctions, or very low moisture film thickness, would be required to retard the microbial activities

**TABLE 5.2**
**Moisture Film Thickness and Moisture Suction Limits for Microorganisms**

| Affected Microbial Activity | Water Film Thickness, m | Suction, Bars | Suction, cm | Value of $A_w$ | Reference |
|---|---|---|---|---|---|
| Movement, of protozoa | 4 | –0.3 | 306 | 0.999 | Griffin, 1972 |
| Movement, zoospores | 1.5 | –1.0 | 1020 | 0.999 | Griffin, 1972 |
| Movement, bacteria | 0.5 | –5.0 | 5100 | 0.996 | Griffin, 1972 |
| Nitrification, sulfur oxidation | 0.003 | –15 | 15,300 | 0.99 | Griffin, 1972 |
| Phycomycete fungal growth | 0.003 | –15 | 15,300 | 0.99 | Griffin, 1972 |
| Nitrification, sulfur oxidation, bacillus growth, competitive bacterial and actinomycetes growth, Phycomycete fungal growth | < 0.003 (< 10 $H_2O$) | –40 | 40,800 | 0.97 | Dubey, 1968; Moser and Olsen, 1953, 1953; Wilson and Griffin, 1975b; Griffin, 1972; Kouyeas, 1964 |
| Fungal growth | < 0.0015 (< 5 $H_2O$) | –100 | 100,200 | 0.93 | Griffin, 1972; Kouyeas, 1968; Wilson and Griffin, 1975b. |
| Fungal growth | < 0.0009 (< 3 $H_2O$) | –400 | 408,000 | 0.75 | Griffin, 1972; Kouyeas, 1968. |

**TABLE 5.3**
**Moisture Availability Limits for Specific Soil and Waste Organisms**

| Organism | Species | Landfill | Minimum $a_w$ | Max. bars | Reference |
|---|---|---|---|---|---|
| Protozoa | NA | Yes | 0.999 | –0.3 | Griffin, 1972 |
| Zoospores | NA | Yes | 0.999 | –1.0 | Griffin, 1972 |
| Bacteria | NA | Yes | 0.996 | –5.0 | Griffin, 1972 |
| Bacteria | Rhizobium, G–, rods | NA | 0.99 | –15 | Steinborn and Roughley, 1975 |
| Bacteria | Clostridium, G+, rods | | 0.97 | –40 | Brown, 1978 |
| Bacteria | Escherichia spp., G–, rods | Yes | 0.95 | –70 | Leistner and Rodel, 1976 |
| Bacteria | Clostridium, G+, rods | Yes | 0.95 | –70 | Brown, 1978 |
| Bacteria | Pseudomonas, G–, rods | Yes | 0.93 | –100 | Prior, 1978 |
| Bacteria | Micrococcus, G+, cocci | Yes | 0.93 | –100 | Measures, 1975 |
| Bacteria | Lactobacillus, G+, rods | Yes | 0.93 | –100 | Brown, 1978 |
| Bacteria | Aerococcus, G+, cocci | NA | 0.93 | –100 | Buchanan and Gibbons, 1975 |
| Bacteria | Bacillus, G+, rods | NA | 0.90 | –150 | Rose, 1976 |
| Bacteria | Lactobacillus, G+, rods | Yes | 0.90 | –150 | Leistner and Rodel, 1976 |
| Bacteria | Micrococcus, G+, cocci | Yes | 0.90 | –150 | Leistner and Rodel, 1976 |
| Bacteria | Micrococcus, G+, cocci | Yes | 0.86 | –200 | Brown, 1978 |
| Bacteria | Micrococcus, G+, cocci | Yes | 0.83 | –250 | Rose, 1976 |
| Actinomycetes | | Yes | NA | NA | |
| Fungi | Basidiomycetes | Yes | 0.97 | –40 | Brown, 1978 |
| Fungi | Basidiomycetes | Yes | 0.95 | –70 | Brown, 1978 |
| Fungi | Candida, Torula | NA | 0.93 | –100 | Rose, 1976 |
| Fungi | Saccharomyces | NA | 0.90 | –150 | Brown, 1978 |
| Fungi | Rhodotorula | NA | 0.90 | –150 | Leistner and Rodel, 1976 |
| Fungi | Pichia | NA | 0.90 | –150 | Leistner and Rodel, 1976 |
| Fungi | Hansenula | NA | 0.90 | –150 | Leistner and Rodel, 1976 |
| Fungi | Saccharomyces | NA | 0.86 | –200 | Brown, 1978 |
| Fungi | Candida | NA | 0.86 | –200 | Leistner and Rodel, 1976 |

| | | | | | |
|---|---|---|---|---|---|
| Fungi | Hanseniaspora | NA | 0.86 | –200 | Leistner and Rodel, 1976 |
| Fungi | Debaromyces | NA | 0.83 | –250 | Brown, 1978 |
| Fungi | Saccharomyces | NA | 0.76 | –300 | Brown, 1978 |
| Fungi | Phycomycetes | NA | 0.95 | –70 | Griffin, 1972; Adebayo et al., 1971 |
| Fungi | Basidiomycetes | NA | 0.95 | –70 | Griffin, 1972 |
| Fungi | Rhizopus | NA | 0.93 | –100 | Rose, 1976 |
| Fungi | Mucor | NA | 0.93 | –100 | Rose, 1976 |
| Fungi | Botrytis | NA | 0.93 | –100 | Rose, 1976 |
| Fungi | Fusarium | NA | 0.93 | –100 | Wilson and Griffin, 1979 |
| Fungi | Fusarium | NA | 0.90 | –150 | Wilson and Griffin, 1979 |
| Fungi | Penicillium | NA | 0.90 | –150 | Wilson and Griffin, 1979 |
| Fungi | Geastrum | NA | 0.90 | –150 | Wilson and Griffin, 1979 |
| Fungi | Cladosporium | NA | 0.86 | –200 | Leistner and Rodel, 1976 |
| Fungi | Paecilomyces | NA | 0.86 | –200 | Leistner and Rodel, 1976 |
| Fungi | Aspergillus | NA | 0.86 | –200 | Griffin, 1972 |
| Fungi | Aspergillus | NA | 0.83 | –250 | Rose, 1976 |
| Fungi | Alternaria | NA | 0.83 | –250 | Rose, 1976 |
| Fungi | Penicillium | NA | 0.83 | –250 | Griffin, 1972 |
| Fungi | Aspergillus | NA | 0.76 | –350 | Rose, 1976 |
| Fungi | Penicillium | NA | 0.76 | –350 | Rose, 1976 |
| Fungi | Aspergillus | NA | 0.69 | –500 | Rose, 1976 |
| Fungi | Chrysporium | NA | 0.69 | –500 | Pitt, 1977 |
| Fungi | Wallemia | NA | 0.69 | –500 | Pitt, 1977 |
| Fungi | Xeromyces | NA | 0.62 | –650 | Brown, 1978 |

The table of selected organisms above includes the major soil decomposers in landfills.

that are associated with wastes in landfills. It is unlikely that such high bulk moisture suctions would be reached in waste sites. Blight and Blight (1992) report from investigations of moisture suction at landfills in arid and semi-arid areas (Australia) that, while suctions near the surface with dry crusted soil covers can be as high as 5300 kPa (−53 bars or 54,060 cm), suctions in the interior of the landfills never exceeded 1500 kPa (−15 bars). The latter is considered the wilting point of plants and has been used as a lower limit for soil biological growth of organisms. The authors also note that the suction considered the limit for bacterial activity, 5500 kPa, corresponding to a relative humidity of 96% ($a_w = 0.96$), was not found for the borehole samples tested. This indicates that suction or matric potential in waste disposal sites might not be limiting, except for arid conditions. However, to maintain biologically favorable matric suction within a waste site, water input must be sufficient to replace loss through evaporation or gas outflow. This imbalance helps explain slowed degradation when waste site landfills are capped.

## WATER AVAILABILITY TO ORGANISMS

As water content of porous, absorptive materials can be represented in terms of sorption forces such as matric potential, it is possible to relate matric or water potential to water content and activity — and this to modeling approaches involving moisture content. Depending on water content, it is possible to consider suitability of the material for degradation by various microbial groups. If the solid material to be degraded is above the suction limit or relative humidity restricting growth, biodegradation should occur. Such constraints can be included as control features of a bioreactor model.

A fraction of water in the condensation range of humidity of the material can be available to the environment, including to microorganisms, dependent on the relative humidity of the environment. For instance, if the GAB model moisture sorption isotherm shows that equilibrium water content of cardboard is at 60% RH while it should be 13 g/100 g — but the actual water content of the cardboard is 15 g/100 g — only 2.0 grams would be available to microorganisms or to evaporation. Thus water content as compared to equilibrium humidity determines water available to organisms.

Some reported environmental limits in terms of relative humidity, moisture potential or suction (in negative bars of capillary pressure head) for organisms are tabulated below. The negative pressure values generally indicate the pressure that must be exerted by an organism to obtain water from a material. Small changes in water content are indicated to involve relatively significant changes in suction head; thus, relatively small water inputs for a particular soil gas humidity could significantly affect matric potential of a waste, increasing water availability.

The waste site microorganisms of greatest importance are influenced by moisture suction level present in the soil or waste materials. Fungi, actinomycetes and bacteria are likely to be directly affected by the available moisture at material surfaces, either as a result of the thickness of the water film left by drying and wetting cycles or by the suction required for the organism cells to obtain water.

The moisture content of a biodegradable material is thus a key indicator of susceptibility to decomposition in soil. The moisture isotherm model can provide the equilibrium moisture content of the waste material in the soil, but actual water vapor pressure vs. actual moisture content of individual wastes can determine water availability and thus rates of microorganism growth.

## HYDRAULIC CONDUCTIVITY

The conductivity of liquid (water, leachate) media can be expressed in terms of the fluid flow velocity, relative saturation, water content and porosity. Though the focus of this work is the decomposition processes occurring in a waste site, it is useful to discuss hydraulic conductivity, as it is often the foundation of many waste site models.

Campbell (1974) noted that an empirical relationship between water potential and water content existed and has the form:

$$K_{unsat} = K_{sat}\left(\frac{\psi_e}{\psi}\right)^{2+\frac{2}{b}} \tag{5.24}$$

where $\theta$ = water content of medium in volume/volume, $\theta_s$ = saturated water content, $\psi_e$ = water potential at pressure where air enters the pores of the material, $\psi$ = unsaturated water potential and $b$ is some constant that describes the material.

It is reasonable to adopt the saturated water content of landfilled material as around 0.28 (Harris and Cook, 1992) or 0.29 (moderate to high compaction, EPA HELP model, Schroeder et al., 1988). $\theta_s$ was taken as 0.29. The dry water content of landfills has been reported as low as 0.0128 (air and oven-dried MSW samples, Zeiss and Major, 1992). Use of these two values in the Hillel (1971) relationship noted above results in:

$$a = -1.6642 \times 10^{-5}, b2 = 3.3088$$

Substituting in the relationship provided by Campbell (1974) provides that for these values for saturated and dry water content and $a$ and $b$, the value of the air-entry water potential $\psi_e$ was estimated as −0.001 bars or −1.02 cm moisture suction, or near to saturation and b1 = 3.3088 = b2. This is slightly wetter than the −15 cm value estimated by Zeiss and Major (1992) or the −20.2 cm value used in the EPA HELP model.

Papendick and Campbell (1981) have noted that an approximate relation between relative humidity and water potential is:

$$\psi = 1350 \ln (a_w)$$

where $a_w$ has a value equal to that of relative humidity and is usually described as the water activity of a material. Brooks and Corey (1966) state that the *relative saturation* or *effective saturation* of a porous material can be defined as $S_e$, where:

$$S_e = \frac{\theta - \theta_r}{\theta_s - \theta_r} = \left(\frac{\psi_e}{\psi}\right)^{\lambda} \tag{5.25}$$

where $\gamma$ = pore size distribution, $\theta_r$ = residual saturation (air dry water content) and other parameters are as previously described. Comparing with the previous relationships:

$$S_e \frac{1}{\lambda} = \left(\frac{\theta - \theta_r}{\theta_s - \theta_r}\right)^{\frac{1}{\lambda}} = \left(\frac{\psi_e}{\psi}\right)^{\lambda} = \left(\frac{\theta}{\theta_s}\right)^{b} \tag{5.26}$$

This indicates that for any value of material moisture content $\theta$ between air dry and saturation, the value of the pore size distribution index $\lambda$ can be found. The value of the pore size distribution index $\lambda$ for landfilled waste has been presented by various sources, for instance:

$\lambda = 0.21$ EPA HELP model

$\lambda = 0.67$ Zeiss and Major (1992), for laboratory refuse columns.

This is well within the range for soils and close to values for sandy soils. By comparison, use of the values:

$\theta_s = 0.12$ (Bagchi, 1970)

$\theta_s = 0.366$ (compacted MSW, EPA HELP Model, Schroeder et al., 1988)

$\theta_s = 0.272$ (moderate compaction, EPA HELP Model, Schroeder et al., 1988)

imply that $\lambda$ at 0.358–0.322. Campbell (1974) noted that combining the relation for $\psi$:

$$\psi = \psi_e \left(\frac{\theta}{\theta_s}\right)^{-b} \tag{5.27}$$

with his integrated relationship for the relation between saturated and unsaturated conductivity:

$$K_{unsat} = K_{sat} \left(\frac{\psi_e}{\psi}\right)^{2b+2} \tag{5.28}$$

gives:

$$K_{unsat} = K_{sat} \left(\frac{\psi_e}{\psi}\right)^{2+\frac{2}{b}} \tag{5.29}$$

where parameters are as described earlier. Comparing and rearranging:

$$K_{unsat} = K_{sat} \left(\frac{\theta - \theta_r}{\theta_s - \theta_r}\right)^{\frac{2b+2}{b\lambda}}$$

$$\psi = \psi_e \left(\frac{\theta - \theta_r}{\theta_s - \theta_r}\right)^{-\frac{1}{\lambda}} \tag{5.30}$$

| Saturated Conductivity $K_s$ | Meters/day | Source | Conditions |
|---|---|---|---|
| 0.0323 cm/sec | 27.91 m/day | Zeiss and Major, 1992 | Channeled flow, pilot scale refuse columns |
| 22.86–36.86 ft/day | 6.97–11.2 m/day | Korfiatis, 1984 | Constant head refuse samples |
| $1 \times 10^{-2}$ cm/sec | 8.64 m/day | Hughes, 1971 | Landfill |
| $1 \times 10^{-3}$ cm/sec | 0.864 m/day | Oweis et al., 1990 | Landfill |
| 0.57 ft/day | 0.1737 m/day | EPA HELP Model, 1988 | Landfill |
| $1 \times 10^{-5}$ cm/sec | 0.0084 m/day | Fungaroli and Steiner, 1979 | Landfill |

where $(2b+2)/\lambda = 8.088$ and $1/\lambda = 3.1056$. Comparison with the unsaturated conductivity relationship developed by Campbell (1974):

$$K_{unsat} = K_{sat}\phi^B \tag{5.31}$$

where $B = 2b + 3$, $\Phi = \theta/\theta_s$ indicates $B = 9.618$. This is close to the value $B = 11, b = 4$, recommended for this relationship for landfills by Korfiatis (1984).

The value of $K_s$ is, however, unique to a site and thus any chosen value cannot be truly representative. Care has to be taken to choose a reasonable value. The saturated conductivity varies with waste–soil composition, waste density, degree of compaction and degree of waste decomposition (older waste is likely to be less porous to flow). Various values of $K_s$ have been reported for MSW in landfills. Some are listed below.

The values of 0.864 m/day provided from studies by Oweis et al. (1990) is considered reasonable for bulk saturated hydraulic conductivity of landfilled waste materials. The higher value from Zeiss and Major's (1992) studies of channeled flow is interesting, however, since channeling could dominate at most inactive landfills, especially those where fully saturated flow is rare and trickle or fingered flow is more probable.

## CAPILLARY EFFECTS IN WASTE SITES

While wetter waste site conditions promote degradation, waste site wetness can vary with wetting and drying (moisture flow) cycles. Some portions of the waste site are likely to remain wet enough that soil organisms and associated waste biodegradation rates could remain unaffected. The degree of dryness or wetness of the soil-waste material can be generally described in terms of its moisture suction or water potential. The latter is considered a more fundamental measure of water availability to organisms than water content; and values of the limiting water potential (wilting point), 1500 kPa, have been included in waste site moisture flow models such as the EPA HELP model referred to in this study. With environmental water variation in water availability, waste degradation rates could be affected. Blight and Blight (1992) state that it appears that bacterial activity would not be affected unless moisture suction or potential is above 1500 kPa; but even at the semi-arid and dry (Australian) landfills examined, the moisture suctions were above those that would cause plant wilting. Thus, bacterial activity in the landfills should proceed uninhibited except for dry, crusted cover soils.

It is well recognized, however, that the better landfill soil and membrane covers reduce biodegradation rates. If it is recognized that actual moisture content is not a good indicator of potential bacterial activity, another approach to analyzing the effect of moisture must be considered. Alternatives are consideration of the moisture suction or water potential of the wastes, if sorption/drying behavior of selected wastes are known to any reasonable extent, and consideration of the limiting effects of drying on key decomposer organisms in the landfill. Moisture presence, volume and movement in the soil and on waste surfaces has an obvious effect on the transport of leachate nutrients, especially the movement of organisms on wastes; for instance, bacteria are considered aquatic organisms.

While description of moisture effects in terms of water content is unfruitful, actual moisture content and water potential (moisture suction) effects are readily expressed in terms of capillary effects. The various landfill models described have not addressed capillary effects, although various researchers have mentioned its importance in relative landfill degradation rates. Rees (1983) reported increased biological (bacterial) activity in the capillary zone near the water table of landfills studied; and Noble and Nair (1990) state that capillarity may play a large role in waste biodegradation and gas formation.

An implication of moisture capillarity in landfills is that wetted zones around regions of landfill flow may actually be the major zones of biodegradation in a landfill. Particularly for closed or inactive landfills and landfill sections, such capillary zones become important in long-term simulations, as models estimating long-term biodegradation could overestimate the volume of the landfill involved. Zeiss and Major (1992), with flow column studies, estimated that saturated moisture movement in landfills is likely to occur through established channels (channeled flow) that are a small fraction of the cross-sectional area, in the range of 23–30% of area. This has implications for inactive landfills with limited inflow. Description of this type of flow and the volumes of a landfill affected is limited. An approximation of waste site areas and volumes affected by channeled flow adapts more readily to consideration of wetting front characteristics during wetting and drying cycles. If degradation is considered optimal within these zones, an approximation of rates of waste removal can be made for inactive or closed landfills.

## THEORY

One of the commonly used relations describing single-phase, vertical unsaturated flow (used in the HELP model, which does not consider capillarity) of an incompressible fluid in a soil column is the Richards (1931) relationship:

$$\frac{\delta\theta}{\delta_t} + \frac{\delta q}{\delta z} = 0 \tag{5.32}$$

where, in terms of the Campbell (1974) relations for unsaturated flow:

$$q = K(\theta) - D(\theta)\left(\frac{\delta\theta}{\delta z}\right)$$

$$K(\theta) = K_s\left(\frac{\theta}{\theta_s}\right)^B$$

$$D(\theta) = -K(\theta)\frac{\delta\psi}{\delta\theta}$$

$$\psi = \psi_s\left(\frac{\theta}{\theta_s}\right)^{-b} \tag{5.33}$$

as previously stated, where $\theta$, $\theta_s$ are the unsaturated and saturated moisture contents, respectively; $\psi$, $\psi$ are the unsaturated and saturated moisture suction or water potential, respectively, as described elsewhere in this section; $\delta t$ = time unit and $\delta z$ = length (vertical) of the soil column under flow. Referring to the above relations:

$$q = K(\theta) + K(\theta)\left(\frac{\delta\psi}{\delta\theta}\right)\left(\frac{\delta\theta}{\delta z}\right) = K(\theta)\left[1 + \frac{\delta\psi}{\delta z}\right];\ \left(\frac{\delta\psi}{\delta z} = \frac{\delta\theta}{\delta z}\frac{\delta\psi}{\delta\theta}\right)$$

$$\frac{\delta\psi}{\delta\theta} = \left(\frac{-b\psi_s}{\theta_s}\right)\left(\frac{\theta}{\theta_s}\right)^{(-b-1)}$$

$$\frac{\delta\theta}{\delta z} = \frac{(K(\theta) - q)}{D(\theta)} = \frac{(q - K(\theta))}{K(\theta)\frac{\delta\psi}{\delta\theta}}$$

$$\frac{\delta z}{\delta\theta} = \frac{K(\theta)\frac{\delta\psi}{\delta\theta}}{(K(\theta) - q)}$$

$$\delta z = \left[\left(\frac{-b\psi_s}{\theta_s}\right)\left(\frac{\theta}{\theta_s}\right)^{(-b-1)}\frac{(K(\theta))}{(K(\theta) - q)}\right]\delta\theta$$

$$\delta z = \left[\left(\frac{-b\psi_s}{\theta_s}\right)K_s\left(\frac{\theta}{\theta_s}\right)^{(B-b-1)}\frac{1}{(K_s(\frac{\theta^B}{\theta_s}) - q)}\right]\delta\theta \tag{5.34}$$

where $K_s$ = saturated conductivity, $b$ = exponent of the water potential relationship (found for this study by substituting for $\psi_{dry} = 0.0128$ vol/vol, $\theta_{wet} = 0.29$ vol/vol) in the relations:

$$\psi = a(\theta)^{-b} = 1350 \quad \ln(a_w) = \psi_{sat}(\theta/\theta_{sat})^{-b}$$

The value found for the exponent b was b = 3.3088. The Campbell relation with (B = 2b + 3) as an exponent for the unsaturated flow conductivity relationship is thus B = 9.618.

$$x = S_e = \text{effective saturation} = (\theta - \theta_r)/(\theta_s - \theta_r),$$
$$dx/d\theta = x'(\theta) = 1/(\theta_s - \theta_r)$$
$$dz/d\theta = (dz/dx)x(dx/d\theta)$$
$$a = (\alpha\psi_{\text{max}})/(\theta_s - \theta_r)$$
$$b = \alpha$$
$$c = K_s e^{-\gamma}$$
$$d = \gamma$$
$$\gamma = B = 9.618$$
$$b = 3.3308$$
$$\alpha = 5.0 \text{ (Korfiatis, 1984)}$$
$$\psi_{\text{max}} = 29 \text{ ft} = 8.84 \text{ meters (Korfiatis, 1984)}$$

A version of this relationship was used by Noble and Nair (1990) to estimate the capillary fringe height rise above the bottom of the landfill and it was indicated that biodegradation within this fringe might be enhanced due to the greater availability of moisture than elsewhere in the landfill.

The discussion above highlights the value of use of models that are able to handle unusual conditions. One such model, based on capillary moisture uptake between saturation and dryness, was developed by Rossi and Nimmo (1994). This model is discussed in detail in the following section.

## WASTE SITE MOISTURE RETENTION CHARACTERISTICS

Given the importance of moisture content of the material to most biodegradation, flow rate and distribution, porosity, tortuosity and mass and heat transfer, it is useful to consider the bulk moisture sorption characteristics of the landfill layers, since these should affect intrinsic bioreactor properties such as available surface area and supply of oxygen or nutrients to microorganisms. Literature review indicates this topic had not been fully addressed with application to landfills.

The effect of moisture content of soils on biological processes is not the direct result of the amount of water retained by a soil, but of the force with which it is held, i.e., its capillary pressure or moisture suction. Noble and Arnold (1991) and Noble and Nair (1990) point out that capillary fringes of landfills, more wetted than the bulk of the soil system, are likely to support greater decomposition activity. Thus characterization of capillary distribution in landfills would be a major step toward determining biodegradation.

## FULL RANGE MOISTURE CAPILLARITY

A major criticism of the application of mathematical models of moisture movement or hydraulic flow to practical situations (such as landfills) and the EPA HELP Model, is that (1) they generally neglect capillarity (Noble and Nair, 1992) or that (2) they only represent part of the porous media moisture content vs. flow conductivity relationship (Rossi and Nimmo, 1994). The full moisture content curve is typically represented by flow conductivity $K(\theta)$ vs. moisture content $\theta$ and ranges from a maximum value of flow conductivity at media saturation, or $K(\theta) = K_s$, to a minimum value of flow velocity, $K(\theta) \approx 0$ at minimum or essentially dry moisture content $\theta = \theta_{dry}$. The dry moisture content of soil media $\theta_{dry}$ is typically defined as that measured after the media has been oven-dried for 24 hours, while the saturated moisture content is defined as that at which the pore space is filled.

Regions of this moisture retention curve for which the practical hydraulic flow models generally do not hold include (1) in the wet region, i.e., between or complete saturation, $\theta = \theta_s$ and $\theta = \theta_{fc}$ or field capacity; and (2) in the dry region, i.e., that between the dry end of the Brooks-Corey (1966) mathematical relationship for porous media hydraulic flow:

$$\frac{\phi}{\phi_s} = \left(\frac{\psi_s}{\psi}\right)^{-b} \tag{5.35}$$

and the lower end of the complete moisture content curve, i.e., at $\theta = 0$. The Brooks-Corey (1966) relationship is considered a power law function of moisture potential or pressure *head* vs. moisture content. Moisture potential, which is also called the suction pressure of a porous medium or its capacity to hold water against gravity, affects flow rate, with drainage occurring when potential falls below that for maximum moisture retention, defined as when moisture equals *field capacity* $\theta = \theta_{fc}$. The moisture potential at this point is typically the bubbling pressure ($\psi = \psi_b$) of the medium. The Brooks-Corey (1966) does not hold in the dry and wet regions of the moisture concentration curve, either because the differential does not exist or because the relationship is no longer a power function (Rossi and Nimmo, 1994).

If a goal is to model moisture content and flow behavior in porous media more accurately, more exact representation of the flow is useful. This has particular value for representation of porous media systems where moisture content and moisture distribution affect the rate of reaction, such as in soil systems or biofilm and trickling filter bioreactors. These systems must often be modeled to examine how treatment through manipulation of bioreactors can be effected. Landfills, for example, can fall into one or more of the above variable moisture systems; and with the restricted flow resulting from the placement of thick or low-permeability covers, they can develop variable moisture saturation and distribution and subsequent difficult-to-define overall reaction rates, unless moisture retention can be more fully defined. Formulation of moisture retention curve over saturation regions of interest could reduce the modeling errors associated with the typical assumption that an average moisture content represents the porous media system.

A straightforward approach to development of a full range of moisture content vs. flow relationship is to develop mathematical formulations covering the whole curve, from saturated to dry conditions. The middle or moist-to-wet section of this curve has long been represented by mathematical relationships such as developed by Richards (1931), Brooks and Corey (1966) and Campbell (1974). In all regions of the curve, including the wetter and drier regions, moisture content vs. moisture potential relationships can be formulated. Considering that at each junction of the three sections of the curves, expressions for abutting curve sections and derivatives are matched, examination of these junctions can allow development of water potential over the whole range of water content. For smooth, continuous transitions between the curve regions involved, constants or exponents that allow the curves to draw smoothly into each other must then be found. The value of these can be approximated based either on actual data end-point and curve data.

For landfilled material, it has been reported (Oweis et al., 1992) that the saturated water content, $\theta_s$, under moderate to high site compaction, is approximately 0.42 volume water/volume solid. Zeiss and Major (1993) have reported that the oven-air dried moisture content, $\theta = \theta_{dry}$, is 0.012 vol/vol. The EPA HELP Models (1988, 1993) also indicate the field capacity moisture content for moderately compacted landfilled material, $\theta_{fc}$, is 0.29. These values can be used in development of the extended moisture retention curve once satisfactory formulations are found.

Recent examinations of ways of formulating full-range moisture retention curves have been carried out by Kosugi (1994), who developed a general moisture retention mathematical model to cover the whole range of soil moisture retention vs. soil capillary pressure, and Rossi and Nimmo (1994), who developed a full moisture range moisture retention relationship that could be used with other models to characterize moisture retention and hydraulic conductivity.

The Rossi-Nimmo (1994) model considers soil capillary pressure or moisture potential vs. moisture content and is useful for consideration as a relatively simple mathematical approach, free of the statistical considerations of the Kosugi (1994). The formulation can be reconsidered in terms of capillary pressure or moisture potential and mathematical solutions developed that apply to drainage and capillary flow or distribution. In the following sections, water potential vs. water content formulations are explored to allow insight into moisture retention or flow under a full range of capillary uptake forces.

## MIDDLE MOISTURE CONTENT RANGE

The Rossi-Nimmo (1994) mathematical model assumes that the central portion of a dimensionless curve describing moisture retention vs. moisture potential in soils is adequately described by a power law function of the type adopted by the Brooks-Corey (1966) relation, i.e.,

$$\theta/\theta_s = \theta_{\mathrm{II}} = (\psi_0/\psi)^{\lambda}; \text{ for } \psi_i \leq \psi \leq \psi_i \tag{5.36}$$

where $\theta$ = moisture content in volume/volume, $\theta_s$ = saturation moisture content, $\psi_0$ = bubbling or air entry pressure head, $\psi$ = actual pressure head at soil moisture content $\theta$, $\lambda$ = constant related to the pore size distribution (but not necessarily the same), $\psi_i$ = pressure at the wet end of the Brooks-Corey (1966) curve, i.e., where the wet-to-saturation curve section begins and $\psi_j$ = capillary pressure head at the dry end, i.e., where it joins the curve describing moisture retention as it falls toward zero.

## MOIST TO SATURATION OR WET MOISTURE CONTENT SECTION OF CURVE

Rossi and Nimmo assumed that this portion of the curve is parabolic in form, as had been proposed by Hutson and Cass (1987). By this modification the sharp upper corner of the Brooks-Corey (1966) curve, i.e., where the derivative becomes discontinuous or infinite, is replaced by a smooth curve tending toward saturation moisture content. The form of the parabolic curve was given as:

$$\theta/\theta_s = \theta_{\mathrm{I}} = 1 - c(\psi/\psi_0)^2;\ \text{for } 0 \le \psi \le \psi_i \tag{5.37}$$

where $c$ is a constant representative of the medium and values of the other parameters are as previously defined.

## MOISTURE RETENTION CURVE IN THE DRY RANGE FOR LANDFILLED WASTE

Ross et al. (1991) had also proposed that the dry range of the moisture retention curve can be represented by a function that makes the water content $\theta = 0$ at the point where the dry pressure head $\psi_d$ approaches some final, maximum value in meters or feet. Zeiss and Major (1993) have shown that $\psi_d$ has a value of 31,090 cm for municipal solid waste. However, Blight and Blight (1990) found dry range suction pressures as high as 54,060 cm at landfills in arid and semi-arid climates. Rossi and Nimmo (1994) also reported dry end suction potentials for sands, clays and other soils. These authors proposed that moisture content vs. moisture potential in the dry range could be represented by a logarithmic function:

$$\theta/\theta_s = \theta_{\mathrm{III}} = \alpha \ln(\psi_d/\psi);\ \text{for } \psi_j \le \psi \le \psi_d \tag{5.38}$$

Rearrangment of the relations (i)–(iii) results in:

$$\begin{aligned}
\theta_{\mathrm{I}} &: \psi = (\psi_0/c)(1-\theta/\theta_s)^{0.5};\ \text{for } \theta_s \ge \theta \ge \theta_i \\
\theta_{\mathrm{II}} &: \psi = \psi_0(\theta/\theta_s)^{-1/\lambda};\ \text{for } \theta_i \ge \theta \ge \theta_j \\
\theta_{\mathrm{III}} &: \psi = \psi_d \exp\,(\theta/\theta_s)^{-1/\alpha}\ \text{for } \theta_j \ge \theta \ge 0
\end{aligned} \tag{5.39}$$

Differentiating:

$$\begin{aligned}
\theta_{\mathrm{I}} &: \psi'(\theta) = [(-0.5\psi_0)/(\theta_s c)](1-\theta/\theta_s)^{-0.5}\ \text{for } \theta_s \ge \theta \ge \theta_i \\
\theta_{\mathrm{II}} &: \psi'(\theta) = [-\psi_0/\lambda\theta_s](\theta/\theta_s)^{-1/\lambda}\ \text{for } \theta_i \ge \theta \ge \theta_j \\
\theta_{\mathrm{III}} &: \psi'(\theta) = [-\psi_d/\alpha\theta_s]\exp[(-1/\alpha)(\theta/\theta_s)]\ \text{for } \theta_j \ge \theta \ge 0
\end{aligned} \tag{5.40}$$

## BOUNDARY CONDITIONS

As noted by Rossi-Nimmo (1994), at the two-curve junction $\theta = \theta_i$:

$$\theta_{\mathrm{I}}(\psi_i) = \theta_{\mathrm{II}}(\psi_i)$$
$$\theta'_{\mathrm{I}}(\psi_i) = \theta'_{\mathrm{II}}(\psi_i)$$

where:

$$\theta_{\mathrm{I}}(\psi_i) = 1 - c(\psi_i/\psi_0)^2$$
$$\theta'_{\mathrm{I}}(\psi_i) = [(-2c)/(\psi_0^2)]\psi_i$$
$$\theta_{\mathrm{II}}(\psi_i) = (\psi_0/\psi_i)^\lambda$$
$$\theta'_{\mathrm{II}}(\psi_i) = (-\lambda/\psi_i)(\psi_0/\psi_i)^\lambda$$

It is implied that

$$\begin{aligned} c &= (\lambda/2)(\psi_0/\psi_i)^{\lambda+2} \\ \ln(\psi_i) &= [(\lambda - 2)\ln(\psi_0) + \ln(c)]/(2 - \lambda) \end{aligned} \tag{5.41}$$

Additionally, at the two-curve junction $\theta = \theta_j$:

$$\theta_{\mathrm{II}}(\psi_j) = \theta_{\mathrm{III}}(\psi_j)$$
$$\theta'_{\mathrm{II}}(\psi_j) = \theta'_{\mathrm{III}}(\psi_j)$$
$$\theta'_{\mathrm{II}}(\psi_j) = \theta'_{\mathrm{III}}(\psi_j)$$

where:

$$\theta_{\mathrm{II}}(\psi_j) = (\psi_0/\psi_j)^\lambda$$
$$\theta'_{\mathrm{II}}(\psi_j) = (-\lambda/\psi_j)(\psi_0/\psi_j)^\lambda$$
$$\theta_{\mathrm{III}}(\psi_j) = \alpha \ln(\psi_d/\psi_j)$$
$$\theta'_{\mathrm{III}}(\psi_j) = -\alpha/\psi_j$$

From the above:

$$\psi_j = (\alpha/\lambda)^{-1/\lambda} \tag{5.42}$$

## ESTIMATION OF CONSTANTS FULL-RANGE (WET TO DRY) MOISTURE CAPILLARITY RELATIONS

It is apparent that the value of the fitting constants of the full range moisture content curve, $\alpha, c, \psi_i, \psi_j, \theta_i, \theta_j$ and $\lambda$, can be found by varying the range of q between a suitable range, i.e., between the saturated value $\theta = \theta_s$ and the dry value $\theta = \theta_d$, such that the points $\theta_i$ and $\theta_j$ are covered. The curve points $\theta_i$ and $\theta_j$ would coincide with values of moisture where curve section I = curve section II and their differentials also match.

The values for moisture suction at the dry and saturated content must be known in order to do this estimation. In most cases, experimental values of $\psi_0$ and $\psi_{dry}$ exist for the types of soils (MSW and sand) to be used in the present simulation. For instance, for municipal solid waste, these are given as $\psi_0 = 15$ cm (Zeiss and Major, 1993) or 20.2 cm (EPA HELP Model, 1988) and $\psi_{dry} = 31{,}090$ cm (Zeiss and Major, 1993). Similar values appear in literature for other media. A MATHCAD program was thus developed to determine the constants listed.

In the middle to dry range, the middle range curve:

$$\psi = \psi_0(\theta/\theta_s)^{-1/\lambda}$$

meets the dry range curve $\psi = \psi_d \exp(\theta/\theta_s)^{-1/\alpha}$ at junction point $\theta_j, \psi_j$. At this point derivatives are the same:

$$(\theta/\theta_s)(\psi)' = (-\lambda/\psi_j)(\psi_0/\psi_j)^{\lambda} = -\alpha/\psi_j$$

From earlier analytic comparison of the junction points:

$$\psi_j = \psi_d \exp(-1/\lambda)$$

Using the values of $\theta_s$ and $\psi_d$, and $\psi_0$ from literature, and a range of moisture content $0.18 \geq \theta \geq 0.002$, the estimated values of dry range parameters were found to be:

$$\theta_s = 0.41 \text{(Oweis et al., 1990)}$$

$$\psi_d = 310.90 \text{ m (Zeiss and Major, 1993)}$$

$$\psi_0 = 15 \text{ cm (Zeiss and Major, 1993)}$$

Estimates:

$$\alpha = 0.062, \lambda = 0.425, \psi_j = 2950 \text{ cm}, \theta_j = 0.06$$

Testing with $\theta_s = 0.52$ and $\psi_0 = 20.2$ cm (0.202 m) EPA HELP Model, default values, 1988, landfilled MSW:

$$\alpha = 0.055, \lambda = 0.468, \psi_j = 3664 \text{ cm}, \theta_j = 0.061$$

Testing with $\theta_s = 0.671$ and $\psi_0 = 20.2$ cm (0.202 m) EPA HELP Model, default values, 1994, MSW:

$$\alpha = 0.047, \lambda = 0.495, \psi_j = 4130 \text{ cm}, \theta_j = 0.063$$

Testing with $\theta_s = 0.168$ and $\psi_0 = 15$ cm (0.202 m) EPA HELP Model, default values, 1994, MSW channeling:

$$\alpha = 0.107, \lambda = 0.331, \psi_j = 1520 \text{ cm}, \theta_j = 0.054$$

To estimate constants in the wet moisture content range including $\theta_i$, $\theta$ was varied between $\theta_s = 0.40$ and 0.15. It was assumed this range would cover the upper (wet range) two-curve junction. The junction point can be used to estimate the constants:

$$\theta_{\mathrm{I}}(\psi_i) = \theta_{\mathrm{II}}(\psi_i) \Rightarrow \psi = (\psi_0/c)(1 - \theta/\theta_s)^{0.5} = \psi_0(\theta/\theta_s)^{-1/\lambda}$$

$$\theta_{\mathrm{I}}(\psi_i) = \theta_{\mathrm{II}}(\psi_i) \Rightarrow (\theta/\theta_s)(\psi)' = [(-2c)/(\psi_0^2)]\psi_i = (-\lambda/\psi_i)(\psi_0/\psi_i)^{\lambda}$$

Equating and rearranging the above:

$$\ln(1 - \theta/\theta_s) = 2\ln(c) + (-2/\lambda)\ln(\theta/\theta_s)$$

This gives a slope of $(-2/\lambda)$ and an intercept of 2 ln(c). Rearranging and equating for other factors:

$$\psi_i = \psi_0^{2\lambda/(\lambda+1)} = \text{moisture suction head at upper curves junction}$$

$$c = (\lambda/2)(\psi_0/\psi_i)^{(\lambda+2)} = \text{constant for wet range curve}$$

$$\theta_i = (\theta_s)(\psi_0/\psi_i)^{\lambda} \text{ moisture content at upper curves junction}$$

Assuming the values shown below:

$$\psi_0 = 15\,\text{cm (Zeiss and Major, 1993)}$$

$$\theta_s = 0.41\text{(Oweis et al., 1990)}$$

Estimates of capillarity sorption constants are:

$$\lambda = 0.431\text{(landfilled MSW)}$$

$$\psi_i = 0.2954 = 29.54\,\text{cm} = \text{MSW suction head, upper curves junction}$$

$$\theta_i = 0.306 = \text{water content at upper curves junction}$$

$$c = 0.193 \text{ (landfilled MSW)}$$

Testing for the EPA HELP Model, 1988, default values of $\psi_0 = 0.202$m, $\theta_s = 0.52$ and $\lambda = 0.468$ as calculated earlier from the dry-range to mid-range curves junction $\psi_j$:

$$\lambda = 0.468 \text{ (landfilled MSW)}$$

$$\psi_i = 0.3607 = 36.07\,\text{cm.} = \text{MSW suction head, upper curves junction}$$

$$\theta_i = 0.3964 = \text{water content at upper curves junction}$$

$$c = 0.06 \text{ (landfilled MSW)}$$

Testing for the EPA HELP Model, 1994, default values of $\psi_0 = 0.202\,\text{m}$, $\theta_s = 0.671$ and $\lambda = 0.495$ as calculated earlier from the dry-range to mid-range curves junction $\psi_j$:

$$\lambda = 0.495 \text{ (landfilled MSW)}$$
$$\psi_i = 0.347 = 36.07\,\text{cm.} = \text{MSW suction head, upper curves junction}$$
$$\theta_i = 0.513 = \text{water content at upper curves junction}$$
$$c = 0.064 \text{ (landfilled MSW)}$$

Testing for the EPA HELP Model, 1994, default values of $\psi_0 = 0.15\,\text{m}, \theta_s = 0.168$, which apply to the case of MSW with channeling, and $\lambda = 0.331$ as calculated earlier from the dry-range to mid-range curves junction $\psi_j$:

$$\lambda = 0.331 \text{(landfilled MSW, channeling)}$$
$$\psi_i = 0.390 = 39.0\,\text{cm.} = \text{MSW suction head, upper curves junction}$$
$$\theta_i = 0.1225 = \text{water content at upper curves junction}$$
$$c = 0.108 \text{ (landfilled MSW, channeled flow)}$$

## RELIABILITY OF ESTIMATED VALUES

The assumption is made that a single value of $\lambda$ satisfies the mid-range of the three curves, two-junction full range moisture retention mathematical model.

From statistical examination, it is believed that $\lambda$ values found at this junction would be more accurate than found at the upper curve junction. The values of the important general curve parameter $\lambda$ were thus taken from values also estimated for the dry–moist range curves junction. Rossi and Nimmo (1994) have also noted the greater accuracy of the junction point $\psi_j$ found at this two-curve junction point. If $\lambda$ is reliable, incorporation into and reliability of the other parameters, $\alpha, c, \psi_i, \psi_j, \theta_i$ and $\theta_j$, are improved.

If the values estimated above are sufficiently reliable, they can obviously be used in simulation of a full-range moisture retention for the landfilled material, assuming overall bulk material consistency with regard to hydraulic flow. The values tabulated below include parameters estimated by Rossi and Nimmo (1994) for relevant soils and default values of saturation moisture content and suction head from the EPA HELP Model (1988, 1994) and from literature, as indicated.

## RELEVANCE OF THE LOWER CURVE JUNCTION TO BIOREACTOR SIMULATION

The EPA HELP Model (1994) lists the wilting point moisture content, i.e., the bulk moisture content at which biological activity is considered to cease, at $\theta = 0.077$ for MSW and 0.019 for MSW with channeling. It is obvious, from comparison with the estimated values of $\theta_j$ estimated for various cases of landfilled MSW in Table 5.4, that the logarithmic region of the curve, or the dry range curve, should fall outside the range of bioreactor simulation interest, even though a considerable fraction of the landfilled waste–soil material might have moisture contents in the range. Thus the two upper range moisture retention curves should be sufficient to

**TABLE 5.4**
**Estimated Moisture Capillarity Properties of Landfill Layer Materials**

| Soil Description | Moisture Content Range Used | $\theta_s$ | $\psi_0$ | $\lambda$ | $\alpha$ | c | $\psi_i$ | $\psi_j$ | $\theta_i$ | $\theta_j$ | Reference |
|---|---|---|---|---|---|---|---|---|---|---|---|
| MSW, Landfilled | 0.41–0.002 | 0.41 | 15 cm | 0.431 | 0.062 | | 29.54 | 2950 | 0.306 | 0.06 | Estimated |
| MSW, Landfilled | 0.52–0.002 | 0.52 | 20.2 cm | 0.468 | 0.052 | | 36.07 | 3664 | 0.3964 | 0.061 | Estimated |
| MSW, Landfilled | 0.671–0.002 | 0.671 | 20.2 cm | 0.495 | 0.047 | | 34.7 | 4130 | 0.513 | 0.063 | Estimated |
| MSW, Channeling | 0.168–0.0 | 0.168 | 15 cm | 0.33 | 0.107 | | 39.0 | 1520 | 0.1225 | 0.054 | Estimated |
| L-Soil, Sand | 0.18–0.0 | 0.18 | 32.7 | 1.83 | | | 57.1 | 20,000 | | | Rossi & Nimmo, 1994 |
| Paisleys B, Silty Clay | 0.55–0.0 | 0.55 | 1.1 | 0.15 | | | 19.6 | 470,000 | | | Rossi & Nimmo, 1994 |

simulate landfill biological activity. Simulation of bioreactor moisture content and variation with volume can thus be simulated without the dry-end moisture curve. The Brooks-Corey (1966) mathematical representation co-joined to a wet-range parabolic curve correction should be sufficient for simulation.

## DEVELOPMENT OF MOISTURE CAPILLARITY–HYDRAULIC CONDUCTIVITY RELATIONSHIPS

To use the extended moisture range approach in hydraulic conductivity simulations, mathematical formulations more amenable than the Campbell (1974) relationships used in other landfill models should be developed. As described in this section, the general or full-range moisture retention curve consists of (1) an upper, saturated-to-wet moisture range parabolic curve, joined smoothly at a junction with coordinates $(\psi_i, \theta_i)$ to (2) a middle range moisture content power law curve, also joined smoothly to (3) a dry-end logarithmic moisture retention curve.

Rossi and Nimmo (1994) state that the obtainable, integrated forms of sections (1)–(3) of the general moisture retention curve are suited to use in a soil hydraulic conductivity model such as proposed by Mualem (1976):

$$K_r(\theta) = \sqrt{\frac{\theta}{\theta_s}}\frac{I^2(\theta)}{I^2(\theta_s)} = \frac{K(\theta)}{K_s} \tag{5.43}$$

Rossi and Nimmo (1994) provided integrated forms of curve sections (1)–(3).

### Dry Range Logarithmic Curve Section, for $\theta_j \geq \theta \geq 0$

For this range, Rossi and Nimmo (1994) provide that:

$$I(\theta) = I_{\mathrm{III}}(\theta) = \frac{\alpha}{\psi_d}\left[\exp\left(\frac{1\theta}{\alpha\theta_s}\right) - 1\right] \tag{5.44}$$

Squaring this relationship:

$$I^2(\theta) = I^2_{\mathrm{III}}(\theta) = \frac{\alpha^2}{\psi_d^2}\left[\exp\left(\frac{2\theta}{\alpha\theta_s}\right) - 2\exp\left(\frac{1\theta}{\alpha\theta_s}\right) + 1\right] \tag{5.45}$$

Thus considering the Mualem (1976) ratio relationship and the case where $\theta = \theta_s =$ saturated moisture content:

$$\frac{I^2(\theta)}{I^2(\theta_s)} = \frac{I^2_{\mathrm{III}}(\theta)}{I^2_{\mathrm{III}}(\theta_s)} = \frac{\frac{\alpha^2}{\psi_d^2}[\exp(\frac{2\theta}{\alpha\theta_s}) - 2\exp(\frac{1\theta}{\alpha\theta_s}) + 1]}{\frac{\alpha^2}{\psi_d^2}[\exp(\frac{2}{\alpha}) - 2\exp(\frac{1}{\alpha}) + 1]} \tag{5.46}$$

Considering this ratio and constants as follows:

$$A_3 = \left[\exp\left(\frac{2}{\alpha}\right) - 2\exp\left(\frac{1}{\alpha}\right) + 1\right]^{-1};\ B_3 = \frac{2}{\alpha}\frac{1}{\theta_s};\ C_3 = \frac{1}{\alpha}\frac{1}{\theta_s} \tag{5.47}$$

The Mualem (1976) relationship for soil relative hydraulic conductivity, for the curve dry range, is thus:

$$K_3(\theta) = K_{\mathrm{III}}(\theta) = K_s A_3[\exp(B_3\theta) - 2\exp(C_3\theta) + 1] \tag{5.48}$$

where constants $\alpha$ and $\theta_s$ are as described in this section.

## Medium Moisture Range, Power Law Curve, for $\theta_i \geq \theta \geq \theta_j$

For this soil moisture retention curve section, Rossi and Nimmo (1994) provide that:

$$I_{\mathrm{II}}(\theta) = I_{\mathrm{III}}(\theta_j) + \frac{I}{\psi_0}\frac{\lambda}{\lambda+1}\left[\left(\frac{\theta}{\theta_s}\right)^{\frac{\lambda+I}{\lambda}} - \left(\frac{\theta_J}{\theta_s}\right)^{\frac{\lambda+1}{\lambda}}\right] \tag{5.49}$$

Considering constants and the point $I_{\mathrm{III}}(\theta) = I_{\mathrm{III}}(\theta j)$:

$$I_{\mathrm{III}}(\theta) = I_{\mathrm{III}}(\theta_j) = \frac{\alpha}{\psi_d}\left[\exp\left(\frac{1}{\alpha}\frac{\theta_j}{\theta_s}\right) - 1\right] = \beta_1;\ \frac{\lambda}{\lambda+1} = \beta_2;\ \left(\frac{\theta_J}{\theta_s}\right)^{\frac{\lambda}{\lambda-1}} = \beta_3 \tag{5.50}$$

Thus:

$$I_{\mathrm{III}}(\theta) = \beta_1 + \frac{\beta_2}{\psi_0}\left[\left(\frac{\theta}{\theta_s}\right)^{\frac{I}{\beta_2}} - \beta_3\right] = \left[\beta_1 + \frac{\beta_2}{\psi_0}\left(\frac{\theta}{\theta_s}\right)^{\frac{I}{\beta_2}} - \frac{\beta_2\beta_3}{\psi_0}\right] \tag{5.51}$$

Squaring this relationship:

$$\begin{aligned} I_{\mathrm{II}}^2(\theta) = {} & \left[\beta_1^2 - \frac{2\beta_1\beta_2\beta_3}{\psi_0} + \beta_2^2\beta_3^2\right] \\ & + \left[\frac{2\beta_1\beta_2}{\psi_0} - \frac{2\beta_2^2\beta_3}{\psi_0}\right]\left(\frac{\theta}{\theta_s}\right)^{\frac{1}{\beta_2}} + \left[\left(\frac{\beta_2}{\psi_0}\right)^2\right]\left(\frac{\theta}{\theta_s}\right)^{\frac{2}{\beta_2}} \end{aligned} \tag{5.52}$$

Considering the case where $\theta_2 = \theta_s$ and groups of constants in the above relationship:

$$A_2 = \left[\beta_1^2 - \frac{2\beta_1\beta_2\beta_3}{\psi_0} + \beta_2^2\beta_3^2\right];\ B_2 = \left[\frac{2\beta_1\beta_2}{\psi_0} - \frac{2\beta_2^2\beta_3}{\psi_0}\right];\ C_2 = \left[\left(\frac{\beta_2}{\psi_0}\right)^2\right]$$

$$I_{\mathrm{II}}^2(\theta) = A_2 + B_2\left(\frac{\theta}{\theta_s}\right)^{\frac{1}{\beta_2}} + C_2\left(\frac{\theta}{\theta_s}\right)^{\frac{2}{\beta_2}}$$

$$I_{\mathrm{II}}^2(\theta_s) = A_2 + B_2 + C_2 = \frac{1}{D_2} \tag{5.53}$$

Thus the unsaturated conductivity in the middle general moisture retention curve can be represented as:

$$\underline{K_2(\theta) = K_s D_2\left[\left(\frac{\theta}{\theta_s}\right)^{0.5}\right]\left[A_2 + B_2\left(\frac{\theta}{\theta_s}\right)^{\frac{1}{\beta_2}} + C_2\left(\frac{\theta}{\theta_s}\right)^{\frac{2}{\beta_2}}\right]} \tag{5.54}$$

where the values of $\alpha$, $\lambda$, $\theta_j$ and $\theta_s$ are as previously described; thus $A_2$, $B_2$, $C_2$ and $D_2$ can be readily estimated.

## Saturated-to-Mid Range (Parabolic) Curve, $\theta_i \geq \theta \geq \theta_j$

For this range of moisture content, Rossi and Nimmo (1994) provide that the integrated form of the parabolic curve representing soil moisture retention is provided by:

$$\underline{I_{\mathrm{I}}(\theta) = I_{\mathrm{II}}(\theta_i) + \frac{2C^{0.5}}{\psi_0}\left[\left(1 - \frac{\theta_i}{\theta_s}\right)^{0.5} - \left(1 - \frac{\theta}{\theta_s}\right)^{0.5}\right]} \tag{5.55}$$

Because the term $I_{\mathrm{II}}(\theta)$ is as described earlier:

$$\underline{I_{\mathrm{II}}(\theta) = I_{\mathrm{III}}(\theta_j) + \frac{1}{\psi_0}\frac{\lambda}{\lambda+1}\left[\left(\frac{\theta}{\theta_s}\right)^{\frac{\lambda+1}{\lambda}} - \left(\frac{\theta_j}{\theta_s}\right)^{\frac{\lambda+1}{\lambda}}\right]} \tag{5.56}$$

the term $I_{\mathrm{II}}(\theta_i)$, which is now a constant, is thus:

$$\underline{I_{\mathrm{II}}(\theta_i) = \tfrac{\alpha}{\psi_d}\left[\exp\left(\frac{1}{\alpha}\frac{\theta_j}{\theta_s}\right) - 1\right] + \frac{1}{\psi_0}\frac{\lambda}{\lambda+1}\left[\left(\frac{\theta_i}{\theta_s}\right)^{\frac{\lambda+1}{\lambda}} - \left(\frac{\theta_j}{\theta_s}\right)^{\frac{\lambda+1}{\lambda}}\right] = \beta_4} \tag{5.57}$$

With the other constant groups in $I_{\mathrm{I}}(\theta)$ defined as:

$$\underline{\frac{2c^{0.5}}{\psi_0} = \beta_5;\ \left(1 - \frac{\theta_i}{\theta_s}\right)^{0.5} = \beta_6} \tag{5.58}$$

the integral $I_{\rm I}(\theta)$ can be redefined as:

$$I_{\rm I}(\theta) = \beta_4 + \beta_5\left[\beta_6 - \left(1 - \frac{\theta}{\theta_s}\right)^{0.5}\right] = \beta_4 + \beta_5\beta_6 - \beta_5\left[\left(1 - \frac{\theta}{\theta_s}\right)^{0.5}\right] \tag{5.59}$$

Squaring this relationship and grouping constants:

$$\begin{aligned} I_{\rm I}^2(\theta) &= [\beta_4^2 + 2\beta_4\beta_5\beta_6 + \beta_5^2\beta_6^2] \\ &\quad -2[\beta_4\beta_5 + 2\beta_5^2\beta_6]\left(1 - \frac{\theta}{\theta_s}\right)^{0.5} + [\beta_5^2]\left(1 - \frac{\theta}{\theta_s}\right) \end{aligned} \tag{5.60}$$

The constant terms in the relation can also be defined as:

$$A_1 = [\beta_4^2 + 2\beta_4\beta_5\beta_6 + \beta_5^2\beta_6^2];\ B_1 = 2[\beta_4\beta_5 + 2\beta_5^2\beta_6];\ C_1 = [\beta_5^2] \tag{5.61}$$

Thus the squared relation is:

$$\begin{gathered} I_{\rm I}^2(\theta) = A_1 - B_1\left(1 - \frac{\theta}{\theta_s}\right)^{0.5} + C_1\left(1 - \frac{\theta}{\theta_s}\right) \\ I_{\rm I}^2(\theta_s) = A_1 = \frac{1}{D_1} \\ \frac{I_{\rm I}^2(\theta)}{I_{\rm I}^2(\theta_s)} = \frac{A_1 - B_1\left(1 - \frac{\theta}{\theta_s}\right)^{0.5} + C_1\left(1 - \frac{\theta}{\theta_s}\right)}{A_1} \end{gathered} \tag{5.62}$$

and the unsaturated conductivity in the saturated-to-moist range is given by:

$$K_1(\theta) = K_s D_1\left[A_1 - B_1\left(1 - \frac{\theta}{\theta_s}\right)^{0.5} + C_1\left(1 - \frac{\theta}{\theta_s}\right)\right] \tag{5.63}$$

## SUMMARY OF EXTENDED RANGE CONDUCTIVITY RELATIONSHIPS

The relations found for $K_1(\theta)$, $K_2(\theta)$ and $K_3(\theta)$ can obviously be used once appropriate values of A, B, C and D have been estimated. These relations allow extension of the Brooks-Corey model for soil moisture content into the wet and dry range and they remove the limitations of the Brooks-Corey model in these moisture retention regions.

## MOISTURE INFLOW AND MOISTURE BALANCE

To solve the relations in $dz/d_\theta$, $z$, $A_{fp}$ and $V_{fp}$, the value of theoretical or actual value of the inflow $q$ has to be formulated. Noble and Nair (1990) addressed this issue by assuming that the value of $q$ is the average inflow over a long period for the landfill site and by using two HELP models based on data for annual rainfalls in Alaska (very wet, 102 in/year) and Nevada (very dry, 3.8 in/year). By using long enough times, the capillary fringe so generated becomes an average between the waste wetting and drying cycles. This approach is useful, as it well recognized that the moisture retention curves that can be developed to represent wetting and drying cycles differ significantly in capillary pressure for soil and other materials that might be present in landfills (Bear, 1988 (1972); Stamm, 1974; Marshall and Holmes, 1979; Johnson and Duckworth, 1985). Some values of waste site flows with site top cover types I–VII and estimated for this study using the USEPA HELP model, are shown in Table 5.5.

## LOCATIONS USED FOR LANDFILL COVER MOISTURE IMPACT SIMULATIONS

For this study, as nine locations and a variety of covers were used in simulation of flows through covers, it is possible to consider the average inflow $q$ and thus the theoretical *capillary fringe* in terms of the locations chosen for the final analyses. These are shown in Table 5.5.

**TABLE 5.5**
**Landfill Cover-Liner Combinations Used for Simulations**

| Case | Location | Cover Type | Linear Type | Estimated Annual Inflow q Past Cover, m/year |
|---|---|---|---|---|
| I | Ithaca, NY | VII (permeable, sand) | VII (no liner) | 0.382 |
| II | Ithaca, NY | II (permeable) | II (good liner) | 0.43 |
| III | Ithaca, NY | IV (modern, impermeable cap) | VI (no liner) | 0.0002 |
| IV | Ithaca, NY | V (modern cap) | V (modern liner) | 0.0002 |
| I | Denver, CO | VII | VII | 0.0404 |
| II | Denver, CO | II | II | 0.063 |
| III | Denver, CO | IV | VI | 0.000015 |
| IV | Denver, CO | V | V | 0.000005 |
| I | San Juan, PR | VII | VII | 0.3683 |
| II | San Juan, PR | II | II | 0.412 |
| III | San Juan, PR | IV | VI | 0.00097 |
| IV | San Juan, PR | V | V | 0.000097 |

Monthly inflows and mean temperatures at all waste site locations also exhibit cyclic variations. These values can be regressed with statistical methods to develop algorithms for moisture input to a site or temperatures at ground level.

## MICROORGANISM RATE VS. WATER CONTENT AND WATER ACTIVITY

As discussed in Chapter 2, water potential is a property unifying organisms and their environment. Overall water potential, $\psi$, of a system can be defined as:

$$\psi = \psi_\pi + \psi_m + \psi_g - \psi_p + \psi_\Omega \tag{5.64}$$

where $\psi_\pi$ = osmotic potential, $\psi_m$ = matric potential, $\psi_g$ = gravitational potential, $\psi_p$ = pressure potential and $\psi_\Omega$ = overburden potential. Combined osmotic and matric potential can be defined as:

$$\psi = \psi_\pi + \psi_m \tag{5.65}$$

For plant and microorganism cells, water potential is similarly defined by:

$$\psi = \psi_\pi + \psi_p \tag{5.66}$$

Microorganisms are close enough to their immediate surroundings to be in equilibrium; thus, their water potential matches that of their location. Water potential is also the chemical potential of water in soils, organics or cells, compared to free water:

$$\psi = \frac{\mu_w - \mu_{w,o}}{V_w} \tag{5.67}$$

where $V_w$ = molal volume of water, $\mu_w$ = chemical potential of water in the system and $\mu_{w,o}$ = chemical potential of free, pure water. Water potential can also be defined in terms of equilibrium between liquid and vapor (water) phases:

$$\text{Water potential} = \psi = \frac{RT}{V_w}\ln(a_w) \tag{5.68}$$

and $R$ = universal gas constant, $a_w$ = *water activity*, $T$ = °K and $V_w$ = molal volume of water as in Eq. 5.31. The term $a_w$ or water activity has the value of *equilibrium humidity* and ranges from 0.0 or bone dry to 1.0 or saturated condition. Metabolic activities of organisms are within certain ranges of $a_w$. Eq. 5.33 is thus a direct link between equilibrium water potential and $a_w$ value. Water activity ($a_w$) has been more widely used than water potential to define water relations of soil microorganisms (Brown, 1978). Reduced water activity, or a related property, is also indicated to be the factor limiting and stopping growth (Christian, 1978). Water vs. microbial growth should thus be modeled in terms of $a_w$ rather than water content ($\theta$).

Eqs. 5.24, 5.27, 5.28, 5.31 and 5.33 combined with Eq. 5.68, enable modeling of water potential ($\psi$) vs. water activity ($a_w$) but require determination of water potential ($\psi$). While water activity ($a_w$) around organisms relates to water potential established for that location in the unsaturated medium, difficulty of measuring water potential limits use of Eq. 5.33. A more efficient approach is to link water potential to water retention of unsaturated soils or solid materials. Water retention vs. humidity determinations have been made for the whole moisture range from saturated to bone-dry for soils. These also exist for food processing and construction industry materials (Iglesias and Chirife; Hansen, 1982), but do not yet exist for many solid organic wastes, though some have been developed from published data (Miller, 1998).

An important relation for water retention vs. water activity ($a_w$) of organic materials is the Guggenheim-Anderson-Boer (GAB) sorption model discussed in Weisser (1985), based on multi-layer molecular water coverage of pore structure, and, as discussed in Chapter 2, is

$$\frac{w}{w_m} = \frac{cka_w}{(1 - ka_w)(1 - ka_w + cka_w)} \tag{5.69}$$

This reduces to

$$\frac{a_w}{w} = A + Ba_w + Ca_w^2 \tag{5.70}$$

Constants $A$, $B$ and $C$ can be estimated data for $w$ and $a_w$ as suggested in Chapter 2 and are algebraic combinations of $w_m$, $c$ and $k$, constants related to thermodynamic equilibrium; $w$ = water content, dry weight basis; $w_m$ = water content if the porous material is covered by a water layer one molecule thick; and $a_w$ = water activity.

Models of soil water potential vs. water content developed and mentioned include:

$$\text{Soils (Gardner et al., 1971; Hillel, 1971)}: \psi = a(\theta)^{-b} \tag{5.71}$$

$$\text{Soils (Campbell, 1974)}: \psi = \psi_e(\theta/\theta_s^{-b},$$

$$K_{\text{unsat}} = K_{\text{sat}}(\psi_e/\psi)^{(2b+2)} \tag{5.72}$$

$$\text{Soils ( Brooks and Corey, 1966)}: \theta/\theta_s = (\psi_e/\psi)^{-b}$$

$$\text{Soils ( Rossi and Nimmo, 1976)}: \theta/\theta_s = 1 - c(\psi_e/\psi)^2,$$

$$\theta/\theta_s = (\psi_e/\psi)^{\lambda}, \theta/\theta_s) = \alpha \ln(\psi_d/\psi) \tag{5.73}$$

Constants $a$ and $b$ in Eqs. 5.71 and 5.72–5.74 are also soil-specific and $\theta$ refers to water content, as in models by Gardner et al. (1971) and Hillel (1971). In the soil water models of Brooks and Corey (1996), Campbell (1974) and Nimmo and Rossi (1994), $\theta_s$ refers to saturation water content and $\psi_e$ is the water potential (or bubbling pressure) at which drainage begins. $K$ is hydraulic conductivity (Gardner et al., 1971). In Eq. 5.74, separate expressions describe $\psi$ vs. $\theta$, in the near-saturation, moist and near-dry-to-dry ranges; the constants $c$, $\lambda$ and $\alpha$ are soil-specific; and $\psi_d$ refers to water potential at the end of the dry range (bone-dry). Eqs. 5.71 and 5.72–5.74 are essentially similar. Eqs. 5.72 and 5.73 have been widely used for water drainage and moisture characteristics models, with the advantage of simplicity.

Eqs. 5.68 to 5.74 link water content ($\theta$), water potential ($\psi$) and water activity ($a_w$). Combining these expressions, however, requires that (1) water content and water retention be similarly defined and cover the same range and (2) biodegradation modeling of waste sites requires that water activity estimation is sufficiently accurate for microbial growth models.

## LIMITATIONS OF APPLYING WATER POTENTIAL CONCEPTS

### MODELS OF WATER CONTENT VS. WATER POTENTIAL

The models quoted above are often unreliable for porous media in near-saturated or near-dry condition. Except for the Rossi and Nimmo (1994) approach, simpler water content or hydraulic flow models such as have generally not incorporated near-saturation and near-dry water potential data in development. However, microorganisms are sensitive to small variations in water potential in the wet range. Bohn and Bohn (1999) reported that for biofilters at waste facilities, bacterial rates are high in the range $-0.1$ to $-3$ bars ($a_w = 0.99$ to $0.97$), and slow down at about $-3$ bars and stop at about $-20$ bars matric potential. The range $-0.1$ bars (moist) to $-20$ bars representing water availability in low-moisture to dry surfaces is relatively wide, compared to the 0 to $-3$ bars range for high microbial activity rates. The latter range is associated with enhanced bioremediation or biodegradation of solid organic waste in unsaturated media (composting, landfilling, fermentation, anaerobic digestion). The wide range also contrasts with the limited range for soil drainage and hydraulic flow models. For example, water drains from soils or organic material at about $-0.2$ bars or higher pressure values. Soil water flow vs. $\psi$-models must thus be used with caution. Improved accuracy also requires consideration of near-saturated and near-dry conditions.

Noble and Nair (1990) examined the role of moisture in biodegradation of municipal waste in a landfill, a water potential vs. water content model was combined with a one-dimensional hydraulic flow model (Richards, 1931) to predict capillary fringe in a landfill with bottom-layer saturation. The capillary fringe, or the rise of water in soil media above the saturated layer, was indicated to be a landfill zone with enhanced biodegradation. This model did not incorporate a biodegradation rate. The researchers (Noble and Nair) noted that, despite the ability to model the capillary fringe, the $\psi$ vs. $\theta$ relationship (Campbell, 1974) and associated hydraulic conductivity expression (Richards, 1931) produced unrealistic predictions of soil water potential ($\psi$) for drier conditions and finite flow if water content was zero. This is obviously erroneous, and model shows the potential for error if a hydraulic flow model is combined with water potential for only the range of water content where flow is involved. This point is critical, as errors, especially in the high-biodegradation range for microorganisms, can invalidate estimation of moisture effect on decomposition.

## LIMITATIONS OF MODELS OF WATER RETENTION VS. HUMIDITY

Water retention vs. humidity data for organic materials can be developed into sorption isotherms. These isotherms, or $\theta$ vs. $a_w$ (Eqs. 5.44 and 5.46) are sigmoid curves (Christian, 1978). Plots of water potential vs. water content ($\psi$ vs. $\theta$), e.g., Ragab et al. (1972), also result in sigmoid relations on data analysis, e.g., Fink and Jackson (1951):

$$\text{Ln}\,(\theta) = A + B\ln\left[\left(\frac{\psi}{\psi_s}\right)^{-c} - 1\right] \tag{5.74}$$

$A$, $B$ and the exponent $c$ are found by statistical analysis of ($\psi$ vs. $\theta$) data. Both ($\psi$ vs. $\theta$) and ($\theta$ vs. $a_w$) relations represent *unique* pore size distribution of the material involved. Pore size distribution of media undergoing sorption and percolation is given by the Kelvin relation for condensation pressure (Sahimi, 1995):

$$\frac{\psi_s}{\psi} = \exp\left[\frac{2\,\sigma\,V_L}{RT(r-t)}\right] \tag{5.75}$$

where $\sigma$ = the liquid-to-vapor surface tension, $V_L$ = molar volume of the liquid (= $V_w$), $R$ and $T$ are as before in Eq. 5.68, $r$ = pore radius corresponding to the value of $\psi$ and $t$ is the thickness of the moisture film on the pore surface. The value of $t$ increases or decreases with water content, as does open pore volume. If pores are tortuous, water may be present as discontinuous films, thus film thickness estimation based on total surface area can be subject to error.

Researchers of styrene degradation in a biofilter (Cox et al., 1996), concluded that, while styrene removal correlated with water activity ($a_w$) rather than water content $a_w$ as estimated from humidity studies, water activity average did not reflect correlate with microorganisms in biofilms. Microorganism distribution is however, more likely to correlate imperfectly with water activity ($a_w$) than with water content ($\theta$). Discontinuous water films and associated microbial activity under drier conditions can be defined in terms of specific pore size distribution vs. surface tension. A link between pore size distribution and $a_w$ is defined by Eq. 5.75. This suggests that mathematical models incorporating specific pore size and pore water distribution, rather than average $a_w$ or ($\theta$), can improve the accuracy of biodegradation models.

For microorganisms, size limits access to water held in pores. Motile organisms also cannot move in small pores. Flux of nutrients, air and metabolic wastes depends on pore water status (Brown, 1978). As pore size decreases, matric force increases, or water is held more strongly, as Eq. 5.75 suggests. Sorption and capillary (matric potential) forces dominate at low water content, retaining water against gravity and sorption by organisms (Bohn and Bohn, 1999). *Water film thickness* is also affected with drier conditions causing thinner water films (in which bacteria must live), low nutrient flow and increased organism energy use to obtain water. For example, at water film thickness reduced to 6 to 8 times the thickness of a molecule of water (2 to 3 nm), plants can no longer extract water (Griffin, 1972). Increased energy use to obtain water reduces metabolic activity or slows biodegradation rates. Effect of water potential, water activity and water film thickness is described in data reported by Richards (1960).

**TABLE 5.6**
**Water Potential Effect on Microorganisms**

| Limitation (Cox et al., 1996) | Water Potential $\psi$, bars | Equivalent $a_w$ | Water Film Thickness, $\mu$m |
|---|---|---|---|
| Protozoa movement | –0.3 (1 bar = 1020 cm $H_2O$) | 0.999 | 4 |
| Yeast movement | –1.0 | 0.998 | 1.5 |
| Bacteria movement | –5 to –15 | 0.998–0.999 | 0.5–0.003 |
| Growth: yeast, bacteria | –15 to –40 | 0.997 | 0.003 |
| Growth, fungi | –100 to –400 | 0.93–0.75 | 0.001–0.0009 |

The limits (Table 5.6) for water potential ($\psi$) and water activity ($a_w$) show that water availability to microorganisms is limited by pore size and sorption-desorption and that optimal metabolic activity is limited to a narrow range of water content ($\theta$), e.g., moist and decreases with conditions reducing water content or film thickness (such as aggressive aeration and heating).

The effect on microbial activity of soil water film thickness is especially significant to modeling biodegradation with Eq. 5.75. For various soils, water begins to act as a solvent at moisture content corresponding to a water film thickness of 4 to 5 molecular layers (implying the wet range) (Petersen et al., 1995; Petersen et al., 1996) at which thickness Henry's Law applies and sorbed VOCs enter the water film (Holden et al., 1995). Thus porous media water potentials corresponding to thinner moisture films may be limiting for bioremediation. Microorganism colonies are likely to be attached to soil or solid wastes rather than in solution; thus moisture film thickness, as it affects amount of organic contaminants and nutrients in solution, also affects biodegradation rate. The specificity of water content ($\theta$) vs. water potential ($\psi$) to the porous medium has been noted (Holden et al., 1995; Taylor and Ashcroft, 1972). Determination of specific ($\theta$) and ($\psi$) and use of relations such as Eq. 5.75 for water film thickness could thus improve accuracy of biodegradation models incorporating water limitation and efficiencies of treatment systems based on these models.

The need to incorporate $a_w$ into biological process kinetics was stated by Moser (1988), who showed the dependency between organics destruction rate and $a_w$:

$$k = k_\infty(a_w)\exp\left[\frac{E_a(a_w)}{RT}\right] \tag{5.76}$$

where $k$ = biokinetic rate, $k_\infty$ = maximum rate, $k_\infty(a_w)$ describes $k$ vs. $a_w$, $E_a$ $(a_w)$ describes activation energy $E_a$ vs. $a_w$, $a_w = [\psi_{\text{material},T}/\psi_{envir.,T}]$, as previously described and $R$ and $T$ are as before. The values of ($E_a$ vs. $a_w$) and ($k_\infty$ vs. $a_w$) in Moser (1988) can be statistically determined from experimental data. Application of Eq. 5.76 to biodegradation in soils or unsaturated media would require (1) suitable laboratory determined values of ($a_w$) vs. $k$ and $E$; and (2) measurement or availability of ($a_w$) and $\psi$ values for the media to be used for biodegradation.

An empirical model of $(a_w)$ vs. microbial growth (Eq. 5.77), based on $E_a$ and $T$, covering 50 years of data and various microorganisms on various substrates (Davey, 1989) showed excellent agreement with results from an additional study:

$$\text{Ln } k = C_0 + \frac{C_1}{T} + \frac{C_2}{T^2} + C_3 a_w + C_4 a_w^2 \tag{5.77}$$

where $k = G \ln 2$ and $G$ = microbial divisions/hr, $T$ = °K and the $C$ constants are microorganism-specific. It was noted (Davey, 1989) that Eq. 5.77 was suitable for $a_w$ values between 0.85 and 0.998 and $T$ values between −1 °C and 42 °C but could not predict $T$ and $(a_w)$ limits for high and low bacterial growth rates. From the previous discussion of $(a_w)$, it is clear that unpredictability of Eq. 5.77 for growth rates is related to unreliability of models for predicting $(a_w)$ at near-saturation and near-dry conditions without supporting data. The values of $\psi$-values are rarely (closely) measured for these conditions for porous media on which organisms are grown; rather, they are extrapolated. Values of $\psi$ are also indicated to be nonlinear in the near-saturation and dry regions of the moisture curve (Holden et al., 1995). This suggests that full-range $\psi$ vs. $(a_w)$ models could be more accurate in estimating effect of water availability on microbial growth.

Biodegradation of solid organic wastes in fermentation, composting and landfilling involve aerobic decomposition, with fungi in a fundamental role. Modeling how water availability affects fungi is thus useful. Table 5.6 indicates fungi tolerate drier conditions (lower $a_w$ and water film thickness) than other organisms. Limits of $a_w$ for fungal growth has long interested the food and manufactured goods industries for mold damage control (Wilson and Griffin, 1979; Troller, 1980); however, the effect of $a_w$ on fungi should also be of interest to fermentation or composting, as soil fungi have a primary role in cellulose hydrolysis. Models for prediction of the effect of $a_w$ (water activity) on fungal growth (Gervais et al., 1988) include Eq. 5.56:

$$v = \left[A \ln\left(\frac{a_w}{a_{w,0}}\right)\right] + v_m \tag{5.78}$$

where $v$ = radial extension/second, $v_m$ = maximum radial extension/second, $a_w$ = water activity as in Eq. 5.33 and $a_{w,0}$ = water activity for optimal growth. Eq. 5.76 indicated weak fungal growth and long lag time for low values of water activity $(a_w)$, which matches findings of other studies (Snow et al., 1944; Pitt and Hocking, 1977). Fungal growth (Lamare and Legoy, 1995) at controlled $(a_w)$ was shown to correspond, for $a_w$ = 0.4 to 0.7, to transition between unbound and free water in the medium and associated enzymes in solution. The $(a_w)$ range for fungal activity is much wider than for bacteria, as Table 5.6 indicates. Eq. 5.76 also indicated that water activity (aw) is a fundamental parameter of mass transfer of nutrients or contaminants (e.g., odorous methyl ketone) between water and fungi or yeast (Gervais, 1990) and that variation of $a_w$ greatly affects microbial cell permeability, even at high moisture. This shows microorganisms respond to small changes at high moisture content and suggests the need for accuracy in this range for modeling growth vs. $a_w$.

The combined effect of water activity, temperature and solute on microorganism growth has also been modeled (McMeekin et al., 1987; Ratkowsky and Rose, 1995) as shown in Eqs. 5.79 and 5.80:

$$\sqrt{r} = 0.0205[\sqrt{a_w} - 0.838][T - 275.9] \quad (5.79)$$

$$\sqrt{k} = c[T - T_{\min}][\sqrt{pH - pH_{\min}}][\sqrt{a_w - a_{w_{\min}}}] \quad (5.80)$$

For Eqs. 5.79 and 5.80, growth rate $= r = k$, bacterial growth stops at $a_{w(\min)}$, $pH_{\min}$ and $T_{\min}$ are similar limits and $T$ and $a_w$ are as previously defined. Validation of Eqs. 5.76 and 5.77 showed (Gervais et al., 1988; McMeekin et al., 1987) an inverse relationship between decreasing $a_w$ and $T_{\min}$, or $f(a_w) = f[1/T_{\min}]$. This nonlinear growth response may be organism adjustment to low temperatures, but it also suggests the value of careful modeling of microbial growth for $T$ and $a_w$ extremes and limitations. Eq. 5.80 includes pH limits and discussion by Ratkowsky and Rose (1995) states that Eq. 5.79 is easily modified for logistic microbial growth affected by multiple limitations.

## DISCUSSION

The scarcity of water vs. biodegradation modeling hints at difficulties and assumed limited practical value, of measuring useful corollaries of water content in unsaturated materials: water potential and water activity. Numerous biodegradation topics reviewed show conclusively that biodegradation rate is strongly influenced by water availability to microorganisms and, subsequently, nutrients or dissolved substrates. Models examined show links between measurable parameters such as moisture content and those less measurable — water potential and water availability — through soil and water physics. Water potential is a central link between the water content of a medium and metabolic activity of biodegradative organisms. Models of water content vs. water potential ($\theta$ vs. $\psi$) examined indicate, however, unsuitability for biodegradation predictions due to poor predictability in the narrow range associated with enhanced biodegradation, i.e., high moisture content. Water retention vs. water activity ($\theta$ vs. $a_w$) models examined link water activity, a fundamental variable of organism growth, to water content, and, if water loss or gain in unsaturated materials is incorporated, the models are applicable for assessing biodegradation in porous media. Microbial growth vs. water activity models examined were shown to be powerfully influenced by moisture variation but were nonlinear for wetter or drier conditions (Miller, 2000).

No models of water effect on degradation of solid organics were discussed; however, earlier models showed good to excellent correlation with soil water content or potential for agricultural residuals, sewage sludge and forest litter (Sommers et al., 1981). Organic materials aerobic decomposition is for unsaturated condition, water potentials from $-0.1$ to $-15$ bars, optimal water potential from $-0.2$ to $-0.5$ bars, is reduced up to 50% by anaerobic condition (limited air flow); and involves bacteria ($-0.1$ to $-15$ bars), actinomycetes (to $-100$ bars) and soil microfauna

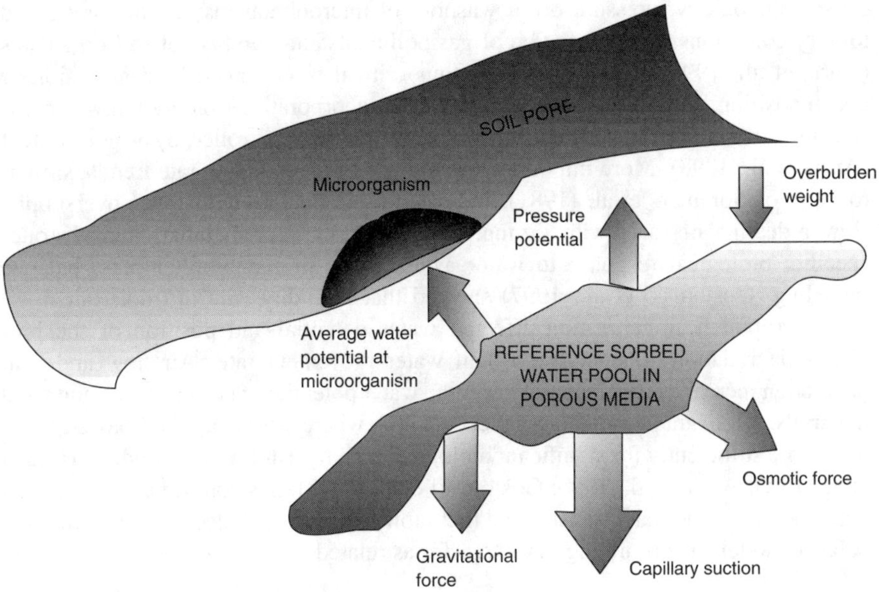


**FIGURE 5.1** Forces acting on moisture absorbed by soil and waste pores.

(Sommers et al., 1981). Aerobic biodegradation rapidly decreases as the media nears saturation, but more slowly as it nears dryness (Sommers et al., 1981). These biodegradation modes probably reflect slower natural water loss than saturation rate, suggesting the value of water balance models incorporating water potential vs. growth. Similar respiration rate models indicate rate of oxygen flow and mass transfer in water film. Water film thickness, as a potential modeling parameter, is also implied by several of the growth or biological activity models and studies.

In biodegradation, water potential affects sorption or desorption and thus mass transfer (Petersen et al., 1995; Holden et al., 1995) of compounds of interest to biological treatment in unsaturated media. These include enzymes (Lamare and Legoy, 1995), pesticides (Sommers et al., 1981), gases (Boeckx and van Cleemput, 1996), alcohol vapors, aromatic and aliphatic hydrocarbons, or VOCs (Petersen et al., 1995; Petersen et al., 1996; Holden et al., 1995; Tahraoui and Rho). The influence of mass transfer limitation on biodegradation of gaseous compounds was shown by Holden et al. (1995) and Boeckx and van Cleemput (1996). In Boeckx and van Cleemput, moisture content influenced methane removal in landfill cover soil, which improved with moist-to-unsaturated conditions. This indicates better mass transfer to organisms oxidizing methane (methanotrophs), with unblocked landfill gas flow in soil and soil water film thick enough to support significant methane-to-water transfer and subsequent metabolic activity.

Biofiltration, which involves the use of bioactive aerobic soil columns or beds to strip and degrade gas compounds, is seriously affected by media water content. Water content affects partitioning of compounds from air or soil to water; too-wet conditions cause anaerobicity, pressure drop, washout of microorganisms and nutrients; and too-dry conditions reduce sorption of gas pollutants and numbers of soil organisms (Auria et al., 1998). Water control is thus critical to enhanced biodegradation or decomposition. A biofilter mass balance model incorporating evaporation water loss (weight difference) showed that ethanol removal was controlled by water content (Auria et al., 1998). More important, the study's biodegradation rate trends, similar to those of Sommers et al. (1981), indicated rapid decline with water oversupply, slower decline of rates with drying. This suggests a predictable, nonsubstrate-specific, biological response to water availability. Similar biofilter mass balance modeling (Gostomski et al., 1997) showed that humidity control (moist air flow) and microbial heat generation affected water potential and position of the biofilter's degradation zone more so than water movement rate. Air flow and heat generation increase water loss and lower the water potential of porous media and solid materials. Biodegradation zones reflect regions where water supplied has adjusted water loss sufficiently for significant biological activity. Biofiltration studies (Holden et al., 1995; Auria et al., 1998; Gostomski et al., 1997) thus confirm a link between water potential and mass transfer and that biofiltration would also benefit from mass balance models incorporating mass transfer as related to water potential.

# 6 Heat Generation and Transport

## INTRODUCTION

Heat concentration and distribution in a landfill are likely to reflect the influence of several factors absent from other porous media systems. These include a lack of effective insulation from the outside environment, making the landfill subject to heat inflows from and heat losses to the atmosphere. Other factors include heat conduction and advection related to gas and liquid flows through the pore structure; heat diffusion through the waste–soil–water–gas mixture; and internal heat generation from microbial metabolic activity such as waste decomposition (an exothermic process). Heat content of the system is additionally influenced by its moisture content, temporal and spatial variation of heat input with climatic season and heat properties of the types of solid materials present.

Analysis of landfill heat generation and flows thus requires an adequate mathematical heat balance model incorporating the most influential of the factors mentioned. As a moisture-unsaturated system, the effect of gas filling unsaturated pore space and responsible in part for heat transport must be incorporated. Heat models for unsaturated flow have been developed and employed by various authors and are often modifications of the classic Fickian model for one-dimensional heat diffusion through a conductive medium:

$$\frac{\partial T}{\partial t} = \kappa \frac{\partial^2 T}{\partial z^2} \tag{6.1}$$

where $T$ = soil system temperature, $\kappa$ = thermal diffusivity and $z$ = soil column depth, with the soil upper surface usually taken as a zero reference point.

In more general terms, for heat transport through a porous medium, the transport equation can be written according to:

Heat accumulation + heat convection = heat diffusion + heat sources − heat sinks

$$\frac{\partial T}{\partial t} + \Delta \bullet (uT) = \Delta \bullet (\kappa T) + \sum R_a \tag{6.2}$$

which is the same as:

$$\frac{\partial T}{\partial t} = -\Delta \bullet (uT) + \Delta \bullet (\kappa T) + \sum R_a \tag{6.3}$$

Forms of this one-dimensional expression have been widely used for heat flow modeling. Restating the expression in one-dimensional Cartesian coordinates, where heat transport is vertical:

$$\frac{\partial T}{\partial t} = \frac{\partial (uT)}{\partial z} + \frac{\partial}{\partial z}\left(\kappa \frac{\partial T}{\partial z}\right) + \sum R_a = -u\frac{\partial t}{\partial z} + \kappa \frac{\partial^2 T}{\partial z^2} + \sum R_a \tag{6.4}$$

The term $u$ represents the velocity of fluid passing through the medium; the sources/sinks term $\sum(R_a)$ is the sum of all heat sources and sinks internally generated; and $\kappa$, the *thermal diffusivity*, is the ratio of the system heat conductivity ($k$) to its heat capacity ($c_p$):

$$\kappa = \frac{K_m}{\rho c_p} \tag{6.5}$$

where $K_m$ = thermal conductivity, $\rho$ = density and $c_p$ = heat capacity of the system. Inclusion of the above expression for the thermal diffusivity $\kappa$ leads to:

$$(\rho c_p)_{sys}\frac{\partial T}{\partial t} = -u(\rho c_p)_{gas}\frac{\partial T}{\partial z} + \kappa \frac{\partial^2 T}{\partial z^2} + (\rho c_p)_{hs}\sum R_a \tag{6.6}$$

Multiplication throughout by the heat capacity term ($\rho c_p$) leads to $(\rho c_p)_s$ for the system (liquids plus solids) heat capacity, $(\rho c_p)_f$ for the heat capacity of the fluid (gas or liquid) in motion and $(\rho c_p)_{hs}$ for the source or sink heat capacity when the heat exchange above is stated in terms of the mass and other properties.

In one of the few simulations of landfill heat generation and transport, a heat balance model, similar to the above expression, was recently used with a numerical solution by El-Fadel et al. (1996). It was presented as:

$$\frac{\partial(\rho c_p)T}{\partial t} = -\frac{\partial(u_g(\rho c_p)_g T)}{\partial z} + \frac{\partial}{\partial z}\left(\kappa \frac{\partial T}{\partial z}\right) + a \tag{6.7}$$

This model assumed landfill gas to be the major fluid transporting heat out of the medium and that a single lumped term ($\rho c_p$) represented $(\rho c_p)_s$. The term $a$ represented heat generation during decomposition of the organic substrates. Heat generation from landfilled waste decomposition was developed by using a reaction enthalpy balance in a stoichiometric relation for MSW bio-oxidation, with stoichiometry based upon a glucose equivalent for each substrate. Three substrate classes were used: biodegradable, slowly biodegradable and nonbiodegradable. Reasonable hydrolysis rates based upon previous findings were assigned to these substrate classes.

While the El-Fadel (1996) model is adequate for landfill heat generation modeling, it neglects some factors that could be important in more comprehensive bioreactor modeling. Factors not considered for the El-Fadel model include:

- The contribution of leachate or water input to heat loss or gain
- The effect of system moisture content on heat capacity

- External atmospheric temperature variation
- Specific variations and chemical/physical properties of the wastes

The types of MSW materials considered for this model were also limited.

This approach could affect accuracy, as a wide range of materials undergoes simultaneous decomposition and releases energy at highly specific rates. Also, a stoichiometric model based upon glucose equivalents of wastes types may not be applicable to all materials present.

However, the stated purpose of the El-Fadel et al. (1996) model was incorporation of the effect of temperature on gas generation capacity; so the factors not considered are not serious shortcomings. Additionally, El-Fadel models various kinetic factors to landfill temperature.

Consideration of a more comprehensive variety of waste materials for a waste site bioreactor, of which the landfill is representative, and the effect of heat and moisture for improved predictability of atmosphere and natural process on the reactor as a treatment system, suggests a different type of model and, specifically, a different heat model.

The model presented in this section, similar to that of El-Fadel et al. (1996), can be written in the general heat conservation form as:

$$(\rho c_p)_{sys}\frac{\partial T}{\partial t} = -u_{gas}(\rho c_p)_{gas}\frac{\partial T}{\partial z} + \kappa\frac{\partial^2 T}{\partial z^2} + (\rho c_p)_{hs}\sum R_a \tag{6.8}$$

## THE HEAT MODEL

When mass, energy, fluid species and momentum conservation are incorporated into energy balance, the energy conservation expression becomes:

$$(\rho c_p)\left[\alpha\frac{\delta T}{\delta t} + (u \bullet \Delta T)\right] = k\Delta^2 T + q''' + \left(\frac{\mu}{k}\right)u^2 \tag{6.9}$$

where $k$ represents the thermal conductivity of the porous medium as before, $q'''$ = heat generation sources and sinks within the system volume, $a$ = overall heat capacity ratio for the medium, $u$ = fluid velocity and $(\rho c_p)$ = heat capacity.

### VISCOUS ENERGY DISSIPATION

The right-hand term $\{(\mu/k)u^2\}$ represents *viscous energy dissipation*. Viscous energy dissipation can be considered negligible if it is assumed that viscosity changes in the fluids in the system (gas and liquid) are minimal, which is likely for the temperature ranges encountered inside a landfill or a compost pile (generally above freezing or 0°C and less than 100°C). However, fluid viscosity changes with temperature; for instance, gas viscosity increases with temperature, while liquid viscosity decreases as temperature increases. The relation between viscosity and a specific base temperature (20°C) is:

$$\mu = (\mu_{20°C})\exp\left[\left(\frac{d(\ln\mu)}{dT}\right)(T°C - 20°C)\right] \tag{6.10}$$

**TABLE 6.1**
**Viscosity Change vs. Temperature of Some Fluid Phases**

| Fluid | Viscosity 20°C | d(ln $\mu$)/dT | Viscosity 0°C | Viscosity 100°C |
|---|---|---|---|---|
| Water | 0.001 | –0.0284 | 0.00177 | 0.000131 |
| Water vapor | 0.00000957 | 0.00367 | 0.0000089 | 0.0000128 |
| Air | 0.0000182 | 0.00256 | 0.0000173 | 0.0000223 |
| Oxygen | 0.0000203 | 0.00256 | 0.0000193 | 0.0000249 |
| Nitrogen | 0.0000176 | 0.00250 | 0.0000167 | 0.0000215 |
| Carbon dioxide | 0.0000147 | 0.00307 | 0.0000138 | 0.0000188 |
| Carbon monoxide | 0.0000176 | 0.00262 | 0.0000167 | 0.0000217 |
| Benzene | 0.000652 | –0.0157 | 0.000893 | 0.000186 |
| Methyl alcohol | 0.000584 | –0.0157 | 0.000800 | 0.000166 |
| Landfill gas | 0.0000119 | 0.006082 | 0.0000121 | 0.0000145 |
| Methane | NA | NA | 0.0000103 | NA |

If, for instance, the temperature change from landfill interior to exterior (or vice versa of the gas air or landfill gas) or liquid (leachate, or water recharge) is known, the change in viscosity can be calculated. Values of the term $d(\ln\mu)/dT$ must be derived beforehand to estimate how the viscosity changes with temperature. A few of these values are listed below.

The fluid flows of concern would be gas flow (generally upward) and leachate flow (downward under gravity). Both flows would affect heat transfer, thus the term $\{(\mu/k)u^2\}$ can be modified to account for viscosity, thermal conductivity and velocity of landfill gas and leachate. For instance, the total viscous energy dissipation due to landfill gas and leachate flows would be:

$$\sum\left(\frac{\mu}{k}\right)u^2 = \left(\frac{\mu_{gas}}{k_{gas}}\right)u^2_{gas} + \left(\frac{\mu_{leach}}{k_{leach}}\right)\mu^2_{leach} \tag{6.11}$$

Trevisan and Bejan (1990) presented a heat model of the above type, where the term $q'''$ was neglected because internal heat generation was assumed to be small. This is obviously inappropriate for landfills. Neglect of the viscous heat dissipation term, if considered acceptable for the temperature and velocity ranges involved with the landfill bioreactor, leads to:

$$(\rho c_p)\left(\alpha\frac{\delta t}{\delta t} + (\mu \bullet \Delta T)\right) = k\Delta^2 T + q''' \tag{6.12}$$

## DEFINITION OF WASTE SITE SYSTEM HEAT CAPACITY

Trevisan and Bejan (1990) defined the heat capacity ratio term $\alpha$ as:

$$\alpha = \phi + (1-\phi)\frac{(\rho c_p)_s}{(\rho c_p)_f} \tag{6.13}$$

where $\phi$ = system porosity, $(\rho c_p)_s$ = heat capacity of the medium and $(\rho c_p)_f$ = heat capacity of the fluid. This definition of $\alpha$ assumes saturation by the fluid. In the unsaturated landfill environment, gas (landfill gas or air) is the saturating fluid since it would fill any unoccupied pore. El-Fadel et al. (1996) made this assumption, although $(\rho c_p)_s$ was presented as a single lumped value for the medium. In Cheng's (1978) model of heat generation in a geothermal reservoir, upon which the El-Fadel et al. (1996) landfill heat generation model was based, saturation by liquid was assumed. The landfill is neither saturated by liquid nor by gas; thus a redefinition of $\alpha$ would be useful.

The heat capacity ratio term $\alpha$ is a volumetrically averaged term. It is understood as the ratio of system heat capacity to that of the fluid filling the porous structure of the medium. For an unsaturated medium and assuming that gas is the saturating fluid as argued above and by analogy in the Cheng (1978) model, total heat capacity of the medium must include that of the water held against gravity, of the liquid flow and of the solids present. The heat capacity is also a summed property. According to Kirkham and Powers (1979) and Marshall and Holmes (1979), the total heat capacity can be stated as:

$$\rho c_p = (\rho c_p)_{total} = \sum_i (V_i(\rho c_p)_i) = \sum_i (M_i(c_p)_i) \qquad (6.14)$$

The product of the material volume $V_i$ and $\rho_i$ the material density, for solid materials present, is the mass $M_i$. Stating the heat capacity in terms of mass $M$ is useful, because it eliminates the need to know beforehand both the volume $V$ and density $\rho$ of materials (solid or liquid) in the system.

Assuming a landfill or waste site with many material types present, the solid materials can be classed by the subscript $j$, where the $j$th material could be *waste paper or gravel* and particle sizes are classed according to the subscript $i$, where the $i$th material size can be 32 mm screen size and the overall heat capacity of the material can be stated as a function of $M_j$, $V_j$, $r_j$ and $c_{p,j}$. Additionally, kinetic equations for biological processes are usually stated in terms of substrate mass $S$, which has the same meaning as $M$. The overall heat capacity of the solid materials (wastes, soils) can thus be stated in mass units as:

$$(\rho c_p)_{solids} = (\rho c_p)_{total,solids} = \sum [M_j(c_p)_j] \qquad (6.15)$$

The specific heat capacity term $c_p$ has units kcal/(kg°C); and if mass of the material (solid, liquid or gas) is stated in kg, the product $(M_i c_p)$ obviously has the correct dimensions (kcal/°C). This definition is only useful for the part of the system containing a mixture of solid materials.

Taking into account volumetrically averaged properties, segments or phases of the landfill system are described according to their volumetric fractions. When the system contains a moisture fraction $\theta$ (volume/volume), the open pore fraction of the unsaturated medium is:

$(\theta_s - \theta)$, where $\theta_s$ = saturated moisture content, or system porosity, which represents the total volume of saturated pores. The volumetric fraction $\theta$ has been described as the moisture or liquid holdup. Its maximum value, called the *field capacity*, is a constant for each porous system (landfilled MSW, soils, sands or clays). The overall volume fraction of solids is $(1 - \theta_s)$, or simply system volume minus porosity. If individual masses of solid materials present are known, total heat capacity for solid materials can be stated as described by $\sum[M_i(c_p)_i]$. Heat capacities for the combined system can thus be stated as:

Gas saturating open pores of landfill:

$$\text{Heat capacity } = (\rho c_p)_{gas}(\theta_{sat,water} - \theta_{water}) \tag{6.16}$$

Moisture held in the system against gravity, or instantaneous:

$$\text{Heat capacity } = (\rho c_p)_{water}(\theta_{water}) \tag{6.17}$$

Solids fraction, including MSW and soils:

$$\text{Heat capacity} = \sum V_j(\rho_j c_{p,j}) = \sum(M_j c_{p,j}) \tag{6.18}$$

where $V \times \rho = M =$ mass and $j =$ waste type (class). With gas assumed to be the saturating fluid, the total heat capacity ratio thus becomes:

$$\alpha = \left((\theta_{sat,\,w} - \theta_w)\frac{(\rho c_p)_{gas}}{(\rho c_p)_{gas}} + \theta_w\frac{(\rho c_p)_w}{(\rho c_p)_{gas}} + \frac{\sum[M_j(c_p)_j]}{(\rho c_p)_{gas}}\right) \tag{6.19}$$

This is the same as:

$$\alpha = ((\theta_{sat,w} - \theta_w) + \theta_w(\rho c_p)_w + \sum[M_j(c_p)_j])\left(\frac{1}{(\rho c_p)_{gas}}\right) \tag{6.20}$$

Notably, if a layer of soil in the landfill, e.g., sand or clay, is considered, the solids heat capacity term in the formulation above would be incorrect. For example, the heat capacity ratio for a partially saturated layer of soil (or landfill cover or liner material) would be:

$$\alpha_{layer} = ((\theta_{s,w} - \theta_w) + \theta_w(\rho c_p)_w + (V_{soil}\,\rho_{soil}\,c_{p,soil}))\left(\frac{1}{(\rho c_p)_{gas}}\right) \tag{6.21}$$

with reference to the water content and the mass of the soil in the layer.

## HEAT CONTENT OF SYSTEM: LANDFILL GAS OR AIR AS SATURATING FLUID

The previous definition of the velocity term $u \bullet \Delta T$ in the general one-dimensional heat flow relation for motion of a single saturating fluid (gas) is the same as $u_g \bullet \Delta T$ where $u_g$ is the volume-averaged flow velocity of the gas. Assuming one-dimensional (vertical) flow:

$$u \bullet \Delta T = u_g \bullet \Delta T = u_g(dT/dz)$$

The energy conservation expression can be restated as the one-dimensional partial differential equation:

$$(\rho c_p)_{gas}\left[a\frac{\partial T}{\partial t}+u_{gas}\frac{dT}{dz}\right]=k\frac{d^2T}{dz^2}+q''' \tag{6.22}$$

The volume-averaged system properties $(\rho c_p)_g$, $u_g$ and $k$ (or $K$), which are average gas heat capacity, gas flow velocity and overall system thermal conductivity, respectively, may be considered constants for analysis of *instantaneous* heat change for purposes of analysis. Substitution for $\alpha$ would thus provide a complete expression. However, little information is available on the velocities of landfill gas in actual landfills. McBean et al. (1995) suggested that a value of about 1 cm/sec for gas might be appropriate, which can be used for simpler analyses. More relevant gas flow velocities should be used in simulations of an actual system.

## THE VOLUMETRIC HEAT GENERATION TERM q‴

The term q‴ represents cumulative effect of various other environmental factors or heat sources and sinks. Major sources of heat supply to the landfill soil-waste system include heat energy evolved from microorganism activity during the decomposition process and from atmospheric heat input. Heat losses or sinks result from moisture entering or leaving the system. For instance, heat losses or sinks can result from moisture evaporation, moisture uptake, or from liquid absorbing and transporting heat out of the system. Changes in water content of the medium generally act as a heat sink, i.e., lower the temperature of the system.

Heat distribution profiles inside landfills, as in composting heaps, typically show higher temperatures at a depth and location coincident with the center of highest decomposition activity; and temperatures decrease in upward, downward and outward directions from this center. The distribution of heat inside this system will reflect the concentration of microorganisms and type and amount of organic wastes undergoing decomposition, at the center of highest heat production (Figure 6.1). Relations between unit decomposition rates and heat generation inside layers and heat dissipation can be developed to allow analysis of heat variation with depth and initial values can be determined with sensors or probes.

## HEAT IMPACT OF MOISTURE UPTAKE AND FLOWS

Injection of a liquid of volume $q_{in}$ into a segment of a landfill or waste site, as precipitation, as leachate being recycled, as leachate inflow through gravity drainage between landfill layers, or as nutrient supply in liquid form would result in an increase in heat content, according to:

$$\text{Heat in} = \{q_{in}(\varrho c_p)_w\}$$

where $(\varrho c_p)_w$ is the product of the fluid's specific heat capacity and its density.

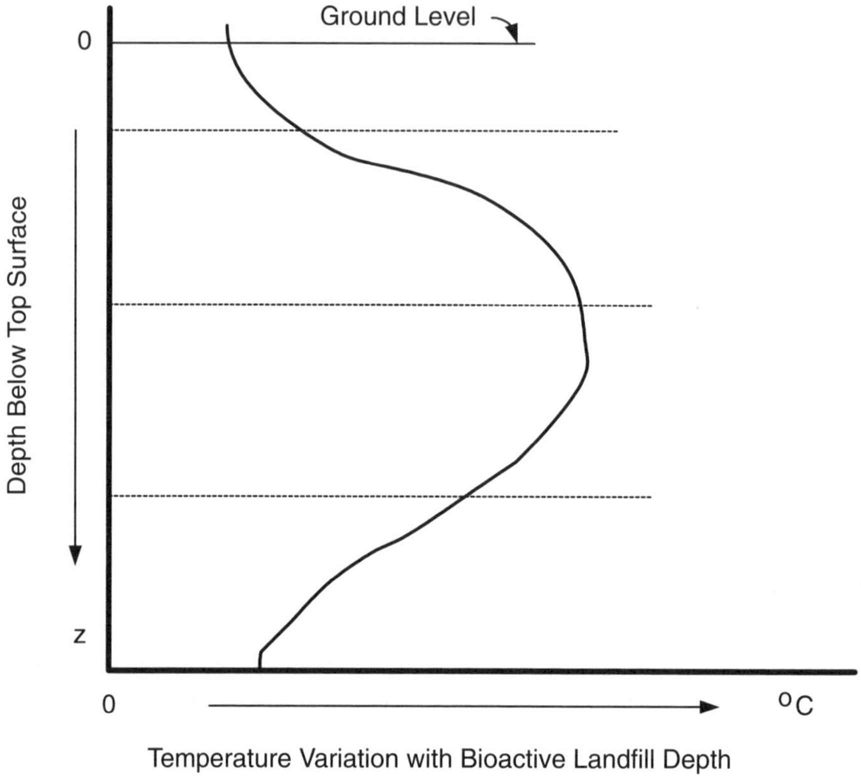


**FIGURE 6.1** Heat distribution in bioreactor

Similarly, drainage of liquid of volume $q_{out}$ and heat capacity $(\varrho c_p)_w$ from a landfill layer or out of the system would result in a heat content decrease of $\{q_{out}(\varrho c_p)_w\}$. These terms are associated with flow vectors and must be incorporated into the general expression for velocity terms. Heat conduction associated with the moving water phase can be stated as:

$$\{q(\varrho c_p)_w\}\, T$$

where $q$ = water input or output to the control volume, $(\varrho c_p)_w$ = heat capacity of the moving water phase and $T_{iw}$ = temperature of the input water. Thus, for an input liquid flow of $q_{in}$ into the landfill, heat input would be:

$$\text{mobile liquid phase – heat input: } + q_{liq,in}(\rho c_p)_{liq}\, T_{liq,in} \tag{6.23}$$

Assuming that, in an open, large system such as a landfill bioreactor there would be liquid outflow, heat loss with outflow would be:

$$\text{mobile liquid phase – heat input: } -q_{liq,out}(\rho c_p)_{liq}\, T_{liq,out} \tag{6.24}$$

The sorbed moisture phase (moisture uptake per inflow-outflow cycle) affects the system heat content by conduction of heat into the particles. Assuming this is associated with a thermal conductivity, this contribution may be stated as:

$$\text{heat input, moisture uptake: } + (q_{liq,in} - q_{liq,out})(\rho c_p)_{liq} \left(\frac{T_{liq,in} + T_{liq,out}}{2}\right) \tag{6.25}$$

## HEAT EFFECT OF MOISTURE EVAPORATION

Moisture evaporation and condensation are exchanges also occurring between the liquid and gas or gas and solids systems, respectively and can be expected to be normal phenomena for the soils and materials in the landfill. Evaporation flux is indicated as a significant source of heat transfer out of a moist porous medium (Cahill and Parlange, 1998).

As water is evaporated, a loss of heat occurs related to the latent heat of vaporization, which can be stated for an unsaturated medium as (Piver and Lindstrom, 1991):

$$-[\zeta \alpha_{tort} D_{atm} \delta_\varrho / \delta z]$$

In this expression, $\alpha_{tort}$ represents the overall tortuosity of the porous structure of the medium, $D_{atm}$ represents moisture diffusivity between water and air and $\delta\varrho/\delta z$ represents the change in density of water vapor with depth. The term $\zeta$ is the latent heat of vaporization for moisture, which varies with temperature as:

$$\zeta = 598.88 - 0.547(T - 273.16) \text{ cals/g or kcal/kg}$$

However, in general terms, an alternative to the above expression for heat loss associated with evaporation is the relation between mass of water lost via evaporation ($q_{evap}$) and the latent heat of vaporization $\zeta$:

$$q_{heat,\ evap} = \zeta q_{evap}$$

This is the same as for Fourier's law applied to water vapor heat density flux in soil (Cahill and Parlange, 1998):

$$\text{water vapor flux – heat transport: } \rho_w \, L q_{vapor} \tag{6.26}$$

The evaporation amount $q_{vapor}$ can be stated in terms of the moisture diffusivity and the change of water potential with system moisture content:

$$q_{vapor} = D_v \frac{\partial \theta}{\partial \psi} (\Delta z) = D_v \frac{\partial \theta}{\partial \psi} (dz)\, \alpha_{tortuosity} \tag{6.27}$$

The above expression for vapor flux is stated in terms of the moisture gradient (matric vs. water content). Examination of water vapor flux vs. temperature by Philip and de Vries (1957) has noted that vapor flux density is the result of thermal gradient as well:

$$\begin{aligned} q_{vapor} &= D_{v,\theta}\nabla\theta + D_{v,T}\nabla T \\ &= \frac{(a\alpha_{tort}D_a v g\rho_{vap})}{RT\rho_{liq}}\frac{\partial\theta}{\partial\psi} + (\eta a\alpha_{tort}D_a v)\frac{d\rho_{vap,\ sat}}{dT} \end{aligned} \tag{6.28}$$

In the above expression by Cahill and Parlange (1998), $\tau$ = average medium porous structure length/actual length, accounting for the fact the path traveled by gas through the system will be longer than its depth in either vertical or horizontal directions. The term $a$ represents the volumetric gas content of the medium, or porosity minus water content $(\theta_{sat} - \theta)$. The term $D_a$ is (De Vries, 1975):

$$D_a = \text{water vapor diffusivity in air} = (2e^{-7})\left(\frac{T}{273.15}\right)^{1.88} \tag{6.29}$$

The mass flow factor is given by $v = 1.0$, g = gravitation constant (9.81 m/s$^2$), $\psi$ = matric potential, $\theta$ = moisture content (vol/vol) and $\eta$ = a temperature gradient enhancement factor. The saturated vapor density term $\rho_{vap,sat}$ is related to matric potential and temperature by (Edlefson and Anderson, 1943):

$$\rho_{vap} = \rho_{vap,sat}\exp\left(\frac{\psi g}{RT}\right) \tag{6.30}$$

## EVAPORATION ENHANCEMENT DUE TO THERMAL GRADIENT IN PORE STRUCTURE

In the Philip and de Vries (1957) expression for vapor flux due to evaporation, the enhancement factor $\eta$ defines the combined effect of water vapor flow that condenses on water film on the pores but evaporates at the film surface due to a temperature gradient between the open volume of the pore and bulk water temperature. Evaporation vapor flux is indicated to be a major source of heat loss from soil systems; thus, in a waste site bioreactor such as a compost operation, landfill or bioreactive soil system, heat loss through water evaporation is worthy of separate discussion. Philip and de Vries (1957) state $\eta$ as:

$$\eta = \frac{a + f(\theta)}{\alpha_{tort}a}\frac{(\nabla T)_a}{\nabla T} \tag{6.31}$$

where $a$ = volume of gas in pore, $f(\theta)$ = relation between moisture content and open pore volume, $(\Delta T_a)$ = temperature gradient from between pore center (moving gas) and pore wall and $\alpha$ represents tortuosity. Cahill and Parlange (1998) note that the value of $(\nabla T_a)$ cannot be measured and results from models for $\eta$ have not matched actual measurements. These authors also state that an alternative formulation for vapor flux formulation by Milly (1982) avoids the issue of $\eta$ and also states flux in terms of the moisture potential ($\psi$) — removing some of the uncertainties associated with water uptake hysteresis and soil inhomogeniety:

$$q_v = \frac{D_a \alpha_{tort} \theta_a}{\rho_l} \frac{\partial \rho}{\partial z} \tag{6.32}$$

In this expression, $\theta_a$ = volumetric gas content of the porous medium = medium porosity–volume of liquid present = $(\varepsilon - \theta)$, $\alpha_{tort}$ = tortuosity (Chapter 1), $\rho_1$ = density of liquid and $D_a$ = water vapor molecular diffusivity. Saravanapavan and Salvucci (2000) show that at the drying front or vapor-dominant zone in a porous medium, where water vapor flux should occur, the matric potential should have reached a critical value. Notably, all of the above variables except $\partial \rho / \partial z$ are defined elsewhere in this work; thus, the Milly (1982) expression can be used for vapor flux. For instance, Milly (1982) defines the evaporation flux in terms of matrix potential as:

$$\frac{D_a \alpha_{tort} \theta_a}{\rho_l} \frac{\partial \rho_{vapor}}{\partial z} \frac{\partial z}{\partial \psi} = \frac{D_a \alpha_{tort} \theta_a}{\rho_l} \frac{\partial \rho_{vapor}}{\partial \psi} = K(\psi) \tag{6.33}$$

where $\psi = \psi_{critical}$ = matric potential value at which water and vapor transport are equal (boundary between saturated and unsaturated phase in medium) and the effective hydraulic conductivity is $K(\psi) = K(\psi_{critical})$ at that value of $\psi$.

However, the relation for $\eta$ in vapor flux is assumed to be correct (Liu et al., 1998; Saravanapavan and Sarvucci, 2000); thus a reformulation using the same terms can be considered. Haraiwa and Kasabuchi (2000) reexamined the enhancement factor $\eta$ as introduced by Philip and deVries (1957) in their expression for effect on thermal conductivity of latent heat transfer:

$$\text{Thermal conductivity (latent heat transfer)} = \lambda_{vapor} = \eta \theta_a \, \alpha_{tort} DH \left( \frac{d\rho_{vapor}}{dT} \right) \tag{6.34}$$

where $\eta$ = enhancement factor, $\alpha_{tort}$ = tortuosity, $\theta_a$ = air volume in (soil) medium, $D$ = water vapor diffusivity in air and $H = \zeta$ = latent heat of evaporation water in air. Haraiwa and Kasabuchi (2000) show that the constants ($\eta \alpha_{tort}\, a$) in the Philip and deVries (1957) expression can be replaced by the phenomenological enhancement constant $\beta$, where $\beta = f(1/\text{matric potential}) = f(1/\psi)$, or $\beta = f(\text{air volume in pore}) = f(\theta_s - \theta)$:

$$\begin{aligned} q_v &= \frac{(a\alpha_{tort} D_a v g \rho_{vap})}{RT\rho_{liq}} \frac{\partial \theta}{\partial \psi} + (\eta a \alpha_{tort} D_a v) \frac{d\rho_{vap,\,sat}}{dT} \\ &= \frac{(a\alpha_{tort} D_a v g \rho_{vap})}{RT\rho_{liq}} \frac{\partial \theta}{\partial \psi} + (\beta D_a v) \frac{d\rho_{vap,\,sat}}{dT} \end{aligned} \tag{6.35}$$

Two soils — Ando (clay loam) and Red Yellow (light clay) — were used in this study for laboratory examination of variation of $\beta$ vs. $1/\psi$ and of $\beta$ vs. $\theta_{ait}$. The data plot values for these variables indicate that the phenomenological enhancement factor $\beta$ has the following pattern:

- $\beta$ increases when water content reaches a maximum at a humidity ($a_w$ value) near 1.0 for the medium (soil). Above this value the soil saturation increases and $\beta$ decreases in value, reaching zero at effective saturation of the medium.
- For $\beta$ vs. water potential ($1/\psi$), $\beta$ increases *linearly* from a dry condition of the medium to water vapor saturation ($a_w \approx 1.0$) and decreases *logistically* thereafter. This is interpreted as the value of $\beta$ influenced by water potential, up to the critical threshold (humidity $= a_w \approx 1.0$) and the influence of decreasing air volume as the pore structure fills up.

Behavior of the phenomenological enhancement factor $\beta$ vs. water potential is likely to relate to two factors: available wetted surface of the medium and potential energy of the water sorbed. The drier the material, the farther the drying front will have retreated away from its surface. Higher potential energy (as thermal energy) would be required to remove water from inside the material to its surface and to vapor flux. Additionally, water in fine pores is reduced under drier conditions, meaning severe reduction of wetted surface and thus less evaporation of water. As the material becomes wetter, the potential energy required for evaporation becomes smaller; thus evaporation flux becomes higher with a thermal gradient. The inverse relation between $\beta$ and the water potential $\psi$ (and humidity $a_w$) reflects this. Note that the relation between $\psi$ and $a_w$ is linear:

$$\psi_{water\ vapor} = \frac{RT_{air}}{M_w} \ln(h_{air}) = \frac{RT_{air}}{M_w} \ln(a_w) \tag{6.36}$$

In the above, $M_w$ = molecular weight of water (18.0), $R$ = gas constant for air (8.314 Joules/mol.°K), $T_{\text{air}}$ = °Kelvin = 273.16+ °C and $h_{air}$ = relative humidity of the air. The water potential relationship also assumes equilibrium between water *at the sample surface* and in the air (or gas), thus at equilibrium:

$$h_{air} = h_{material} = a_{w,material} \tag{6.37}$$

The point of maximum evaporation enhancement ($\beta_{\text{max}}$) also reflects the point of maximum coverage of pore surface by water film, also coincident with low potential energy and thus sustained supply *from* the surface. A temperature gradient at between the pore interior and bulk water would thus result in the maximum transport to vapor flux for this value of $\beta$ and humidity. Enhanced evaporation due to thermal gradient in the porous structure can thus be said to reflect available wetted pore surface.

## TEMPERATURE VS. WATER VAPOR DIFFUSION, LATENT HEAT AND DENSITY VARIATION

### WATER VAPOR DIFFUSION

The molecular diffusion coefficient of water vapor in air (at 25°C) is reported at $D_a = 0.256\ \text{cm}^2/\text{sec}$. In unsaturated soil the diffusivity of water vapor is corrected for the mean path of travel of water molecules by use of the average pore tortuosity $\alpha_{tort}$, volume of air or open pore (porosity minus liquid content) and water content:

$$D_{diff} = D_a \alpha_{tort} \theta_a \tag{6.38}$$

The variation of diffusion coefficient ($D_a$) of water vapor in still air with temperature (°K) is provided by (Boulet et al., 1997):

$$D_a = (2.7e^{-7}) \left( \frac{T}{273.15} \right)^{1.88} \tag{6.39}$$

According to Philip and de Vries (1957) the effect of soil moisture on diffusion coefficient of water vapor diffusivity $D_{diff}$ is provided by:

$$D_{diff} = D_a f(\theta_{air}) \left( \frac{p_{atm}}{p_{\text{atm}} - p_{vapor}} \right)$$

$$f(\theta_{air}) = \theta_{air} \left( 1 + \frac{\theta}{\theta_k} \right) : \theta \leq \theta_k$$

$$f(\theta_{air}) = porosity : \theta \geq \theta_k \tag{6.40}$$

In the above expression, $\theta_k$ represents the minimum water content for continuity of the liquid phase. This value of water content can be assumed to be at or near the field capacity of the soil, or at humidity approximating water vapor saturation. Various data measurements can also be used to determine a modeling algorithm for effect of temperature ($T$) on water vapor diffusivity ($D_{diff}$), or of $T$ vs. $D_a$, moisture diffusion coefficient in still air (see end of this section).

## Latent Heat of Vaporization

Thibodeaux (1996) states that, at the air-water interface, the heat, vaporization or condensation due to movement of other (trace) chemicals does not affect latent heat transfer; thus, only water needs to be considered. Heat transfer via water evaporation from subsurface soil or waste material wetted by leachate should mostly involve the latent heat of vaporization of water. The latent heat of vaporization $\zeta$ has been defined earlier as $2.44 \times 10^6$ Joules/kg (748.91 calories/kg) at 24°C. Its change in value with temperature is likely to be relatively small for the range of temperatures likely to be experienced in waste sites (0−80°C). The variation of $\zeta$ with temperature $T$ can also be assessed from existing data (see end of section).

## Water Vapor Density Variation

The diffusivity relationship developed by Philip and de Vries (1957) and modified by later investigations requires assessment of the term $\partial\rho/\partial T$. Multiplication by the matric potential term as done by Milly (1982) introduces:

Thermal moisture diffusivity

$$D_{Tv} = (\eta a \alpha_{tort} D_a v) \frac{d\rho_{vap,\,sat}}{dT} = \frac{D_a \alpha_{tort} \theta_a \eta(RH)}{\rho_{liquid}} \left[ \frac{d\rho_{vap,sat}}{dT} \right] \left( \frac{p_{atm}}{p_{atm} - p_{vapor}} \right)$$

$$= \frac{D_a [\beta](RH)}{\rho_{liquid}} \left[ \frac{d\rho_{vap,\,sat}}{dT} \right] \left( \frac{p_{atm}}{p_{atm} - p_{vapor}} \right) \tag{6.41}$$

The value of $\partial\rho_v/\partial T$ has been separately defined by Boulet et al. (1997) as:

$$\rho_v = (RH)\frac{p_{vapor,sat}(T)}{RT}$$

$$RH = \exp\left(\frac{g\psi}{RT}\right)$$

$$p_{vapor,sat}(T) = (618.78)\exp\left(\frac{17.27(T - 273.16)}{T - 35.86}\right)$$

$$\rho_v = (618.78)\left(\frac{1}{RT}\right)\exp\left(\frac{g\psi}{RT}\right)\exp\left(\frac{17.27(T - 273.16)}{T - 35.86}\right)$$

$$= (618.78)\left(\frac{1}{RT}\right)\exp\left[\frac{g\psi}{RT} + \frac{17.27(T - 273.16)}{T - 35.86}\right] \quad (6.42)$$

The above expression only needs definition of the matric potential $\psi$ of the medium (soil or soil/waste mixture). The explicitly defined relative humidity (RH) term has the same value as the water activity term $a_w$ used in Chapter 1 for solid materials; $g$ is gravitational constant (9.81 m/sec$^2$), $R$ = gas constant (8.314 joules/mol.°K) and $T$ is in degrees Kelvin (°K). The partial derivative:

$$\partial p_v/\partial T$$

can be determined if values $\psi$ and $T$ are defined elsewhere. However, data can also be used to assess how the water vapor density changes with temperature (see end of section). Philip and de Vries (1957) note that for the range 10°C to30°C,

$$\partial\rho_v/\partial T = 1.05\ \text{g/cm}^3.°\text{C}.$$

Because the above range of temperature would not cover likely ranges in waste sites, the above value of $\partial\rho_v/\partial T$ has limited application.

## OTHER DATA FOR EVALUATING $D_A$, $\zeta$ AND $\partial\rho_V/\partial T$ VS. TEMPERATURE (T)

In their study of the relation of soil thermal conductivity vs. temperature over the range 0°C to 75°C, Haraiwa and Kasubuchi (2000) used data from the National Astronomical Observatory, Tokyo, Japan (1998) to evaluate conductivity. Conductivity was defined as a function of latent heat of vaporization of water, moisture diffusion coefficient of water vapor in air and water vapor density vs. temperature. As this data for water can be applied to the present case, e.g., waste sites, it is used below for simple statistical expressions to benefit modeling over the 0°C to 75°C temperature range, likely to cover several waste site internal temperature scenarios. This data is presented as (Table 6.2, Haraiwa and Kasabuchi, 2000):

**TABLE 6.2**
**Variation of Water Vapor Energy Parameters with Temperature**

| Temperature T (°C) | Latent Heat, $\zeta(\times 10^6$J/kg) | Diffusion Coefficient $D_a$ ($\times 10^{-4}$m$^2$/sec) | Water Vapor Density Variation, $\partial\rho_v/\partial T$ ($\times 10^{-4}$kg/K.m$^3$) |
|---|---|---|---|
| 0 | 2.500 | 0.234 | 2.40 |
| 5 | 2.490 | 0.244 | 3.30 |
| 15 | 2.465 | 0.264 | 5.13 |
| 25 | 2.441 | 0.285 | 7.26 |
| 35 | 2.418 | 0.308 | 9.95 |
| 45 | 2.392 | 0.330 | 13.46 |
| 55 | 2.368 | 0.354 | 18.06 |
| 65 | 2.344 | 0.379 | 24.01 |
| 75 | 2.319 | 0.405 | 31.58 |

Statistical regression with the above data (Table 6.2) and using MathCad 2000 software provides the following:

$T$ = degrees Kelvin = 273.15 + °C

$\Delta T = (T_0 - T_1)$ degrees Centigrade

$F(T) = (T)/(T - \Delta T)$

$\zeta(T) = 3.063 - 0.577F(T)$ $\quad n = 9$, correlation = 0.946

$D_a(T) = -0.373 + 0.606F(T)$ $\quad n = 9$, correlation = 0.963

$[\delta\rho/\delta T).\zeta.D_a] = -0.228 + 1.936[F(T)]^{11.286}$ $\quad n = 9$, correlation = 0.99

## DEFINITIONS OF WASTE SITE SYSTEM TORTUOSITY

The surface area of porous media exposed to gas or liquid flow is called the *dynamic surface area*, to distinguish from stagnant areas mostly associated with fine or unconnected pores or small pores filled with water. The pore in contact with the gas flow will also be involved with heat flow and transfer because the gas flow stream may be the largest heat transfer for heat through and out of the system. In terms of water absorbed by or trickling through the porous media, the area in contact with the gas stream can be described as the total area of the porous structure unoccupied by the wetting fluid (water in this case) and interconnected across the system. This area is called the *interface area* and, under mass and heat transfer considerations, is a function of particle arrangement and size and of tortuosity.

Fluid (liquid or gas) flow through the system establishes preferential pathways. The length and total surface area of these pathways has been found to be dependent upon the particle arrangement. In a mixed or disordered system such as a waste site, particle size distribution along flow path is highly irregular; and system properties such as particle arrangement, interface area and tortuosity have to be deduced through mathematical and statistical approaches.

As fluid (liquid or gas) moves through the porous media, it typically travels along a path more winding or tortuous than any vertical or horizontal length. In a vertical direction, this path is longer than the media depth. The ratio of the pore path length to the actual layer or soil column depth is called the *tortuosity*, referred to above as $\alpha_{tort}$. The tortuosity of a waste–soil system, described earlier as a porous bed, is a function of both moisture content and the bed porosity. As physical arrangement of particles and thus tortuosity affects gas or liquid flow rate and interface area — but is exceedingly difficult to measure for irregular particle systems — various mathematical considerations have been proposed by investigators.

## Tortuosity as a Function of Particle Flatness

In Chapter 1, overall porous system tortuosity was defined according to mass transfer convention as:

$$\zeta^* = \alpha_{tort} = \left[ \frac{(M^*)^2}{N} \frac{2\gamma\eta\varepsilon^3}{(0.0968\rho)^2} \right]^{\frac{1}{4}} \tag{6.43}$$

in which expression $M$ and $N$ were defined, according to the Comiti and Renaud (1989) approach for flow through packed beds, as:

$$M^* = \left[ 1 - 0.0413 \left( \frac{D - d_{particle}}{D} \right) + 0.0968 \left( \frac{D - d_{particle}}{D} \right) \right] \alpha_{tort}^3 \rho a_{vd} \frac{(1-\varepsilon)}{\varepsilon^3}$$

$$N^* = \left[ 2\gamma\alpha_{tort}^2 a_{vd}^2 \left[ 1 + \frac{4}{a_{vd} D(1-\varepsilon)} \right] \frac{(1-\varepsilon)^2}{\varepsilon^3} \right] \tag{6.44}$$

and $D$ and $d_{particle}$ are diffusion coefficient and mean particle size, respectively, and $a_{vd}$ = dynamic surface area (area of bed in contact with the fluid flow). The mean particle size for a mixed particle waste site can only mean a particle or grain of mean size, shape and specific surface area that represents the whole site. As shown in Chapter 1, these average system properties must be deduced through statistical methods.

Comiti and Renaud (1989) also found that tortuosity increases with particle flatness (thickness/length = $e/a$) and that comparison of length/thickness ratios for various particles shapes and packed bed porosities ($\varepsilon$) agreed with the Pech (1984) expression:

$$\alpha_{tort} - 1 = 1.6 \ln \left( \frac{1}{\varepsilon} \right) \tag{6.45}$$

for the porosity range $0.1 < \varepsilon < 0.6$ and thus tortuosity should vary according to:

$$\alpha_{tort} - 1 = P \ln (1/\varepsilon) \tag{6.46}$$

with $P$ values for several particle shapes in packed beds varying according to:

$$P = 0.577\exp[0.18(a/e)] \tag{6.47}$$

and thus average particle bed tortuosity vary according to:

$$\alpha_{tort} = 1 + [0.577\exp(0.18(a/e))]\ln(1/\varepsilon) \tag{6.48}$$

The issue of how flatter particles affect surface area exposed to fluid flow through a packed bed was also addressed by stating that the ratio $X$ of *specific interface area to specific surface area* was:

$$X = \frac{a_{vd}}{a_{vs}} = 0.43 + 0.577(e/a) \tag{6.49}$$

where $a_{vd}$ = specific surface area actually exposed to fluid flow, $a_{vs}$ = geometric specific surface area, $e$ = average particle thickness and $a$ = average particle length. This requires that $e$ or $a$ be known or derived from other values.

However, later work by Mauret (1995) and Sabiri (1995) confirmed that the term $P$ in the media tortuosity expressions:

$$\alpha_{tort} 1 - P\ln(1/\varepsilon) \text{ and}$$

$$P = 0.577\exp[0.18(a/e)]$$

(*Equations* 4.46, 4.47)

depend upon the shapes of particles and their orientation in a porous bed.

This further suggests that caution should be taken when applying $P = f(a/e)$ relationships to porous systems such as waste sites, as $P = f(a/e)$ expressions are unlikely to account explicitly for particle shape and orientation. Particle shape is nevertheless important to mass and heat transfer and thus to tortuosity, in porous granular media modeling as shown by Comiti et al. (2000). Mauret and Renaud (1997) indicate that a $P = 0.49$ value is more applicable to a wide range of particle types than the $P = 0.41$ suggested by Comiti and Renaud (1989). Hence the following expression could be used for a waste site, with the proviso that further consideration should be given to effect of particle shape and surface area:

$$P = f(\textit{mass, volume, shape, area})$$

$$P \simeq 0.49$$

$$\alpha_{tort} = 1 + P\ln\left(\frac{1}{\varepsilon}\right) \tag{6.50}$$

According to Ferguson (1993), waste particles inside a landfill can be characterized as plates of average diametric length-to-thickness ($d_s/t$) ratios of about 5 and of inclination about 30°. This would imply that the *aspect ratio (t/L) of the waste particles is about* 0.2, with a vertical plane length or diameter of about $d_s \cos 30°$

(or 0.866 $d_s$), where $d_s$ = characteristic particle length. Checking this against the Comiti and Renaud (1989) expressions for tortuosity:

$$\{\alpha_{tort} = 1 + P \ln(1/\varepsilon)$$

$$P = 0.577 \exp[0.18(a/e)]\} \text{ Comiti and Renaud, 1989}$$

$$(a/e) = 0.2, \varepsilon = 0.41 \text{ (landfilled waste)}$$

$$\alpha_{tort} = 1.526$$

and by Mauret and Renaud (1995):

$$\{\alpha_{tort} = 1 + P \ln(1/\varepsilon) P = 0.49\} \text{ Mauret and Renaud (1995)}$$

$$\varepsilon = 0.41 \text{ (landfilled waste)}$$

$$\alpha_{tort} = 1.437$$

Seguin et al. (1996) show that the tortuosity of a packed bed of plates of aspect ratio 0.209 is about 2.8(= 2.77). Therefore, for the purpose of analysis, the tortuosity value of a landfilled waste–soil layer could be assumed to be about 2.8. *A different value has to be used for intervening sand and layers (daily cover or final cover material).*

$$\alpha_{tort} = 2.77$$

The above value $\alpha_{tort} = 2.77$ is unlikely to be reliable for waste sites, as the study by Seguin et al. (1996) involved investigation of a relatively limited number of particle types and shapes in packed beds, compared to the great variety of particles likely to be present in a waste site (soil mix, waste–soil mix or composting materials). Use of the Mauret and Renaud (1995) expression for tortuosity and $P$, for a range of waste site porosity ranging from 0.33–0.6, indicates tortuosity decreases as porosity increases from 1.5 to 1.25. Thus a value of $\alpha_{tort}$ based upon system porosity for a landfill-type waste site can be assumed to be approximately:

$$\alpha_{tort} = 1.45$$

With a tortuosity of this value, a correlation of active pore area for the migrating gas can be made. The relation between the tortuosity and packed bed length L is usually given as:

$$\alpha_{tort} = L_e/L$$

which would imply that $L_e$, the true average length of any pore or flow path, would be:

$$L_e = \alpha_{tort} L$$

For practical purposes, the intrinsic permeability $k$ of the medium can be calculated from the conductivity $K$ as:

$$K = k\varrho g/\mu$$

where $K$ is assumed to be 0.864 m/day for landfilled material (Oweis et al., 1992), $\varrho =$ density of water $= 997.1$ kg/m$^3$ at 25°C, $g = 9.81$ m/sec$^2$ and $\mu =$ dynamic viscosity of water $= 8.95 \times 10^{-4}$ m$^2$/sec. This gives a value of the intrinsic permeability of the landfilled material at $k = 0.0068$/day.

According to Du Plessis (1992), for low Reynolds number (laminar) flow through packed (fixed, permeable, granular) beds, use of the representative unit cell (RUC) volume approach (whereby for geometric analysis the solids content of the medium is assumed to be a cubic volume enclosed in an cubic envelope of fluid) leads to the following correlation:

$$k/d_s^2 = [\varepsilon^3/180](1 + 6\varepsilon)$$

where $k$ = permeability, $d_s$ = mean diameter of the representative solid particle and $\varepsilon$ = porosity of the medium. This correlation is based upon modification of the Blake–Kozeny correlation for laminar flow through packed beds or granular materials. Oweis et al. (1992) found the porosity of landfilled material to be approximately 0.41. The above correlation could thus be used to estimate the mean particle diameter $d_s$.

Use of the intrinsic landfill layer permeability value $k = 0.0068$/day and the porosity $\varepsilon$ value of 0.41 leads to an estimation of characteristic landfill waste layer particle size $d_s$ as:

$$d_s = 0.044 \text{ meters}$$

This suggests an average particle size in the landfill as a packed bed of approximately 1.737 inches, which is reasonable considering both the coarser sizes of wastes and the size dilution effect of the fine-sized sand or soil particles.

As discussed by Sahimi (1995), the Blake-Kozeny correlation for flow rates in packings of particles leads to a specific surface area $A_s (m_2/\text{unit volume})$ correlation of:

$$A_s = 6/d_s$$

which, from the value of $d_s$ calculated above, is $A_s = 135.93$ (m$^2$). This value is not unreasonable considering the specific surface areas for potential wastes, estimated and tabulated for this work in Chapter 2, Table 2.8.

According to Seguin et al. (1995), for beds of particles of overlapping plate shapes where the aspect (thickness/length) is 0.209, the ratio $X$ of surface presented by the particles to fluid flow ($A_{vd}$) to actual surface area $A_s$ is:

$$X = A_{vd}/A_s = 0.54$$

For landfilled material the aspect ratio (thickness over length) was estimated by Ferguson (1993) to be 0.2. This would suggest a particle thickness of 0.0088 meters and that the average landfill surface presented to flow would be:

$$A_{vd} = 73.4\,\mathrm{m}^2/\mathrm{m}^3$$

This is also 790.1 square feet per cubic meter. However, this applies to the theoretical area open to (saturating) gas flow. A modification must thus be made to account for degree of saturation. As the estimate of $A_{vd}$ is based upon the assumption that the average particle in a waste site should have an aspect ratio of about 0.2, which is unconfirmed, the above value should not be used when other values of $A_{vd}$ can be determined.

## Tortuosity as a Function of Particle Surface Properties

Mass–volume ($m-v$) considerations for any particle in a landfill or waste site necessitate a strict relationship between these and other particle parameters:

$$\text{Volume} = \frac{M}{\rho} = \mathit{area} \times \mathit{thickness} \tag{6.51}$$

However, thickness is only applicable as an average for particles of predictable geometry. For irregular particles it would have little practical meaning. For instance, a section of plant twig would have a complex outer geometry not easily resolved into thickness and surface area, though its mass and density, and thus its volume, could be otherwise found.

The discussions of specific surface areas of particles in Chapter 1 suggested that the total surface area subject to fluid adsorption can be found from the mathematical models described and thus even the most complex materials could be represented. The surface area involved in sorption and heat or mass transfer is of fundamental importance to reactor considerations, more so than purely geometric area. Tortuosity is additionally an essential parameter when applying transport concepts to granular media reactor theoretical concepts. Thus it could be useful to present tortuosity in terms applicable to particle types and shapes that cannot be considered strictly geometric.

Plots of tortuosity vs. porosity by Mauret and Renaud (1995) suggest that medium tortuosity varies according to:

$$\alpha_{tort} = \alpha_{tort,\max} \exp(-\beta\varepsilon) \tag{6.52}$$

where:

$$\varepsilon_{\max} = 1.0, \text{ at } \alpha_{tort,\max} = 1.0$$

Plots by Mauret and Renaud (1995) for dynamic surface area $a_{vd}$ also suggest a porosity relationship:

$$\alpha_{vd} = a_{vd,\min} \exp(\gamma\varepsilon) \tag{6.53}$$

The hydraulic diameter of a representative pore of a packed bed can be represented as:

$$r_H = \frac{4\varepsilon}{SSA_{bed}} = \frac{2\varepsilon d_{\varnothing}}{3(1-\varepsilon)} \tag{6.54}$$

where $SSA_{bed}$ = specific surface area of the bed, $d_{\varnothing}$ = diameter of a spherical particle equivalent in volume to a representative particle of the medium and $r_H$ = hydraulic diameter of representative pore. Examination of expressions (6.51) and (6.53) leads to:

$$\varepsilon = \left[1 - \frac{1}{6}\,(SSA_{bed})(d_{\varnothing})\right] = \frac{-1}{\beta}\ln\left(\frac{\alpha_{tort}}{\alpha_{tort,\max}}\right) \tag{6.55}$$

Because at porosity $\varepsilon = 1$, the tortuosity $\alpha_{tort} = 1.0$, the above expressions can be restated as:

$$\alpha_{tort} = \exp\left(\beta(1-\varepsilon)\right)$$
$$\varepsilon = \left[1 - \frac{1}{6}(SSA_{bed})(d_{\varnothing})\right] \tag{6.56}$$

In the above (6.55), bed or system porosity is expressed in terms of the specific surface (SSAbed) and equivalent mean spherical diameter ($d\varnothing$) of the media, which, as illustrated in Chapter 1, can be estimated from the individual size distributions and specific surface areas of the types of particles present.

The above expression for porosity $\varepsilon$ can also be applied to the Comiti and Renaud (1989) and Mauret and Renaud (1995) expression for tortuosity:

$$\alpha_{tort} = 1 + P\ln(1/\varepsilon)$$

## ENERGY BALANCE AT ATMOSPHERIC BOUNDARY OF BIOREACTOR

Energy heat generated within bioreactor systems such as landfills and compost piles is likely to diffuse outward; some will be lost to the atmosphere at the upper soil–atmosphere boundary, the soil surface. At this boundary, important heat energy interchanges result from sunlight (incident solar radiation, called shortwave), long-wave radiation from the atmosphere and clouds, latent heat flux from evaporation (discussed elsewhere in this section) and sensible heat flux (loss of soil heat flux due to convection at the surface). The effect of these energy flows is modified by surface reflectance and emissivity; thus, energy inflows from the atmosphere are reduced at the surface of the bioreactor by an amount reflecting resistance to energy inflow. These energy flows are classified here as:

$$R_{is} = \text{incident solar radiation, short wave}$$

$$R_{is}(1-\alpha) = \text{amount of short wave solar radiation reflected}$$

$$L_{wi} = \text{incident atmospheric radiation, long wave}$$

$$L_{wo} = \text{reflected or absorbed long wave radiation}$$

$$LE = \text{latent heat energy}$$

$$Q_{shf} = \text{sensible heat flux due to convection at the soil surface}$$

Energy conservation requires that the sum of these energy flows be zero.

The resistance to energy inflow (or outflow) is also unique to the site because the soil or material comprising the upper surface of the reactor will be different for each location.

## NET SOLAR RADIATION

For this type of analysis, solar radiation is usually defined as the sum of direct and diffuse radiant shortwave energy from the sun, striking the Earth's surface at a normal angle. The difference between incident sunlight and amount absorbed by the Earth surface is the amount reflected and this difference is stated in terms of the surface albedo or reflectance constant $\alpha$. The amount of solar energy delivered is thus:

$$R_{is,\ net} = R_{is}(1-\alpha) \tag{6.57}$$

The amount of solar energy delivered to the surface is affected by surface elevation above sea level, latitude, solar angle, sunlight absorption and scattering by atmospheric particles and substances. Solar angle, maximum at noon, is provided according to Lambert's law for decrease of radiation density on a horizontal surface with angle of inclination. The difference between the angle of incidence ($\beta$) of solar energy directly overhead (normal) at location and angle of incidence at another time of the day is given by:

$$R_{is} = R_{is,noon}\sin(\beta)$$

$$\text{Solar angle} = \beta$$

$$\beta = \sin^{-1}(\sin(D)\sin(L)\cos(D)\cos(H)) \tag{6.58}$$

In the above, $D$ = declination of the sun in radians, $L$ = latitude of location in radians and $H$ = solar time angle (Sumner, 1999). Sun declination is defined (Rosenberg et al., 1993) as:

$$D = (0.4101)\cos\left(\frac{2\pi(J-172)}{365}\right) \tag{6.59}$$

with $J$ = day of the year.

Solar time angle H in radians is (Sumner, 1999):

$$H = \frac{2\pi(T - T_{SN})}{24}$$

$$T_{SN} = 12 + \frac{4(\text{Longitude} - \text{Local meridian})}{60} + T_{EQ} \tag{6.60}$$

where $T$ = time in hours since the start of the day, $T_{SN}$ = time of solar noon, longitude = local longitude in degrees and local meridian = degrees for which local time is calculated. For instance, in the U.S., local standard time is calculated according to: Pacific Standard Time (PST) = 120°, Mountain Standard Time (MST) = 105°, Central Standard Time (CST) = 90° and Eastern Standard Time (EST) = 75°. The term TEQ represents the equation for time at a particular location based on longitude and latitude and is (Jensen et al., 1990):

$$T_{EQ} = (0.1645)\sin(2b) - 0.1255(b) - 0.025\sin(b) \ \text{(hrs)}$$

$$b = \frac{2\pi(J - 81)}{364} \qquad J = \text{day of the year} \tag{6.61}$$

Direct radiant energy from the sun ($R_{si,0}$) can be stated in terms of the relative distance from the earth to the sun, the solar constant, latitude, declination and solar time angle at the center of the period P (24 hours or one day), or solar time angle at noon ($\omega$), in radians (Duffie and Beckman, 1991):

$$R_{is,atm} = \left(\frac{24(60)}{2\pi}\right) G_{SC} d_{rel} \left[\begin{array}{l}(\cos(L)\cos(D))\,(\sin(\omega_2) - \sin(\omega_1)) \\ +(\omega_1 - \omega_2)\,(\sin(L)\sin(D))\end{array}\right]$$

$$G_{SC} = 0.08202\ \text{MJ/m}^2.\text{min}$$

$$d_{rel} = 1 + (0.033)\cos\left(\frac{2\pi J}{365}\right)$$

$$\omega_1 = \omega - \frac{\pi}{(24/P)}$$

$$\omega_2 = \omega + \frac{\pi}{(24/P)} \tag{6.62}$$

According to Sumner (1999), time period length ($P$) per day is given, in terms of day length, in radians (TD) and the sunset time angle from noon to sunset ($\omega_s$) by:

$$T_D = \frac{24\omega_s}{\pi}$$

$$\omega_s = \cos^{-1}(-\tan(L)\tan(D)) \tag{6.63}$$

These definitions allow substitution into the expression for direct incident radiant energy from the atmosphere ($R_{is,atm}$) for:

$$R_{is,atm} = \left(\frac{24(60)}{\pi}\right) G_{SC} d_{rel} \left[\begin{array}{l}(\cos(L)\cos(D))\,(\sin(\omega_s)) \\ +(\omega_s)\,(\sin(L)\sin(D))\end{array}\right] \tag{6.64}$$

This expression provides the radiant energy in $MJ/m^2$, given the day of the year ($J$) and latitude ($L$) in radians (the term $D$ is stated in terms of $J$ and is also in radians).

However, the radiant energy through the atmosphere is usually corrected for adsorption and scattering by molecules and particles in the atmosphere; so the amount actually striking the Earth's surface is less. According to Jensen et al. (1990), the direct energy is reduced by about 75% maximum by the time it reaches the Earth's surface, for clear sky conditions. Thus direct solar shortwave or radiant energy reaching the Earth surface on a daily basis ($R_{is,atm,s}$) is approximately:

$$R_{is,\ atm,\ s} = 0.75(R_{is,\ atm}) \tag{6.65}$$

## EFFECT OF SURFACE ALBEDO

When incident solar energy strikes a surface, a fraction is reflected. A measure of the degree of reflection is the albedo. Surface albedo varies from 0.0 to 1.0. As soil–water content increases, the albedo decreases; shiny and light-colored surfaces also decrease albedo.

In the expression for net radiant (shortwave) energy from the sun,

$$R_{is,\ net} = R_{is}(1 - \alpha) \tag{6.66}$$

the albedo $\alpha$ is a measure of reflectance at that particular location. For the bioreactor, for example at a landfill, compost pile or similar scenarios, the net radiant energy would be reduced according to the soil or other cover present. For landfills with temporary or final cover, or bioreactor landfill designs, the cover can be topsoil, clay, fine sand, asphalt or concrete. Temporary covers can include gravel, stone, carpet, soil, mulch and other materials. For any site, it is thus useful to know the albedo value. Some values of albedo are listed below in Table 6.3.

## INCOMING LONGWAVE RADIATION

The flux of radiant energy at the Earth's surface also includes reception and emission of longwave radiation. These can be stated in terms of direction as $LW_{in}$ = incoming and $LW_{out}$ = longwave emission at the surface. Incoming longwave radiation is stated as (Evett, 1999):

$$LW_{in} = \in \sigma(T_{air} + 273.16)^4 \tag{6.67}$$

The value of the Stefan-Boltzmann constant $\sigma$ is $5.67 \times 10^{-8}$ W/m$^2$.°K$^4$. The emissivity $\in_a$ of the air is a function of its vapor pressure $e_a$ and temperature ($T_{air}$), where $T_{air}$ is in degrees Centigrade (°C). Vapor pressure is a function of relative humidity (RH) and saturation vapor pressure $e_{sat}$ and is in kilopascals (kPa):

$$e_a = \text{relative humidity} \times e_{sat}$$

$$e_{sat} = 0.60078 \exp\left(\frac{1.7269 T_{air}}{237.3 + T_{air}}\right) \tag{6.68}$$

**TABLE 6.3**
**Surface Albedo and Emissivity of Some Materials**

| Surface Type | Albedo | Emissivity | Reference |
|---|---|---|---|
| Soils: dark/wet, to light/dry | 0.05–0.50 | 0.90–0.98 | Sumner, 1999, Table 5.4, pg. A146 |
| Sandy soil, dry | 0.25–0.45 | | Sumner, 1999, Table 5.4, pg. A146 |
| Sand, wet | 0.09 | 0.98 | Sumner, 1999, Table 5.4, pg. A146 |
| Sand, dry | 0.18 | 0.95 | Sumner, 1999, Table 5.4, pg. A146 |
| Quartz sand | 0.35 | | Sumner, 1999, Table 5.4, pg. A146 |
| Clay, dry | 0.20–0.35 | | Sumner, 1999, Table 5.4, pg. A146 |
| Dark clay, wet | 0.18 | 0.95 | Sumner, 1999, Table 5.4, pg. A146 |
| Dark clay, dry | 0.02–0.08 | 0.97 | Sumner, 1999, Table 5.4, pg. A146 |
| Bare, dark soil | 0.16–0.17 | | Sumner, 1999, Table 5.4, pg. A146 |
| Bare field | 0.16 | | Sumner, 1999, Table 5.4, pg. A146 |
| Desert | | 0.84–0.91 | Jacobson, 1992 |
| Field with stubble | 0.15–0.17 | | Jacobson, 1992 |
| Field crops, latitude 22–52 | 0.22–0.26 | 0.94–0.99 | Jacobson, 1992 |
| Field crops, latitude 7–22 | 0.15–0.21 | 0.94–0.99 | Jacobson, 1992 |
| Most field crops | 0.18–0.30 | | Jacobson, 1992 |
| Leaves of common field crops | | 0.94–0.98 | Jacobson, 1992 |
| Green field crops | 0.20–0.25 | | Jacobson, 1992 |
| Grain crops | 0.10–0.25 | | Jacobson, 1992 |
| Green grass | 0.16–0.27 | 0.96–0.98 | Jacobson, 1992 |
| Dried grass | 0.16–0.19 | | Jacobson, 1992 |
| Grass | 0.24 | | Jacobson, 1992 |
| Deciduous forest, with leaves | 0.20 | 0.98 | Jacobson, 1992 |
| Deciduous forest, bare of leaves | 0.15 | 0.97 | Jacobson, 1992 |
| Coniferous forest | 0.05–0.15 | 0.98–0.99 | Jacobson, 1992 |
| Snow, fresh | 0.40 | 0.82 | Jacobson, 1992 |
| Snow, old | 0.95 | 0.99 | Jacobson, 1992 |
| Mangrove swamp | 0.15 | | Jacobson, 1992 |
| Mulch | 0.23 | | |
| Mulch (grass and leaf) | 0.05 | | Brunel |
| Concrete | 0.3 | 0.71–94 | Collette, 1998; Jacobson, 2002 |
| Tar paper | 0.05 | 0.93 | Collette, 1998; Brunel |
| Gravel | 0.72 | 0.28 | Collette, 1998; Javelin–IRCON |
| Crushed rock | 0.20–0.40 | 0.93 | Javelin–IRCON |
| Limestone | 0.30–0.45 | | Brunel |
| Brick | 0.20–0.50 | 0.45 | Brunel |
| Carpet (cotton, wool, synthetics) | | 0.98 | Javelin–IRCON |
| Fabrics (cotton, wool, synthetics) | | 0.90–0.95 | Javelin–IRCON |
| Linoleum | | 0.96 | Javelin–IRCON |
| Plastics | | 0.95 | Javelin–IRCON |
| Asphalt-macadam | | 0.98 | Javelin–IRCON |
| Asphalt (roofing) | 0.07–0.15 | | Brunel |
| Tar and gravel (roofing) | 0.08–0.18 | | Brunel |

*(Continued)*

**TABLE 6.3**
**Continued**

| Surface Type | Albedo | Emissivity | Reference |
|---|---|---|---|
| Artificial turf | 0.07–0.10 | | Brunel |
| Deciduous plants | 0.20–0.30 | | Brunel |
| Rubber cloth | | 0.96 | Javelin–IRCON |
| Tiles (rubber and plastic) | | 0.95 | Javelin–IRCON |
| Tiles (ceramic) | | 0.80 | Javelin–IRCON |
| Wood (chips) | | 0.90 | Javelin–IRCON |
| Cardboard | | 0.88 | Javelin–IRCON |
| Paper | | 0.80–0.85 | Javelin–IRCON |
| Roofing material | | 0.95 | Javelin–IRCON |

A relatively reliable expression for predicting the emissivity of air and combining air vapor pressure and temperature was developed by Idso (1981):

$$\in_{air} = 0.70 + (0.000595) e_{air} \exp\left(\frac{1500}{T_{air} + 273.1}\right) \tag{6.69}$$

However, Evett (1999) notes that incoming longwave radiation is usually only a small part of the energy flows at the Earth surface.

## OUTGOING LONGWAVE RADIATION

Longwave radiation energy from the Earth's surface is similarly provided in terms of the emissivity and temperature of the surface, by the Stefan-Bolzmann law:

$$LW_{out} = \in_{surface} \sigma (T_{surface})^4 \tag{6.70}$$

Note that the above energy flow is in MJ/m$^2$. Also, compared to the outgoing longwave radiation ($LW_{in}$), the temperature and emissivity of the surface of concern must be known.

To convert to bioreactor system energy flows, the unit energy flow values of this section must be multiplied by the upper surface area ($A$). For instance, if the longwave radiation energy per unit area is $Q_{LW,\,out}$ and the upper surface area is $A_{bioreac}$, the longwave radiation emitted by the Earth's surface is:

$$Q_{LW,out} = A_{bioreac}\, LW_{out} = A_{bioreac}(\in_{surface}\ \sigma (T_{surface})^4) \tag{6.71}$$

## LATENT HEAT FLOW OF A BIOREACTOR SYSTEM

It was shown that the latent heat flow from the reactor was a product of the latent heat of evaporation of water and the evaporative flux from the system. Latent heat transfer to gas would occur at the bioreactor gas-sorbed water film interface in the bioreactor pore structure. Assuming little temperature variation in the gas stream upflow through the porous structure before it reaches the upper surface of the bioreactor, this latent heat flow definition would still apply. Thus latent heat flow (LE) would be:

$$A_{LE,out} = A_{bioreac}\zeta(q_{evap})$$
$$\zeta = \text{latent heat of vaporization of water} \tag{6.72}$$

## TEMPERATURE VARIATION WITH DEPTH

The bioreactor system generating substantial heat in the interior poses a relatively unique problem with regard to heat distribution, as the highest temperatures are likely to be at a depth below the surface and above groundwater. This would mean that temperature variation is concentric and three-dimensional from the point or points of highest decomposition activity. For one-dimensional analysis, it would be of interest to estimate temperature variation with depth, as this is likely to have some impact on type of cover used and could indicate performance as a treatment system (for instance, at a large composting operation or at a bioreactor landfill).

## SENSIBLE HEAT FLOW FROM THE BIOREACTOR SYSTEM

It can be assumed that heat diffusing from the bioreactor and reaching the upper surface would be lost to the atmosphere through convection. *Sensible heat flux* refers to this heat loss from the system. It is influenced by air movement, which is affected by air turbulence near the surface. Turbulence is in turn affected by surface roughness, wind speed and the variation ($\Delta T$) between air and bioreactor (upper) surface temperature. As surface temperature is a function of the bioreactor's internal temperature, it is necessary to first define temperature distribution with depth. Heat lost from the system reaches the atmosphere via conduction and convection through the porous media.

## DEVELOPMENT OF THE HEAT GENERATION MODEL

The general heat relation can be written as:

$$(\rho c_p)_{gas}\left[a\frac{\partial T}{\partial t} + u_{gas}\frac{dT}{dz}\right] = k\frac{d^2T}{dz^2} + q''' \tag{6.73}$$

and rewritten as:

$$\frac{\delta T}{\delta t} = \frac{k}{\alpha(\rho c_p)_{gas}}\frac{d^2T}{dz^2} - \frac{u_{gas}}{\alpha}\frac{dT}{dz} + \frac{1}{\alpha(\rho c_p)_{gas}}q''' \tag{6.74}$$

or as:

$$\frac{\delta T}{\delta t} = a_1\frac{d^2T}{dz^2} + a_2\frac{dT}{dz} + a_3 q''' \tag{6.75}$$

where the constants are:

$$a_1 = \frac{k}{\alpha(\rho c_p)_{gas}}$$
$$a_2 = \frac{-u_{gas}}{\alpha}$$
$$a_3 = \frac{1}{\alpha(\rho c_p)_{gas}} \tag{6.76}$$

and $q'''$ is the sum of heat sources and sinks as previously described.

## SOLUTION TO THE HEAT EQUATION

For modeling purposes, it is useful to develop algorithms for the variation of temperature and heat in a waste site system such as a landfill, large compost pile, or other physical systems involving bioreactive media (remediation sites, biofilters, etc.). There are many approaches to developing heat equation algorithms and the modeler is free to choose the most convenient and reliable for the system. One mathematical approach to the equation is presented below as a possible alternative. However, simpler approaches are preferable.

### HEAT EQUATION

Assuming that the temperature function $T(z, t)$ is a function of a system depth-varying function $f(z)$ and a temperature time-varying function $g(t)$, $T(z, t)$ can be written as:

$$T(z, t) = f(z)g(t)$$

Differentiating according to variables and substituting in the general expression:

$$\delta T/\delta t = f(z)g'(t)$$
$$\delta T/\delta z = f'(z)g(t)$$
$$\delta^2 T/\delta z^2 = f''(z)g(t)$$

This implies that the general expression for $\partial T/\partial t$ is:

$$\frac{f''(z)}{f(z)} + \frac{a_2}{a_1}\frac{f'(z)}{f(z)} + \frac{a_3}{a_1}\frac{1}{f(z)g(t)}q''' = \frac{1}{a_1}\frac{g'(t)}{g(t)} \tag{6.77}$$

In the above expression, the term $q'''$ represents the sum of heat sources and sinks for the system, including:

- Heat input from liquid inflow
- Heat loss from liquid outflow
- Heat loss via evaporation
- Heat generation (input) from biological decomposition

The right-hand side (RHS) function $g'(t)/g(t)$ is a function of t only and $f(z)$ is a function of $z$ only. This indicates the RHS = constant = LHS. If the RHS constant is $\beta$, then:

$$(1/a_1)[g'(t)/g(t)] = \beta$$
$$g'(t) - (a_1\beta)g(t) = 0$$

for which a possible solution is:

$$g(t) = A_1 \exp(a_1\beta t)$$

Assuming $(1/a_1)[g'(t)/g(t)] = \beta$ and multiplying the general expression above by $f(z)$ :

$$f''(z) + \frac{a_2}{a_1} f'(z) + \frac{a_3}{a_1} \frac{1}{(g(t))} q''' = \beta f(z) \tag{6.78}$$

Substituting from the previous expressions, this is the same as:

$$f''(z) + \frac{a_2}{a_1} f'(z) - \beta f(z) = \left(-\frac{a_3}{a_1} q'''\right) A_1 \exp(-a_1\beta t) \tag{6.79}$$

Distributing terms as follows:

$$(a_2/a_1) = k_1$$
$$(-\beta) = k_2$$
$$\{A_1(a_3/a_1)q'''\} = k_4$$

The general expression becomes:

$$f''(z) + k_1 f'(z) + k_2 f(z) = k_4 \ \exp\,(-a_4\beta t) \tag{6.80}$$
$$(D^2 + k_1 D + k_2)z = k_4 \ \exp\,(-a_1\beta t) \tag{6.81}$$

The complementary function of the above relationship is:

$$y_c = c_1 \exp\,(r_1 z) + c_2 \ \exp\,(r_2 z)$$
$$r_1 = \{[-k_1 + \surd(k_1^2 - 4k_2)]/2\}$$
$$r_2 = \{[-k_1 - \surd(k_1^2 - 4k_2)]/2\}$$

If it is assumed that $T(z, t)$ has the form:

$$T(z, t) = zA(t) + B(t)$$
$$\delta T/\delta z = A(t)$$
$$\delta^2 T/\delta z^2 = 0$$

Substituting in the expression:

$$0 + k_1 A(t) + k_2[zA(t) + B(t)] = k_4 \exp(-a_1\beta t)$$
$$A(t)[k_2 z + k_1] + k_2(B(t) = k_4 \exp(-a_1\beta t)$$

for $k_2$, $z$ and $k_1 \not\equiv 0$; $A(t) = 0$.
Thus

$$B(t) = (k_4/k_2)\ \exp\,(-a_1\beta t) = T(z, t)$$

which is the *particular solution* $y_p$ of the expression. The general solution $y$ for the expression is given by definition, by:

$$y = y_c + y_p = f(z)$$

Thus the general solution for $f(z)$ can be stated as:

$$\begin{aligned} f(z) = {} & c_1 \exp\left(\left(\frac{-k_1}{2} + \frac{1}{2}(k_1^2 - 4k_2)^{\frac{1}{2}}\right) z\right) \\ & + c_2 \exp\left(\left(\frac{-k_1}{2} - \frac{1}{2}(k_1^2 - 4k_2)^{\frac{1}{2}}\right) z\right) \\ & + \left(\frac{k_4}{k_2}\right) \exp[-\beta a_1 t] \end{aligned} \tag{6.82}$$

The expression for $f(z)$ can thus be rewritten as:

$$\begin{aligned} T(z, t) = {} & c_1 A_1 \exp\left(\frac{-k_1}{2} z\right) \exp\left(\frac{3k_1^2 z}{8}\right) \exp\left(\frac{3}{2}\beta z\right) \exp(\beta a_1 t) \\ & + c_2 A_1 \exp\left(\frac{-k_1}{2} z\right) \exp\left(-\frac{3k_1^2 z}{8}\right) \exp\left(-\frac{3}{2}\beta z\right) \exp(\beta a_1 t) \\ & + \left(\frac{k_4}{k_2}\right) A_1 \end{aligned} \tag{6.83}$$

As previously defined, the function $T(z, t) = f(z).g(t)$, where $g(t) = A_1 \exp(\beta a_1 t)$, multiplying the expression for f(z) by g(t), where $\beta = (iw/a_1)$ and $k_2 = -\beta$.

## TEMPERATURE AT THE WASTE SITE SURFACE

From the definition for seasonally varying temperature at the landfill surface ($z = 0$), one possible solution is that for g(t) the Eulerian form would suffice:

$$g(t) = A_1 \exp(\beta a_1 t) = A_1 \exp(iwt) = g(t) = A_1(\cos(wt) + i \sin(wt))$$

The term $w$ is the phase constant for the temperature cycle and has the value $2\pi/p$, where $p = 365.25$ (correction for leap year). Rewriting $T(z, t)$ for boundary conditions at the surface:

Boundary condition (1):

(1) At $z = 0$, the term $(k_4 A_1)/k_2$, where $k_4 = k_3 q'''$, does not exist, as the porous volume averaged heat generation term $q'''$ would not exist.
(2) At $z = 0$, one solution for $T(z,t) = a \sin wt$, where $a =$ half the range of $t$ (modulus). Comparing with the expression developed for $T(z,t)$:

$$T(0,t) = A_1(C_1 + C_2)(\cos(wt) + i\sin(wt)) = a\sin(wt)$$

which implies $a = A_1(C_1+C_2)$, for the solution $\sin(wt)$. Also, $T(0,0) = 0$ implies $(C_1 + C_2) = 0$ if $A_1 \not\equiv 0$, implying $C_1 = -C_2$.

Rearranging $T(z,t)$ in terms of the sine, cosine terms for addition:

$$\begin{aligned} T(z,t) = c_1 A_1 [\exp(-k_1 z/2)]\{&\exp(3k_1^2 z/8)[(\cos(3wz/(2a_1)) \\ &+ i\sin(3wz/(2a_1))].[\cos(wt) + i\sin(wt)] \\ &- \exp(-3k_1^2 z/8)[(\cos(3wz/(2a_1)) \\ &- i\sin(3wz/(2a_1))].[\cos(wt) + i\sin(wt)]\} \end{aligned}$$

The products of the terms bracketed thus { } are:

$$\begin{aligned} &+[\exp(3k_1^2 z/8)]\cos(3wz/(2a_1)\cos(wt)) \\ &+i[\exp(3k_1^2 z/8)]\cos(3wz/(2a_1)\sin(wt)) \\ &+i[\exp(3k_1^2 z/8)]\sin(3wz/(2a_1)\cos(wt)) \\ &-\exp(3k_1^2 z/8)]\sin(3wz/(2a_1)\sin(wt)) \\ &-[\exp(-3k_1^2 z/8)]\cos(3wz/(2a_1)\cos(wt)) \\ &-i[\exp(-3k_1^2 z/8)]\cos(3wz/(2a_1)\sin(wt)) \\ &+i[\exp(-3k_1^2 z/8)]\sin(3wz/(2a_1)\cos(wt)) \\ &+[\exp(-3k_1^2 z/8)]\sin(3wz/(2a_1)\sin(wt)) \end{aligned}$$

Arranging, the sum of the above terms is:

$$\begin{aligned} &[\exp(3k_1^2 z/8) - \exp(-3k_1^2 z/8)]\,[\{\cos(3wz/(2a_1))\cos(wt)\} - \{\sin(3wz/(2a_1)) \\ &\quad \sin(wt)\}] - i[\{\sin(3wz/(2a_1))\cos(wt)\} - \{\cos(3wz/(2a_1))\sin(wt)\}] \end{aligned}$$

This is the same as:

$$2\sin h(3k_1^2 z/8)[\cos\{3wz/(2a_1) + wt\}] + 2i\{\cos h(3k_1^2 z/8)\}\sin\{3wz/(2a_1) + wt\}]$$

Thus the general expression is:

$$\begin{aligned} T(z,t) = (2c_1 A_1)&[\exp(-k_1 z/2)]\sin h(3k_1^2 z/8)\;[\cos\{3wz/(2a_1) + wt\}] \\ &+ (2c_1 A_1)\,i\,\{\cos h(3k_1^2 z/8)\}\;\sin\{3wz/(2a_1) + wt\}] \end{aligned}$$

By inspection the partial expression in sinh does not exist at $z = 0$, as $\sin h(3k_1^2 z/8) = 0$ at $z = 0$; thus this part of the expression would not apply for the boundary condition of $z = 0, t > 0$. As both the real and imaginary parts of the expressions are solutions for $T(z, t)$, one solution existing at $z = 0$ is:

$$T(z, t) = (2c_1 A_1)[\exp(-k_1 z/2)]\ \{\cos h(3k_1^2 z/8)\}\ \sin\{3wz/(2a_1) + wt\}$$

This has the value $T(z, t) = (2c_1 A_1)\sin\{wt\}$ at $z = 0$, or at the surface. For $t > 0$, the temperature expression has the form:

$$T(z, t) = T_a + a\sin(wt)$$

where $a$ = modulus (half the temperature range) and $T_a$ is the average value of $T$. Thus by inspection:

$$a = 1/2(T_{\max} - T_{\min}) = (2c_1 A_1)$$

and the expression for $T(z, t)$ with a solution at $z = 0$ becomes:

$$T(z, t) = 1/2(T_{\max} - T_{\min})[\exp(-k_1 z/2)]\{\cos h(3k_1^2 z/8)\}\sin\{3wz/(2a_1) + wt\}$$

For the temperature expression to be in phase, it is necessary to add a phase constant $C$ to the temperature expression (Kirkham and Bowers, 1972) such that:

$$T(z, t) = T_a + a = T_a + a\sin(wt + C)$$

which implies:

$$\sin(wt + C) = 1$$
$$(wt + C) = \pi/2$$

Because $w = 2\pi/p$, where $p = 365.25$ days as previously defined, the value of $t$ must be such that the $\sin(wt)$ has its maximum value, i.e., $t$ corresponds to a peak $T$ value. The time $t$ when this occurs can be termed $t_m$ for the time since $t = 0$ to reach a maximum value. It can be found from interpolation between the average temperature values for the landfill geographic locations chosen and were estimated for the locales used in this study. Thus the phase constant $C$ can be defined as:

$$C = \pi/2 - (2\pi/p)t_m = \pi/2[1 - (4t_m)/p]$$

The general expression thus becomes:

$$T(z, t) = T_{ave} + \frac{1}{2}(T_{\max} - T_{\min})\left[\exp\left(\frac{-k_1 z}{2}\right)\right]$$
$$\cos h\left(\frac{3k_1^2 z}{8}\right)\sin\left(\frac{3wz}{2a_1} + wt + \frac{\pi}{2}\left(1 - \frac{4t_m}{p}\right)\right) \tag{6.84}$$

The partial differentials of this expression, where $a = 1/2[T_{\max} - T_{\min}]$, are given by:

$$\frac{\delta T}{\delta z} = \left[\frac{3k_1^2}{8}a\right]\left[\exp\left(\frac{-k_1 z}{2}\right)\sin h\left(\frac{3k_1^2 z}{8}\right)\sin\left(\frac{3wz}{2a_1}+wt+\frac{\pi}{2}\left(1-\frac{4t_m}{p}\right)\right)\right]$$
$$+\left[\frac{3w}{2a_1}a\right]\left[\exp\left(\frac{-k_1 z}{2}\right)\cos h\left(\frac{3k_1^2 z}{8}\right)\cos\left(\frac{3wz}{2a_1}+wt+\frac{\pi}{2}\left(1-\frac{4t_m}{p}\right)\right)\right]$$
$$+\left[\frac{-k_1}{2}a\right]\left[\exp\left(\frac{-k_1 z}{2}\right)\cos h\left(\frac{3k_1^2 z}{8}\right)\sin\left(\frac{3wz}{2a_1}+wt+\frac{\pi}{2}\left(1-\frac{4t_m}{p}\right)\right)\right] \quad (6.85)$$

$$\frac{\delta^2 T}{\delta z^2} = \left[\frac{9k_1^2 w}{16a_1}a\right]\left[\exp\left(\frac{-k_1 z}{2}\right)\right]\sin h\left(\frac{3k_1^2 z}{8}\right)\cos\left(\frac{3wz}{2a_1}+wt+\frac{\pi}{2}\left(1-\frac{4t_m}{p}\right)\right)$$
$$+\left[\frac{9k_1^4}{64}a\right]\left[\exp\left(\frac{-k_1 z}{2}\right)\right]\cos h\left(\frac{3k_1^2 z}{8}\right)\sin\left(\frac{3wz}{2a_1}+wt+\frac{\pi}{2}\left(1-\frac{4t_m}{p}\right)\right)$$
$$-\left[\frac{3k_1^3}{16}a\right]\left[\exp\left(\frac{-k_1 z}{2}\right)\right]\sin h\left(\frac{3k_1^2 z}{8}\right)\sin\left(\frac{3wz}{2a_1}+wt+\frac{\pi}{2}\left(1-\frac{4t_m}{p}\right)\right)$$
$$+\left[\frac{9wk_1^2}{16a_1}a\right]\left[\exp\left(\frac{-k_1 z}{2}\right)\right]\sin h\left(\frac{3k_1^2 z}{8}\right)\cos\left(\frac{3wz}{2a_1}+wt+\frac{\pi}{2}\left(1-\frac{4t_m}{p}\right)\right)$$
$$-\left[\frac{9w^2}{4a_1^2}a\right]\left[\exp\left(\frac{-k_1 z}{2}\right)\right]\cos h\left(\frac{3k_1^2 z}{8}\right)\sin\left(\frac{3wz}{2a_1}+wt+\frac{\pi}{2}\left(1-\frac{4t_m}{p}\right)\right)$$
$$-\left[\frac{3k_1 wa}{4}\right]\left[\exp\left(\frac{-k_1 z}{2}\right)\right]\cos h\left(\frac{3k_1^2 z}{8}\right)\cos\left(\frac{3wz}{2a_1}+wt+\frac{\pi}{2}\left(1-\frac{4t_m}{p}\right)\right)$$
$$+\left[\frac{k_1^2}{4}a\right]\left[\exp\left(\frac{-k_1 z}{2}\right)\right]\cos h\left(\frac{3k_1^2 z}{8}\right)\sin\left(\frac{3wz}{2a_1}+wt+\frac{\pi}{2}\left(1-\frac{4t_m}{p}\right)\right)$$
$$+\left[\frac{3k_1^3}{16}a\right]\left[\exp\left(\frac{-k_1 z}{2}\right)\right]\sin h\left(\frac{3k_1^2 z}{8}\right)\sin\left(\frac{3wz}{2a_1}+wt+\frac{\pi}{2}\left(1-\frac{4t_m}{p}\right)\right)$$
$$-\left[\frac{3k_1}{4a_1}a\right]\left[\exp\left(\frac{-k_1 z}{2}\right)\right]\cos h\left(\frac{3k_1^2 z}{8}\right)\cos\left(\frac{3wz}{2a_1}+wt+\frac{\pi}{2}\left(1-\frac{4t_m}{p}\right)\right) \quad (6.86)$$

The general relation for heat generation was previously defined as:

$$\frac{\delta T}{\delta t} = \frac{k}{\alpha(\rho c_p)_{gas}}\frac{d^2 T}{dz^2} - \frac{u_{gas}}{\alpha}\frac{dT}{dz} + \frac{1}{\alpha(\rho c_p)_{gas}}q''' \quad (6.87)$$

or in terms of the constants:

$$\frac{\delta T}{\delta t} = a_1 \frac{d^2T}{dz^2} + a_2 \frac{dT}{dz} + a_3 q''' \tag{6.88}$$

Substitution for the partial differentials and for the values of $k_1$, $k_2$ and $k_3$ and $q'''$ as defined earlier thus provides a solution for the temperature $T$ after a time $dt$. The final form of the heat equation used is thus:

$$\begin{aligned}\frac{dT}{dt} = &\left[-a_2\frac{a}{2} + \frac{k_1^2 a_1}{4} + \frac{9{k_1}^4 a}{64a_1} - \frac{9w^2 a}{4a_1}\right] \\ &\exp\left(\frac{-k_1 z}{2}\right)\cos h\left(\frac{3k_1^2 z}{8}\right)\sin\left[\frac{3wz}{2a_1} + P\right] \\ &+ \left[3a_2 k_1^2 a - 3a_1\frac{k_1^3}{8}\right]\exp\left(\frac{-k_1 z}{2}\right)\sin h\left(\frac{3k_1^2 z}{8}\right)\sin\left[\frac{3wz}{2a_1} + P\right] \\ &+ \left[\frac{3a_2 wa}{2a_1} - \frac{3k_1 wa}{2}\right]\exp\left(\frac{-k_1 z}{2}\right)\cos h\left(\frac{3k_1^2 z}{8}\right)\cos\left[\frac{3wz}{2a_1} + P\right] \\ &+ \left[\frac{9k_1^2(1+w)a}{8}\right]\exp\left(\frac{-k_1 z}{2}\right)\sin h\left(\frac{3k_1^2 z}{8}\right)\cos\left[\frac{3wz}{2a_1} + P\right] + a_3 Q'''\end{aligned} \tag{6.89}$$

where $P$ represents the expression for phase adjustment developed earlier:

$$\begin{aligned} P &= \left[wt + \pi\left(0.5 - \frac{2t_m}{365.25}\right)\right] \\ w &= \frac{2\pi}{p} \\ p &= 365.25 \end{aligned} \tag{6.90}$$

The time $t$ represents the time since start of the year and $t_m$ = time required to reach the maximum temperature. In this form the relation is integrable for time steps $\Delta t$ and when the starting temperature is known.

The term $Q'''$ represents the biological heat generation and other sources (summed). For instance, reiterating:

$Q'''$ = heat sinks and sources
= atmospheric energy input (shortwave and longwave radiation)
− latent heat of vaporization
+ heat from mobile phase (liquid) inflow
− heat from mobile phase (liquid) outflow
+ heat generation (anaerobic and/or aerobic decomposition)
− heat loss to the atmosphere

These sources and sinks were defined earlier in this section.

Biological heat generation is often assumed to be negligible for waste site segments, except in the case of aerobic decomposition. However, a model for heat generation from decomposition is presented in Chapter 7, Stoichiometry and is based upon enthalpic balance for the anaerobic decomposition and, where based upon the oxygen consumption rate, for aerobic decomposition.

## VARIABLES OF THE HEAT GENERATION MODEL

As described at the beginning of this section, the constants $a_1, a_2$ and $a_3$ are algebraically derived from statement of the energy conservation expression:

$$a_1 = \frac{k}{\alpha(\rho c_p)_{gas}}$$

$$a_2 = \frac{-u_{gas}}{\alpha}$$

$$a_3 = \frac{1}{\alpha(\rho c_p)_{gas}} \tag{6.91}$$

The independent terms of these constants are $k$(or $K_m$), the thermal conductivity of the landfill media (layer); $(\varrho C_p)_g$, the heat capacity of the gas moving through the medium, $u_g$, the average gas velocity value and $\alpha$, the heat capacity ratio of the medium. Separate definitions are as follows.

## LANDFILL THERMAL CONDUCTIVITY $K_m$

The major components of landfill media are soil, water and gas (air or landfill gas) pockets. For soil layers it can be assumed the solid material is homogeneously sand (including organic matter, organisms) and the soil moisture content comprises the liquid fraction. For layers containing wastes and soil mixtures, the solids content heat conductivity must comprise a lumped sum of specific materials. This means an overall conductivity must be established for the MSW fraction contained in the layer.

While thermal conductivity values can be obtained for most of the materials contained in the MSW stream as defined by composition, it is likely a broadly representative classification. Review of values of thermal conductivity reported in literature for various MSW materials indicates that, for the larger bulk fractions such as paper or lignocellulosic materials, variation of thermal conductivity between materials was relatively small.

By comparison, the conductivity of wood varied between 0.114 and 0.186 W/m°K (watts/meter deg. kelvin), while the values for paper varied between 0.14 and 0.16 W/m°K; biological materials were in the range 0.03–1.0, soil materials were in the range 0.9–3 and plastics were in the range 0.15–0.3 W/m°K. However, the metals had sufficiently high values; for example, aluminum has a thermal conductivity

**TABLE 6.4**
**Thermal Conductivity of MSW Components**

| MSW Fraction | Percentage (%) Total Weight | Conductivity K W/m. °K | [K*(%wt)]/100 |
|---|---|---|---|
| Food | 15 | 0.05 | 0.0075 |
| Paper | 40 | 0.14 | 0.056 |
| Cardboard | 4 | 0.14 | 0.0056 |
| Plastics | 3 | 0.16 | 0.0048 |
| Textiles | 2 | 0.33 | 0.0066 |
| Rubber | 0.5 | 0.16 | 0.0008 |
| Leather | 0.5 | 0.05 | 0.00025 |
| Garden waste | 12 | 0.05 | 0.006 |
| Wood | 2 | 0.15 | 0.003 |
| Glass | 8 | 0.77 | 0.0616 |
| Tin cans | 6 | 16 | 0.96 |
| Ferrous metals | 1 | 16 | 0.16 |
| Other metals | 2 | 192 | 3.84 |
| Dirt | 4 | | |

1.0 W/m deg. K = 0.86 kCal/hr*m deg K
1.0 $kCal/m^2 = 4186\ Joules/m^2$
J/kg deg K = $2.39 * 10^{-4}$ kCals/kg deg K

of 202 W/m°K, while the thermal conductivity of steel is W/m°K. The presence of metal wastes in the landfill layers should thus strongly influence overall thermal conductivity.

Information was compiled to establish conductivity properties of materials likely to be landfilled. Some of these values are listed or discussed below.

## THERMAL CONDUCTIVITY AND DIFFUSIVITY VALUES

Niessen et al. (1970) assumed the *thermal diffusivity* of MSW to be equivalent to that of *paper* or wood, at $3.6*10^{-4}\ m^2/hr$ ($0.0086\ m^2/day$). As the thermal conductivity of MSW is of greater interest in the heat equation, one approach might be to assume the thermal conductivity is close to that of its largest volume and mass fraction. According to mass fractions, the thermal conductivities are listed in Table 6.4. Table 6.5 also lists thermal conductivities of various solid, liquid and gaseous substances possible at waste sites.

## ESTIMATING THE MEAN THERMAL CONDUCTIVITY OF MIXED WASTE MATERIALS

Definitions of thermal conductivity for composite materials usually assume the materials are in intimate series or parallel contact, which permits the overall conductivity,

**TABLE 6.5**
**Materials Thermal Conductivity Values**

| Material | Conductivity K W/m. °K | Material | Conductivity K W/m. °K |
|---|---|---|---|
| Water | 0.6 | Stone | 2.56 |
| Water-sat. wood | 0.3203 | Cardboard | 0.14 |
| Corkboard | 0.43 | Earth (dry) | 0.134 |
| Aluminum | 192.17 | Fiber (plastic) | 0.233 |
| Stainless steel | 16.01 | Sand (average) | 0.93 |
| Concrete | 1.6 | Sand (moist) | 1.12 |
| Compost | 0.43 | Gravel | 0.93 |
| Paper | 0.14 | Limestone | 2.21 |
| Paper (hard) | 0.151 | Glass | 0.768 |
| Wood | 0.128–0.186 | Leather | 0.14–0.17 |
| Plywood | 0.114 | Polystyrene | 0.157 |
| Celotex | 0.047 | Rubber (hard, normal) | 0.16 |
| Kapok | 0.074 | Rubber (crepe) | 0.03 |
| Wood (maple) | 0.16 | Rubber (vulcanized) | 0.23 |
| Wood cement | 0.174 | Textiles | 0.33 |
| Ceramics | 1.05–1.57 | Asbestos | 0.163 |
| Aluminum | 202 | Steel | 55 |
| Copper | 386 | Clay | 1.26–2.9 |
| Ice | 2.2 | Air | 0.025 |
| Soil organic matter | 0.25 | Glass | 0.8–1.1 |
| Linoleum | 0.08 | Brick | 0.35–0.7 |
| Natural materials | 0.03–0.07 | Masonry | 0.7–0.93 |
| Steam | 0.188 | Asphalt | 0.7 |
| $H_2$ | 0.182 | $CO_2$ | 0.0166 |

based upon definition for heat conduction, to be stated as for resistors in series or parallel. For example, the overall conductivity $K$ for materials can be stated as:

$$\frac{1}{K_{mean}} = \sum \frac{1}{K_i} \text{ (overall series conductivity, series contact)}$$
$$KA = \sum K_i A_i \text{ (conductivity, materials in parallel contact)} \qquad (6.92)$$

Neither above case is applicable for landfilled MSW materials because random contact is more likely than orderly arrangements.

Another approach is to assume that the conductivity value is a mean. According to Rogers and Mayhew (1967):

*If heat flow is one-dimensional and steady, heat flow is independent of wall thickness and the overall conductivity becomes a mean value.*

**TABLE 6.6**
**Values for Thermal Conductivity Estimation**

| MSW Material | # | Weight | Thermal Conductivity, W/M °K | WT. Fraction × K |
|---|---|---|---|---|
| Food | 1 | 15 | 0.05 | 0.0075 |
| Paper | 2 | 40 | 0.14 | 0.056 |
| Cardboard | 3 | 4 | 0.14 | 0.0056 |
| Plastics | 4 | 3 | 0.16 | 0.0048 |
| Textiles | 5 | 2 | 0.33 | 0.0066 |
| Rubber | 6 | 0.5 | 0.16 | 0.0008 |
| Leather | 7 | 0.5 | 0.05 | 0.00025 |
| Garden Waste | 8 | 12 | 0.05 | 0.006 |
| Wood | 9 | 2 | 0.15 | 0.003 |
| Glass | 10 | 8 | 0.77 | 0.062 |
| Tin cans | 11 | 6 | 192 (Aluminum) | 11.52 |
| Metals: ferrous | 12 | 1 | 53 | 0.53 |
| Metals: non-ferrous | 13 | 2 | 55 (Steel) | 1.1 |
| Dirt | 14 | 4 | 0.134 | 0.0054 |
| TOTAL | 4 | 100 | NA | 13.073 |

Independence from wall thickness reduces the importance of material arrangement; so if one-dimensional heat flow is assumed, a mean value for the MSW layers, or the waste–soil mix, can be derived. The mean thermal conductivity $K_m$ is thus stated as:

$$K_m = \sum \frac{K_i}{n} \tag{6.93}$$

where $n$ = number of materials. However, on a macroscopic scale, the proportions of various materials would affect overall conductivity. This effect could most likely be addressed using a weighted mean value:

$$K_m = \frac{\sum(W_i K_i)}{n W_{total}} \tag{6.94}$$

The following data were used to estimate the mean value of landfill $K_m$ for ambient temperatures as an example of how the value of thermal conductivity can be estimated (see Table 6.6).

For a landfilled waste made up of 14 basic components, as listed above, the mean value of the MSW thermal conductivity would thus be:

$$K_m = 13.073/14 = 0.951 \text{ W/m°K}$$

For a landfill or compost or waste site with a larger or smaller number of liquid and solid components, the thermal conductivity can be estimated using a similar approach.

At an assumed bulk density of approximately 500 kg/m$^3$ and a moisture content of about 20%, this indicates a thermal conductivity of approximately 0.12 kJ/kg deg K, i.e., in the range of wood, paper and biological materials. The strong influence of metals content on landfill layer thermal conductivity suggests that landfill layers with elevated metals waste content could have enhanced temperature distribution, which could be of treatment benefit.

It should be noted that the approach listed above for estimating thermal conductivity of the wastes mixture only addresses the basic components of MSW. For a landfill system, all other materials, including sand, clay, water, air, landfill gas and cover or liner materials, must be included in thermal conductivity estimation. In the case where the landfilled material is contained inside a compacted soil or polymer envelope (closed), overall system heat conductivity affects the rate of heat transfer to the surrounding envelope of the landfill.

# 7 The Kinetics of Decomposition of Wastes

## INTRODUCTION

One of the most important applications of engineering to the management of solid wastes has been the design and control of landfills. The sanitary landfill has evolved in recent decades — in industrialized countries and in developing countries that can afford the costs — from essentially open dumps in convenient locations to systems with modern engineering standards, including sophisticated liner and leachate collection and gas migration controls. These controls have developed from increased understanding of the role of injurious water quality and air quality, public health impacts of the practice of placing large volumes of biodegradable wastes at sites and allowing natural decomposition to proceed. While there is warranted focus on control of the impacts of disposal once wastes are placed in a site, development methods of examination of how natural decomposition proceeds at waste sites can provide engineers with added tools for improved management.

At waste sites and dumps, the latter of which are still used for solid wastes management in many parts of the world, natural decomposition can be either aerobic or anaerobic in nature, or both. Shallow dumps are likely to be more aerobic, due to less compaction and thus higher air porosity. Wastes disposed at larger dumps and landfills are likely to pass through an aerobic phase before deeper burial reduces the air supply and thus increases the likelihood of anaerobic conditions. At compost operations and piles, the practices of loose piling and turning at regular intervals ensures aerobic condition; however, unless large compost piles are turned regularly, the reduced air flow and compaction at the bottom of piles can result in anaerobic conditions, with the potential for odor and other impacts.

## ANAEROBIC AND AEROBIC DECOMPOSITION PATTERNS

For a waste site such as a landfill, decomposition is in the main anaerobic. Oxygen is often depleted within a short distance of the landfill top surface. In this case the decomposition is similar to that in an anaerobic digester. The waste site's heterogeneity, however, increases the likelihood that decomposition is uneven and also aerobic in various portions of the system.

The anaerobic sequence is most often described in the simplest terms as initiated by hydrolysis, digestion of solids, acid production, acid consumption and generation of final products such as methane, carbon dioxide and water:

Organism-mediated process:

Hydrolysis (fungi, bacteria) ⇒ acetogenesis (acetogens) ⇒ acidogenesis (acidogens) ⇒ methanogenesis (methanogens)

Waste conversion process:

Carbohydrates (lignocellulosics, sugars, starches), proteins, fats ⇒ sugars ⇒ alcohol aldehydes ⇒ organics ⇒ methane ($CH_4$), $CO_2$, water, etc. (Reynolds, 1982)

Aerobic decomposition, on the other hand, can be defined as the bio-oxidation of organic material by facultative or obligate organisms:

Organic material + $O_2$ ⇒ new cells + energy + $CO_2$ + $H_2O$ + other end products

Aerobic decomposition occurs in the early stages of landfilled waste disposal; thus all wastes in a landfill would have passed through a brief aerobic degradation process. Aerobic conditions may apply to a relatively thin upper segment of a large landfill, where natural penetration of air is possible; but this segment is likely to be thicker for open or shallow waste sites. When passive or active aeration is carried out, as in an (induced) aerobic landfill, landfill bioreactor, dump site stabilization, a composting heap or pile or contained aerobic composter (tunnel, etc.), aerobic decomposition conditions can be created in all or most of a waste site, as oxygen supplied and dissolved in the soil pore water is sufficient to support the growth of aerobic or facultative organisms. The importance of aerobic decomposition relates to its (1) faster rate of conversion of solid organic material, (2) value to conversion of "green" wastes to soil amendments and (3) avoidance of production of methane (an important greenhouse gas) and odors during decomposition.

## ANAEROBIC DECOMPOSITION

Anaerobic decomposition poses more challenging problems than aerobic decomposition to engineering management of solid wastes because it results from two traditional and thus more hidebound, practices in waste management. The long history of solid waste management shows the practice of site placement and abandonment of large volumes of solid wastes at sites selected for convenience if not environmental propriety and the practice of disposing wastes of indiscriminate nature at these sites. The accumulation and abandonment of large volumes of solid wastes often can create the conditions for anaerobic decomposition, which by its nature sets in motion long-term groundwater and air quality impacts. By comparison, aerobic disposal and subsequent decomposition of wastes is often purposeful (shallow landfills an exception) and can involve selective disposal of highly biodegradable wastes as in yard and food wastes (or biosolids) composting. However, dumps and landfills have long been the major disposal alternative in most countries. This means that most wastes for disposal of materials reach landfills or dumps, weighting management heavily toward control of the impacts of disposal sites undergoing anaerobic decomposition.

## THE ANAEROBIC DECOMPOSITION PROCESS

Anaerobic decomposition of solid, biodegradable organic material proceeds by enzymatic solubilization (hydrolysis) of solid material to sugars and acids, followed by conversion (acidogenesis, acetogenesis) to organic acids, which are in turn converted to gases (methanogenesis, hydrogenotrophy), water (produced and consumed) and heat. These steps in the anaerobic biodegradation process are discussed below.

### Waste Hydrolysis by Soil Organisms

The primary and limiting step in the decomposition process, according to several researchers, is hydrolysis, common to both aerobic and anaerobic conversion of solid organic materials. As a primary step, hydrolysis governs the growth of microbial mass, the removal of solid and liquid substrates from the waste environment and thus the generation of decomposition gaseous and liquid products. It controls the rates of other organic decomposition processes.

Hydrolysis can be described as zero order, first, or second order, depending upon conditions imposed by the reactor. For the conditions imposed by a landfill, first-order hydrolysis is likely, as the concentration of substrate compared to cell mass is relatively large. As degradation of the solid material proceeds, substrate volume decreases, as the remaining organic material is more resistant to attack by microorganisms.

In both aerobic and anaerobic decomposition, the hydrolysis rate may be represented, in terms of Monod (1949) kinetics, as:

$$\frac{dS}{dt} = -\frac{k_H AXS}{K_s + S} \tag{7.1}$$

where $S$ is concentration of the substrate converted, $A$ = unit area of the material in contact with enzyme expressed by microorganisms, $X$ = concentration of organisms in mass units and $K_s$ is the substrate concentration at which the growth rate of an organism is half of its maximum value (or the half saturation constant), or $\mu_{\max}/2$, $X$ is the mass of organisms and $k_H$ = anaerobic or aerobic rate of hydrolysis, in inverse time units ($T^{-1}$). When $X$ is much lower than $K_s$, which is likely to be the case for organism ratio to organic waste in a site such as a landfill or compost pile, the relation has the form:

$$\frac{1}{X}\frac{dS}{dt} = -\frac{k_h S}{K_s} \tag{7.2}$$

Rearrangement and integration of the above provides:

$$\frac{dS}{S} = -(k_h X)dt \tag{7.3}$$

$$S = S_0 \exp(-k_h Xt)$$

In this form, the value of $X$ becomes an average concentration of $X$ in the system, in kg/kg of substrate, or kg/kg of fluid phase, over the period of time $t$, or $\Delta t$.

Waste hydrolysis studies providing data on substrate $S$ vs. time $t$ can be used to determine the likely rates of solid substrate solubilization. In these studies the rate term $kX$ typically has a single (constant) value, or hydrolysis rate $= -k_H$, in units of one/day. This value of the hydrolysis rate should be understood as the overall hydrolysis rate due to hydrolytic organisms (e.g., fungi and bacteria) present in the waste site ecosystem and acting upon the specific waste material. If material-specific rates are available, for instance from studies of conversion of organic materials by actinomycetes or fungi, or specific bacteria groups, these may be used. In this case, the proper form of the hydrolysis relation should be:

$$\frac{dS(j)}{S(j)} = -(k_{h,org} X_{org})dt$$

$$S(j) = S_0(j) \exp(-k_{h,org} X_{org} t) \qquad (7.4)$$

In the above, hydrolysis of the $j$th substrate refers to a single material, for instance waste paper. The approach can be extended to the system if the waste site is considered a representation of an anaerobic or aerobic digester, with an $n = 1 \ldots j$ variety of decomposable substrates (liquid, solid), each of *overall* hydrolysis constant $k_H(j)$ for the specific site conditions, the hydrolysis expression can be written, for the $j$th substrate, as:

$$S(j) = S_0(j) \exp(-k_{\mathrm{H}}(j)t) \qquad (7.5)$$

with the value of $k_h$ specified for either aerobic or anaerobic conditions.

For waste sites, the total hydrolyzer concentration ($X_H$), in biomass weight units, has to be linked to the total hydrolysis rate ($k_H$). Forms of the hydrolysis equation incorporating the hydrolyzer mass concentration are thus more appropriate to decomposition modeling.

Use of this type of relationship in a waste site bioreactor model would also require that the value of the substrate term $S$ be defined in terms of available organic material. In the case of solid waste materials, this would require that the term $S$ be defined as dry weight of volatile organic matter, organic carbon, or *soluble* chemical oxygen demand (COD) concentration of the material $S(i)$. The soluble fraction of an organic solid material is often referred to in terms of its volatile solids ($VS$) content.

## Determination of the Hydrolysis Rates of Organic Solid Materials

For hydrolysis of materials likely to be disposed at waste sites and thus decomposed under aerobic and anaerobic regimes, research studies may not match site conditions. While much data is available, including a growing body of literature involving field studies of composting and landfilling, most hydrolysis rates published apply to relatively controlled or laboratory conditions.

Nevertheless, the range of values published or developed statistically from raw data can provide the modeler with reasonable assumptions of hydrolysis rates for

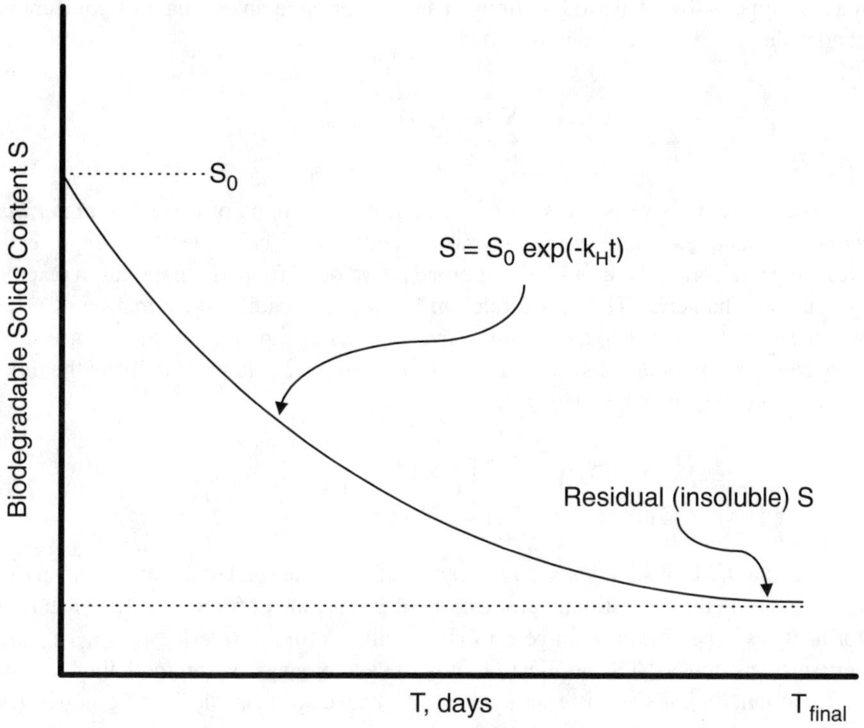


**FIGURE 7.1** Substrate vs. time under hydrolysis

many if not all of solid materials typically disposed at waste sites. Statistical analysis applied to data for substrate concentration vs. time, for instance, provides an estimate of the rate of conversion in terms of the specific hydrolysis rate, which is the slope of the $S$ vs. $t$ curve over the total time for hydrolysis of the substrate (see Figure 7.1):

$$S = S_0 \exp(-k_H X_H t)$$

$$\frac{dS}{dt} = (-k_H) S_0 \exp(-k_H X_H t_{total}) \tag{7.6}$$

## Practical Forms of the Hydrolysis Relationship

For a time step $\Delta t$, the change in substrate $\Delta S(j)$ can be represented by $\Delta S(j) = S_0 - S(j)$, or:

$$\Delta(S(j)) = S_0(j)(1 - \exp(-k_H(j)t) \tag{7.7}$$

For a waste site (compost pile, dump or landfill), estimation of mass and volume changes in total solids content is directly related to estimation of overall change in the mass of organic materials and system porosity. Summation of change in organic mass of type $j$ over a period of time $t$ can thus provide an estimate of solid mass change due to decomposition processes:

$$\sum_{j=1}^{j=n}(\Delta S(j)) = \sum_{j=1}^{j=n}(S_0(j)(1 - \exp(-k_{H,j}(j)t))) \tag{7.8}$$

Note that the above is the sum of changes in mass units of *all of the materials subject to decomposition* within the waste site system. The numerator $j$ represents waste type, in concordance with the approach developed for particulate-based reactor systems in Chapter 2. The above relation lends itself readily to estimation of total volume of materials lost to hydrolytic conversion during a period of time $t$ (since start of decomposition or any shorter period), as the volume $S_j$, in mass units, is the ratio of mass (weight) $M_j$ to density $\rho_j$:

$$Volume = \sum_{j=1}^{j=n}\left(\frac{\Delta S_j}{\rho_j}\right) = \sum_{j=1}^{j=n}\left[\left(\frac{S_0(j)}{\rho_j}\right)(1 - \exp(-k_{H,j}(j)t))\right] \tag{7.9}$$

Once the hydrolysis rates are reasonably known, the total substrate removals in mass units can thus be estimated for a particular substrate or for all substrates chosen for analysis. The volume change estimations allows for *consideration of waste site settlement* as one possible application of waste site biodegradation modeling.

Important findings of waste simulations so far have been that *the* limiting step in gas production from decomposition is not the size of the methanogenic microorganism population but the overall rate of hydrolysis of lignocellulosic material. This suggests that an overall hydrolysis rate $k_H$ taken for any system is a weighted value, as would be expected from the mathematical relationships shown above, when a single substrate is considered representative of the waste site. By contrast, the approach above allows for estimation of specific wastes, once hydrolysis or conversion rates are defined, with as great a degree of accuracy as acceptable for the particular conditions in the waste site. Careful choice or development of hydrolysis factors is necessary.

## Anaerobic and Aerobic Regimes and Lag Time

One of the objectives of a modeling exercise is examination of the effect of operating a waste site (model) as an anaerobic/aerobic-sequencing reactor. This posits a necessary change from anaerobic to aerobic condition during operation, for each period or cycle of operation, with cycle length dependent upon conditions considered most beneficial to the regime needed. For instance, the anaerobic regime might be better suited to simulated dechlorination of particular chemicals in the liquid phase of the system.

The theoretical change between anaerobic and aerobic regimes might merely involve choosing hydrolysis rates representative of the anaerobic condition, though it is more realistic to incorporate a factor representing the response of native soil organisms to the anaerobic or aerobic condition. For instance, aerobic conditions induced through low-pressure aeration would result in decomposition processes mediated by a different consortia of microorganisms — e.g., heterotrophic as well as aerobic organisms — and would involve a wider temperature variation than for an induced anaerobic regime.

Growth responses related to change in anaerobic or aerobic regimes typically involve the incorporation of lag factors into decomposition reactions. In such a case, it takes some time for organisms involved to adapt to the change in regime. This time period for adaptation is typically referred to as the *lag time*. Moisture loss is also likely to become more important, as increased movement of air and increase in landfill temperatures might result in higher rates of moisture removal (evaporation and transport) rates.

Lag times for the aerobic organism consortia could be considered those necessary for optimum development of the most sensitive and important species members. Assuming these would be the primary hydrolyzers (fungi and bacteria), suitable lag times could be incorporated by use of germination lag times, between the start of aeration and initiation of expected hydrolysis rates for each aerobic-to-aerobic cycle of simulation:

$$t_{cycle} = t_{aerobic} + t_{anaerobic} + 2\ (\text{lag time}) \tag{7.10}$$

Aerobic condition development in a landfill under aeration (from anaerobic conditions) has been indicated to require a period of 20 to 60 days (Stone et al., 1975) for aerated landfill cells to reach theft stable or maximum temperatures indicating full growth of the aerobic population (longer times for low air supply) (Stone et al., 1975).

## Hydrolysis Products in Anaerobic Decomposition

As the hydrolysis products are generated, they are converted into microorganism mass and waste products by the hydrolyzers and other dependent organisms in the landfill environment. This conversion of substrates into mass and products has long been modeled in terms of substrate-dependent kinetics.

In many of the substrate-dependent models, Monod (1949) or Michaelis-Menten kinetics have been used to represent anaerobic and aerobic decomposition processes. These models generally follow a mass balance approach, with kinetic relations describing sources, reactions between sources and decompositions and products.

Michaelis-Menten kinetics has been successfully used to model and analyze digestion processes in anaerobic reactors and digesters (Noike et al., 1985; Lawrence and McCarty, 1969; Lin et al., 1989). Monod (1949) kinetics has been used in a number of relatively recent simulations of landfill kinetics. Landfill decomposition models of varying complexity have been developed by Pfeffer (1974), Ghosh et al. (1977), Straub and Lynch (1982), Mata-Alvarez and Martinez-Viturtia (1986), McGowan et al. (1989), El-Fadel et al. (1989), Lee (1993) and Gonullu (1994) and a host of later investigators. The majority of these models simulate waste site organics decomposition as represented by the conversion of a single (lumped) substrate.

Three of these modeling efforts (Findikakis et al., 1988; El-Fadel et al., 1989; Lee et al., 1993) utilize Monod (1949) kinetics to develop rather thorough anaerobic digestion models, all based on a decomposition model proposed by Halvadakis

(1983). This model was based upon first principles and incorporates (simply), by mass balance, some of the basic biological processes known to occur in landfilled waste sites during biodegradation. For the Halvadakis (1983) mathematical (Monod [1949] kinetics) model:

- Solid material is described in terms of its soluble carbon content.
- Hydrolysis rate is defined as first order, as in various other studies.
- Methane generated from combined carbon dioxide-hydrogen ($CO_2 + H_2$) reduction is neglected.
- When there is more than one hydrolyzable solid carbon species (three species used), the various hydrolysis products are summed.
- The landfill void volume is considered initially occupied by nitrogen and carbon dioxide (80% and 20%, respectively).
- Soluble carbon simultaneously exists in solid, liquid, acidogenic biomass, methanogenic biomass, acetate, $CO_2$ and $CH_4$.
- Equal volumes of $CO_2$ and $CH_4$ are generated for each mole of carbon convertible to gas.
- Kinetics for the landfill is considered that of a batch bioreactor.

In the use of the Halvadakis (1983) model by Lee et al. (1993), methane generation from combined carbon dioxide hydrogen ($CO_2 + H_2$) reduction is *included* and four types of biodegradable matter — (1) vinyl, plastic, leather; (2) woods, textile, paper; (3) kitchen wastes; and (4) mixture — were used instead of organic carbon. Also, as in the Halvadakis (1983) model, hydrolysis products were directly used for acidogenic biomass growth and acid product generation. Production of acetic acid was proportional to acidogenic biomass.

Because of its relevance and adaptability to landfill conditions, a modification of Halvakadis' (1983) mathematical description of the landfill biological system as used by Lee et al. (1993) could be used as a mass balance approach. The Halvakadis (1983) model may be described as follows:

- Change in solid carbon = hydrolysis rate × solid material present.
- Amount of liquid carbon = summed hydrolysis products − amount converted to acidogenic biomass.
- Change in amount of acidogenic biomass present is amount present × (growth − death or decay).
- Change in amount of methanogenic biomass present is amount present × (growth − death or decay).
- Amount of acetate carbon present = specific yield× (change in acidogenic mass) − consumption by methanogenic mass.
- Carbon dioxide production is: yield from acidogenic processes + yield from methanogenic processes.
- Methane carbon is: yield from methanogenic processes.

The generation of hydrolytic product to the leachate can be represented according to Monod (1949) kinetics by:

change of hydrolysis product = (yield × refuse fraction mass change) + (product from organism death):

$$\frac{dP_H}{dt} = Y_{P,H}\frac{dS}{dt} + k_{decay}X_{H,0}$$
$$dP_H = (P - P_0) = Y_{P,H}(dS) + (k_{decay}X_{H,0})dt \qquad (7.11)$$

The terms in this relation are substrate- and organism-specific. While modeling with the above expression can be simplified by using a single value of specific yield $Y_{P,H}$ and maintenance factor $k_{\text{decay}}$ to represent either aerobic or anaerobic organisms and a single substrate to represent biodegradable wastes in the site, it is just as easy to apply a multiple substrate and multiple organism approach. For instance, if there are $1 \ldots j$ types of wastes in the site and $1 \ldots k$ types of organisms, the expression becomes, for hydrolysis by organism group $k$:

$$\frac{dP_H(j)}{dt} = Y_{P,H}(j)\frac{dS(j)}{dt} + k_{decay}X_{H,0}$$
$$dP_H(j) = (P - P_0) = Y_{P,H}(j)(dS(j)) + (k_{decay}X_{H,0})dt$$
$$for \quad j = 1 \ldots n \quad waste\ types \qquad (7.12)$$

The time period $dt$ is typically on the order of days. Expressions can be set up to represent the effect of each hydrolytic organism group on the range of wastes present in the site. For instance, hydrolyzer group $k = 2$ may represent the fungi population, and organism group $k = 1$ may represent the hydrolytic bacteria population.

An important note is that the above relation refers to hydrolysis inside a waste segment and excludes inflow or outflow of hydrolytic product. These must be included for modeling any waste site system with liquid phase input and outflows. In these systems, the concentration of hydrolysis product in any segment would be the sum of: production + inflow − outflow, in mass/liquid phase volume units.

## Hydrolysis Products Use for Acidogenic Biomass Growth and Acid Generation

The use of hydrolysis products in acidogenesis can be described as follows:

Change in hydrolysis product = acid generated + incorporation into hydrolyzer/acidogen biomass + input

$$\frac{dP_H}{dt} = -(k_H S) - \frac{1}{Y_{acidogen}}\frac{dX_{acidogen}}{dt} + (1 + R_{leach})\frac{Q_{lpi}}{V_{lp}}(P_{H,l-1} - P_{H,l}) \qquad (7.13)$$

In this relation, $j$ represents the types of waste, $R$ = fraction of throughflow recycled, $Q_{lpi}$ = liquid phase inflow to the segment or system, $V_{lp}$ volume of layer or system involved with the mobile liquid phase; $P_H$ = acid generated and $P_{H,l-1}$, $P_{H,l}$ = acid product inflow into a hypothetical or actual layer $l$ from the layer above $(l-1)$ due to gravity drainage. If there is no recycle of liquid phase, as might be the case for a compost operation, $R = 0$; if for example 15% of leachate outflow is recycled, as is possible in the case of a landfill bioreactor, $R = 0.15$.

## ACID PRODUCTION IN ANAEROBIC OPERATION

Acidogens generate acid as a product from consumption of hydrolysis product for growth processes and maintenance. Methanogens consume a fraction of the acid generated for growth and may contribute a portion to acid through maintenance. The growth and decay of acidogens can be represented as:

$$\frac{dX_a}{dt} = \mu_a X_a - k_{da} X_a = \mu_{a,\max} X_a \left( \frac{P_H}{K_{sa} + P_H} \right) - k_{da} X_a \tag{7.14}$$

The growth and decay of methanogens can be represented as:

$$\frac{dX_m}{dt} = \mu_m X_m - k_{dm} X_m = \mu_{m,\max} X_m \left( \frac{P_a}{K_{sm} + P_a} \right) - k_{dm} X_m \tag{7.15}$$

## ACETIC ACID GENERATION

Acetogenesis as represented by Monod (1949) kinetics is:

Generation of acetic acid = (acetic acid in) + yield from acetogens + yield from death of acetogens – removal of acetic acid by (co-existent) methanogens:

$$\begin{aligned} \frac{dP_{ac}}{dt} &= Y_{ac} \left( \frac{1}{Y_a} \left( \frac{\mu_{a,\max} P_H X_a}{K_{sa} + P_H} \right) \right) + (Y_{ac} k_{da} X_a) + (Y_{ac} k_{dm} X_m) \\ &\quad - \left( \frac{1}{Y_m} \frac{\mu_{m,\max} P_a X_m}{K_{sm} + P_a} \right) \\ &= Y_{ac} \left( \frac{1}{Y_a} \left( \frac{\mu_{a,\max} P_H X_a}{K_{sa} + P_H} \right) + k_{da} X_a + k_{dm} X_m \right) \\ &\quad - \left( \frac{1}{Y_m} \frac{\mu_{m,\max} P_a X_m}{K_{sm} + P_a} \right) \end{aligned} \tag{7.16}$$

and, adjusting for inflow of acetic acid:

$$\begin{aligned} \frac{dP_{ac}}{dt} &= Y_{ac} \left( \frac{1}{Y_a} \left( \frac{\mu_{a,\max} P_H X_a}{K_{sa} + P_H} \right) + k_{da} X_a + k_{dm} X_m \right) \\ &\quad - \left( \frac{1}{Y_m} \frac{\mu_{m,\max} P_a X_m}{K_{sm} + P_a} \right) + (1 + R) \frac{Q_{lpi}}{V_{lp}} (P_{a,l-1} - P_{a,l}) \end{aligned} \tag{7.17}$$

## METHANE GENERATION

Methane ($CH_4$) generation usually takes two pathways: production from acetic acid and production from $H_2$-$CO_2$ complex (Senior, 1991). For the acetic acid □ $CH_4$ - step:

Gas generation rate = gas yield × (acetic acid removal − amount acetic acid converted to biomass) + gas yield × mass of dead acetogens, or:

$$\frac{d(CH_4)_{(1)}}{dt} = Y_{CH_4}\left[(1 - Y_m)\left(\frac{1}{Y_m}\frac{\mu_{m,\max} P_a}{K_{sm} + P_a}\right) + k_{dm}\right] X_m \tag{7.18}$$

$$= CH_4 \text{ from acetic acid conversion by methanogens}$$

By a similar approach, for methanogens able to use hydrogen from the $H_2$-$CO_2$ □ $CH_4$ step, methane gas generation can be represented by:

$$\frac{d(CH_4)_{(2)}}{dt} = Y_{CH_4/CO_2}\left((1 - Y_m)\left(\mu_{m,\max}\frac{P_a}{K_{sm} + P_a}\right) + k_{dm}\right) X_m \tag{7.19}$$

The total methane generation is thus the sum from these two sources:

$$\frac{d(CH)_4)_{\text{total}}}{dt} = \frac{d(CH_4)_{(1)}}{dt} + \frac{d(CH_4)_{(2)}}{dt} = (Y_{CH_4} + Y_{CH_4/CO_2})$$
$$\left((1 - Y_m)\left(\mu_{m,\max}\frac{P_a}{K_{sm} + P_a}\right) + k_{dm}\right) X_m \tag{7.20}$$

## CARBON DIOXIDE ($CO_2$) GENERATION

$CO_2$ is generated throughout the reactor process. For a reactor volume, the total $CO_2$ generated includes the contribution from the hydrolysis-product-using acidogens + that from the acetic acid using methanogen + that from the $H_2$-$CO_2$ utilizing organisms. For the fraction generated by the acidogens:

$CO_2$ = (1 − methane gas yield) × (fraction of hydrolysis products not converted to biomass) × (1 − methane yield) × dead biomass):

$$\frac{d(CO_2)_{acidogens}}{dt} = (1 - Y_{ac})(1 - Y_a)\left(\frac{\mu_{a,\max}}{Y_a}\frac{P_H X_a}{K_{sa} + P_H}\right) + (1 - Y_{ac})(k_{da} X_a)$$
$$= (1 - Y_{ac})\left[(1 - Y_a)\left(\frac{\mu_{a,\max}}{Y_a}\frac{P_H X_a}{K_{sa} + P_H}\right) + (k_{da} X_a)\right] \tag{7.21}$$

Similarly, for $CO_2$ from acidogenesis product and acetic acid using methanogens:

$$\begin{aligned}\frac{d(\mathrm{CO_2})_{methanogens-1}}{dt} &= (1 - Y_{\mathrm{CH_4}})(1 - Y_m)\left(\frac{\mu_{m,\max}}{Y_m}\frac{P_a X_a}{K_{sm} + P_a}\right) \\ &\quad + (1 - Y_{\mathrm{CH_4}})(k_{dm} X_m) \\ &= (1 - Y_{\mathrm{CH_4}})\left[(1 - Y_m)\left(\frac{\mu_{m,\max}}{Y_m}\frac{P_a X_m}{K_{sm} + P_a}\right)\right. \\ &\quad \left. + (k_{dm} X_m)\right] \end{aligned} \tag{7.22}$$

$CO_2$ generation by hydrogen-utilizing organisms can be represented by:

$$\frac{d(\mathrm{CO_2})_{\mathrm{H_2-CH_4}}}{dt} = Y_{\mathrm{CH_4/CO_2}}\left[(1 - Y_m)\frac{\mu_{m,\max}}{Y_m}\frac{P_a X_m}{K_{sm} + P_a} + k_{dm} X_m\right] \tag{7.23}$$

The total $CO_2$ generated, taking into account that a fraction is used by methanogens that can use hydrogen, is thus:

$$\begin{aligned}\frac{d(\mathrm{CO_2})_{tot}}{dt} &= \frac{d(\mathrm{CO_2})_{acidogens}}{dt} + \frac{d(\mathrm{CO_2})_{methanogens-1}}{dt} - \frac{d(\mathrm{CO_2})_{\mathrm{H_2-CH_4}}}{dt} \\ &= (1 - Y_{ac})\left[(1 - Y_a)\left(\frac{\mu_{a,\max}}{Y_a}\frac{P_{\mathrm{H}} X_a}{K_{sa} + P_{\mathrm{H}}}\right) + (k_{da} X_a)\right] \\ &\quad + (1 - Y_{\mathrm{CH_4}})\left[(1 - Y_m)\left(\frac{\mu_{m,\max}}{Y_m}\frac{P_a X_m}{K_{sm} + P_a}\right) + (k_{dm} X_m)\right] \\ &\quad - Y_{\mathrm{CH_4/CO_2}}\left[(1 - Y_m)\frac{\mu_{m,\max}}{Y_m}\frac{P_a X_m}{K_{sm} + P_a} + k_{dm} X_m\right] \end{aligned} \tag{7.24}$$

## Total Gas Output

Anaerobic landfill gas output from landfills and dumps typically reflects percentages of methane ($CH_4$) from 50–70% and carbon dioxide ($CO_2$) from 30–50%, with proportions varying according to which organic materials dominate decomposition (McBean et al., 1995). In most cases, the cellulosic fraction is the largest; for instance, paper and plant material is the largest waste fraction. The discussion of how gas composition is affected by waste fraction by Ham and Barlaz (1987) and McBean et al. (1995) indicates how waste makeup can affect methane percentage in anaerobic gas.

This suggests that, once the relative chemical fraction of the local or regional waste stream has been estimated, the composition of the gas can be determined. For validation of a model, however, the sampling of gas released at a site would provide a more reliable indicator of gas composition.

In any case, the gas released from landfills and dumps is likely to be mainly methane and carbon dioxide, with small percentages of hydrogen, hydrogen sulfides, odorous mercaptans and trace organic substances. According to the mass balance approach

**TABLE 7.1**
**The Effect of Waste Type on Gas Composition**

| Waste Chemical Fraction | Liters $CH_4$/kg Fraction | Anaerobic Gas Composition |
|---|---|---|
| Cellulose | 829 | 50.0 |
| Protein | 988 | 51.5 |
| Fat | 1430 | 71.4 |

discussed above, the (effective) total gas generation can thus be described, in terms of $CO_2$ and $CH_4$, as:

Gas production rate = change in methane gas + change in carbon dioxide

$$\begin{aligned}\frac{d(Gas)}{dt} &= \frac{d(\mathrm{CH_4})_{total}}{dt} + \frac{d(\mathrm{CO_2})_{tot}}{dt}\\ &= (Y_{\mathrm{CH_4}} + Y_{\mathrm{CH_4/CO_2}})\left((1-Y_m)\left(\mu_{m,\max}\frac{P_a}{K_{sm}+P_a}\right)+k_{dm}\right)X_m\\ &\quad + (1-Y_{ac})\left[(1-Y_a)\left(\frac{\mu_{a,\max}}{Y_a}\frac{P_{\mathrm{H}}X_a}{K_{sa}+P_{\mathrm{H}}}\right)+(k_{da}X_a)\right]\\ &\quad + (1-Y_{\mathrm{CH_4}})\left[(1-Y_m)\left(\frac{\mu_{m,\max}}{Y_m}\frac{P_aX_m}{K_{sm}+P_a}\right)+(k_{dm}X_m)\right]\\ &\quad - Y_{\mathrm{CH_4/CO_2}}\left[(1-Y_m)\frac{\mu_{m,\max}}{Y_m}\frac{P_aX_m}{K_{sm}+P_a}+k_{dm}X_m\right]\end{aligned} \tag{7.25}$$

## Gas in Management Scenarios

The *cumulative anaerobic gas output* of a site, in terms of $CO_2$ and $CH_4$, would be the summation of the term $d(Gas)/dt$, for a time period of interest, for instance 50 years. However, the gas production rate declines as materials are decomposed in the site and the $CO_2$ and $CH_4$ rates must subsequently decrease. Choice of a reliable long-term overall or specific hydrolysis rate becomes important to ensuring a degree of accuracy.

Studies at landfills have shown that methanotrophic bacteria resident in the upper aerobic soil layer convert a fraction of the gas released. The amount removed depends upon soil moisture content and gas residence time. While not a major percentage, the removal of this fraction must be considered in modeling.

In many cases, gas at landfills is collected for treatment by flaring or for use in energy production—for instance, use in heating, steam production and electric power. In this case the amount of methane removed via methanotrophs and $CO_2$ uptake by the liquid phase must be considered. $CO_2$ uptake by the liquid phase occurs due to contact between gas moving in the pore structure and sorbed or mobile liquid phase. If gas collection efficiency is $e_{coll}$, typically 50% or higher, gas recovery is:

$$\frac{d(Collection)}{dt} = (1 - e_{coll})\left(\frac{d(Gas)}{dt} - \frac{d(CH_4)_{methanotrophs}}{dt} - \frac{d(CO_2)_{liquid\text{-}phase}}{dt}\right) \tag{7.26}$$

## Decomposition Process Sensitivity to pH

The pH of the liquid phase inside a waste site affects natural decomposition processes such as acidogenesis. Methanogens are sensitive to pH and their growth outside near-neutral range is adversely affected. For instance, van den Berg et al. (1974) show that for methanogens using formic acid, optimum gas production is around a pH of 6.8 to 7.1; and gas production can stop at pH below 6.2 (above pH = 7.2, there is also a rapid decline in methane generation). Grady and Lim (1980) note that the hydrogen-utilizing methanogens are particularly sensitive to pH. Thus, in modeling methane or gas production from a waste site under anaerobic conditions, it is useful to incorporate the effect of pH of the liquid phase on gas production organisms.

The effect of pH on the acidogenic organisms (hydrolyzers) is that with a lower pH the types of organic acid products change. A lower pH can possibly stimulate organic materials hydrolysis (Senior and Talba, 1990) for instance; but it could increase the buildup of acids such as propionic, butyric, valeric, caproic and lactic acids. This results because the growth of hydrogen-consuming methanogens is restricted by a lowered pH, which causes imbalance between metabolic products from the hydrogen-producing acetogenic bacteria and hydrogen-consuming methanogens. This balance is necessary, as with elevated partial pressure of hydrogen (higher concentration), acid products such as proprionate are not converted to acetate and thus accumulate. For example, the hydrogen-using methanogens are essential to removing dissolved hydrogen such that its partial pressure ($p_{H2}$) is less than 0.0001 atm (Senior and Balba, 1990). Impairment of this balance is likely to affect acids accumulation. The effect is noticeable in leachate recycle in landfill bioreactors, where pH improvement increases methane production and reduces the levels of propionic, butyric, valeric and caproic acids in leachate.

### Improvement of Reactor Liquid Phase pH

Improvement of pH of the liquid phase in a waste bioreactor can improve performance by balancing decomposition with decomposition product removal. For a waste site, improvement of the pH is difficult because of the physical dominance of the solid fraction (the liquid volume is relatively small in an unsaturated site). If the production and dissolution of $CO_2$ is insufficient to maintain the system pH in the range of good performance (rapid decomposition or organics conversion rate, usually), the standard practice with biological reactors is to add buffering capacity in the form of solutions that can ionize and increase the hydrogen ion concentration in the liquid phase of the reactor, thereby raising the pH. The main control or buffer capacity of the pH in the liquid phase of the waste environment is the concentration of dissolved carbonic acid–carbon dioxide. The production of $CO_2$ in the liquid phase of the waste site (bacteria and soil microorganisms are essentially aquatic as they must be in contact with the sorbed [film] or mobile [film] liquid phase for transport of nutrients) leads to formation of carbonic acid in the liquid phase through the reaction:

$$CO_2 + H_2O \Rightarrow H_2CO_3 \tag{7.27}$$

The dissolved carbonic acid then dissociates to form bicarbonate and hydrogen ion:

$$H_2CO_3 = HCO_3^- + H^+ \tag{7.28}$$

The pH is a measure of the hydrogen concentration $[H^+]$. It is easy to see from the above relations that reduced carbon dioxide in waste site liquid phase solution would lead to reduced carbonic acid ($H_2CO_3$) in solution, which would result in a lower $[H^+]$ value. A practical implication of this is that it is possible to raise the pH by adding a source of bicarbonate ion (e.g., sodium or calcium carbonates, etc.).

The dissolved carbonic acid–carbon dioxide is also a measure of system liquid phase alkalinity. Grady, Daigger and Lim (1999) note that if the concentration of (organic) wastes (decomposed) is low, less carbon dioxide will be produced, which results in a low alkalinity (of the liquid phase). The implication of this, for lower decomposition rates, is that any site factor that reduces pH level in the bioreactor could also reduce performance level.

One of the most frequently used approaches for improving the alkalinity of the liquid phase in a wastes reactor such as a landfill is to add bicarbonate salts (to liquids injected). Typically, salts such as calcium carbonate ($CaCO_3$) are added through lime addition, although pH control through sodium carbonate ($Na_2CO_3$) is considered more effective though a little more expensive (Kirsch and Sykes, 1971).

Adding salts in this manner impacts pH primarily by increasing the concentration of dissolved carbonic acid ($H_2CO_3$). The carbonic acid goes to ionization, generating excess hydrogen ($H^+$) and bicarbonate ion ($HCO_3^-$). For instance, in adding calcium carbonate to water (solubility is limited to bicarbonate alkalinity of below 500 mg/liter, or pH range 6.9 to 7.5 [Grady, Daigger and Lim, 1999]), care has to be taken that the amount needed is not exceeded to avoid formation of insoluble scale.

For lime (CaO) added to water, the result is slaked or quick lime, ($Ca(OH)_2$), which dissociates in water to:

$$\begin{aligned}
&Ca(OH)_2 \rightarrow Ca^{2+} + 2(OH^-) \quad \text{(ionization)} \\
&Ca^{2+} + 2(OH^-) + 2CO_2 \Leftrightarrow Ca^{2+} + 2(HCO_3^-) \\
&H^+ + HCO_3^- \leftrightarrow H_2CO_3 \quad \text{(carbonic acid)}
\end{aligned} \tag{7.29}$$

Use of the above approach in modeling requires determination of alkalinity provided by adding lime — or in another set of similar reactions, the addition of sodium bicarbonate ($NaHCO_3$) and, subsequently, estimation of pH.

## Approaches to Incorporating the Effect of pH on Decomposition Kinetics

A change in pH affects process rates linked to enzymic reactions, including hydrolysis and growth. This behavior of rate as affected by pH has been the subject of investigation by some researchers and has been referred to as analogous to inhibition kinetics. The analogy to inhibition kinetics creates an opportunity to model the pH effect on rate as similar to other forms of rate inhibition. In most cases the change of rate in response to liquid phase pH can be characterized as (1) reaching a maximum when the pH is at an optimum, which is typically in the neutral range, or around pH = 7.0 for organisms found in waste site, (2) subject to acidic pH-induced inhibition when in the acid range and (3) subject to alkaline pH-induced inhibition, at pHs higher than the optimum pH.

### *Ion Concentration Inhibition*

In this approach, pH effect is modeled according to substrate-limited inhibition. According to Andrews (1971), substrate inhibited growth in chemostat culture can be modeled according to:

$$\mu = \mu_{\max} \frac{s}{K_s + s} \left( \frac{1}{1 + \dfrac{s}{K_{1,s}}} \right) \tag{7.30}$$

where $\mu$ = growth rate, $\mu_{\max}$ = maximum growth rate, $s$ = substrate concentration and $K_s$ = half-saturation coefficient and $K_{1,s}$ = substrate inhibition constant. This relation takes into account that a plot of growth under inhibition followed by an optimum, then by a second inhibition phase, allows for half-saturation values to be found at growth rate $\mu = 1/2\mu_{\max}$ on either side of the optimum growth rate (Figure 7.2).

For modeling the pH effect, the hydrogen ion concentration $[H^+]$, of which the pH is a log value, is assumed as the limiting substrate. The half-saturation rates are termed $K_1$ and $K_2$ and are the 1/2 optimum hydrogen ion concentrations, respectively, for the acid and alkaline phases of growth. With the substrate as hydrogen ion, the expression for pH-inhibited growth becomes:

$$\mu(s, pH) = \mu_{\max} \frac{s}{K_s + s} \left( \frac{1}{1 + \dfrac{K_1}{[H^+]} + \dfrac{[H^+]}{K_2}} \right) \tag{7.31}$$

Because values of $K_1$, $K_2$ and $H+$ concentration can be found to reasonably represent anaerobic digestion in a wastes digest under acid and alkaline phase inhibition, the right-hand side of the above expression becomes a constant upon applying these values; and the expression takes the form:

$$\mu(s, pH) = \mu_{\max} \frac{s}{K_s + s} R_{pH} \tag{7.32}$$

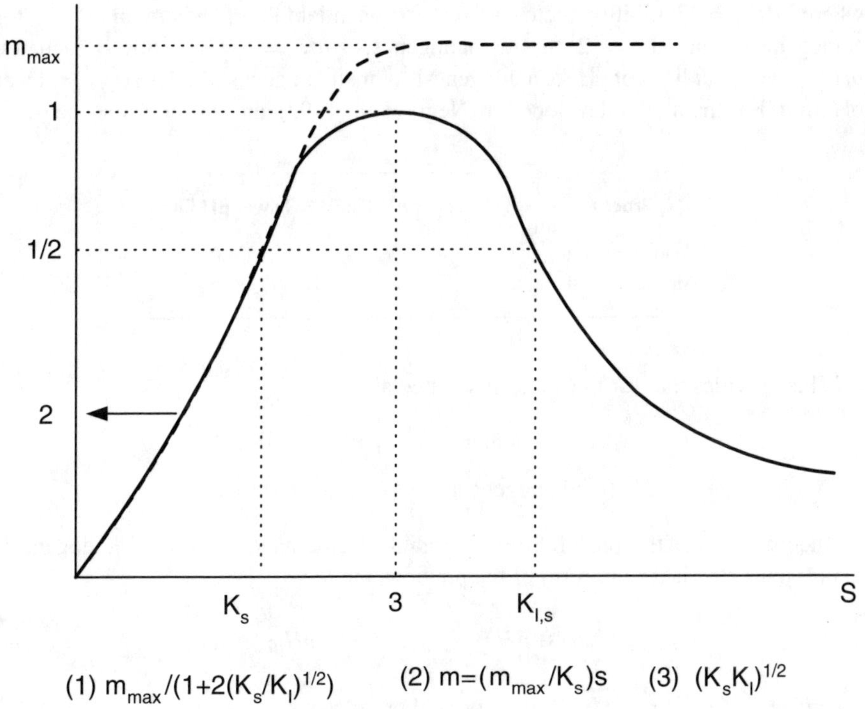


**FIGURE 7.2** Substrate inhibited growth pattern. Source: from Moser, 1988

where $R_{pH}$ represents a constant for pH inhibition of growth. The sensitivity of the hydrogen-consuming (hydrogenotrophic) methanogens to pH makes application of the hydrogen ion concentration inhibition expressions above highly appropriate for predicting growth pattern for this microorganism group.

### *Mechanistic Models of pH Effect*

The inhibition growth pattern influenced by pH as described can be defined by any number of mathematical models. Many of these models simplify the number of variables to be found, for ease of use in reactor performance kinetics, by using statistical and/or probabilistic analyses of data. Other models apply simple mathematical expressions after determining constants appropriate to the growth pattern from experimental studies.

For example, Costello et al. (1991) modeled the effect of pH on growth and product generation by introducing a pH-inhibition factor to terms of the Monod (1949) model. The inhibition factor due to pH is stated as:

$$pH_{if}^{(1/m)} = \frac{pH - pH_{lower}}{pH_{upper} - pH_{lower}} \tag{7.33}$$

where $pH_{if}$ = inhibition factor; $m$ = exponential factor, where $m$ = 3 for acidogenesis and $m$ = 2 for methanogenesis; $pH$ = $pH$ of flow (leachate), $pH_{upper}$ = upper limit of pH for acidogenesis or methanogenesis and $pH_{lower}$ = lower pH limit. For the model developed by Negri et al. (1993) and for the values below:

| Process | M | Upper pH Limit | Lower pH limit |
|---|---|---|---|
| Acidogenesis | 3 | 6.4 | 4.0 |
| Methanogenesis | 2.2 | 7.2 | 5.6 |

This provides that for the values tabulated above,

$$pH_{if}\,(\text{acidogenesis}) = \{(pH - 4.0)/2.4\}^3$$

$$pH_{if}\,(\text{methanogenesis}) = \{(pH - 5.6)/1.6\}^{2.2}$$

Incorporation of this pH inhibition factor into the methanogenesis and acidogenesis expressions also takes the form of the previous model:

$$\mu(s, pH) = \mu_{\max}\frac{s}{K_s + s}pH_{if} \tag{7.34}$$

where $s = P_a$ or $P_H$, the hydrolysis or acid products.

*Models for Effect of Product Inhibition and Incorporating pH Effect*

In many cases, as also applies to waste site decomposition, growth rate of organisms is coupled to rate of product generation — for instance, the concentration in the liquid phase of products such as volatile fatty acids from decomposition. Various investigators have examined this coupling, including the added effect of pH inhibition on organisms. Andreyeva and Biryukov (1973) examined the effect of pH on enzymatic products as secondary metabolites from fermentation with *P. crysogenum*, with a product inhibition model based upon Monod (1949) kinetics and similar to that proposed by Andrews (1968). These authors concluded that substrate concentration does not change the shape of the pH curve and thus the ratio of pHs at different rates (of growth) values are constant values. The effect of noncompetitive inhibition by ($H^+$) and ($OH^-$) ion concentrations present was modeled according to that suggested by Andrews (1968), with the analogy of $H^+$ as a substrate:

$$\mu = \mu_m\frac{s}{K_s + s}\left(\frac{K_{H^+}}{K_{H^+} + H^+}\right)\left(\frac{K_{OH}}{K_{OH} + \dfrac{1}{H^+}}\right) \tag{7.35}$$

where $H^+$ is the concentration of hydrogen ions and $s$ is the product concentration. Rearrangement of the above expression led to the polynomial form:

$$\mu = \mu_m\frac{s}{K_s + s}(\alpha_0 + \alpha_1(pH) + \alpha_2(pH)^2) \tag{7.36}$$

for which the constants are found from the experimental study:

$$\mu = \mu_m \frac{s}{K_s + s}(-919 + 296(pH) - 22.8(pH)^2) \quad (7.37)$$

As the above relation had the goal of finding the optimum pH for a fermentation process, which is essentially anaerobic, it thus applies to strictly anaerobic (hydrolytic and/or fermentative) organisms, likely nonmethanogenic.

Stazak et al. (1994) examined the effect of uncontrolled pH on process kinetics, where product generation rate ($r_P$) resulted in noncompetitive inhibition. The concept of endogenous metabolism (a maintenance or decay factor for growth rate) was included, as well as an exponential factor ($\exp(-k_5 P)$) to correct growth for product (in this case ethanol) inhibition, as proposed for fermentation processes by Aiba and Shoda (1969). The relation between pH and growth ($r_X$) was determined from experimental data to be linear. The basic approach was also classical Monod (1949) growth kinetics. As with Andreyaev and Biryukov (1973), there was no substrate limitation. The authors also noted the importance of stoichiometric relationships for determination of interrelationships and constants and the high importance of the product yield ($Y_{P/S}$) constant in linking growth to product inhibition, e.g., $Y_{P/S} = f(\mu) = Y_{P/S}(\mu)$. The various relationships and associated constants are described below. More familiar equivalents from Monod (1949) kinetics are listed at right.

$$r_X = r_{X,growth} + r_{X,endogenous\ metabolism}$$

$$r_{X,endogenous\ metabolism} = -k_6 X$$

$$r_{X,growth} = \frac{k_6 S}{k_6 F + S} X(\exp(-k_5 P)) = \frac{k_6 S}{k_6 F + S} X(P_{inhib}\ \text{factor})$$

$$\text{Equivalent: } r_X = \frac{dX}{dt} = \mu X - k_d X$$

$$r_{X,growth} = \mu_m \left(\frac{S}{K_s + S} X\right)(P_{inhib}\ \text{factor})$$

$$r_S = -k_3 r_{X,growth}$$

$$\text{Equivalent: } r_S = \frac{dS}{dt} = -(k_h S_0)\frac{dX}{dt} = -(k_h S_0)\left(\frac{\mu_m S X}{K_s + S}\right)(P_{inhib}\text{factor})$$

$$r_P = k_4 r_{X,growth}$$

$$\text{Equivalent: } r_P = \frac{dP}{dt} = -Y_{P/S}\, r_S = (k_h S_0 Y_{P/S})\left(\frac{\mu_m S X}{K_s + S}\right)(P_{inhib}\ \text{factor})$$

$$r_H = k_{11} r_{X,growth}$$

$$\text{Equivalent: } r_H = \frac{dH^+}{dt} = (k_{11})\mu_m \left(\frac{S}{K_s + S} X\right)(P_{inhib}\ \text{factor})$$

$$F = \frac{1 + k_9 H + k_{10} H^{-2}}{1 + k_9 H + k_{10} H_0^{-2}} \quad (7.38)$$

Constants for the above relations were found to be:

$$\begin{aligned} k_1 &= 0.3289 \\ k_2 &= 0.9645 \\ k_3 &= 12.93 \\ k_4 &= 5.705 \\ k_5 &= 0.0453 \\ k_6 &= 0.0288 \\ k_9 &= 10830 \\ k_{10} &= 6E-12 \\ k_{11} &= 0.0023 \end{aligned} \tag{7.39}$$

Values for $P$, $S$ and $X$ were in gm/cubic dm and the value for the hydrogen ion (H) concentration was in mol/dm.

It should be noted that, because the substrate $S$ for this study of fermentation was sucrose and the product $P$ was ethanol, equivalence to anaerobic digestion of waste means adjustments for maximum growth and yield parameters should be made in ethanol and sucrose (or glucose) equivalents.

## Assumptions for Mass Balance Model for Anaerobic Decomposition

The decomposition modeling approach presented above makes some simplifying general assumptions for the hydrolytic and acidogenic process steps. These include:

- Hydrolysis products can be lumped into acetic acid equivalents.
- Methanogenesis proceeds mainly by acetic acid metabolism.
- The acetogenic and methanogenic phases may be modeled using Monod (1949) kinetics.

To solve relations arising from the above, individual or lumped yields for biomass and product are needed for acetogenesis and methanogenesis. Values of this kind are widely reported in solid waste treatment.

Although the mathematical approach is similar to the descriptive landfill anaerobic degradation models employed by Lee et al. (1993) and El-Fadel et al. (1988), variations from these modeling approaches include:

- Incorporation of methane production from $CO_2 + H_2$ reduction (also done by Lee et al., 1993)
- Liquid phase flow recycle considerations
- Waste-specific hydrolysis consideration
- Organism-specific consideration
- Effect of pH variation considered

## Leachate or Gas Recycle as Anaerobic Bioreactor Options

Recycle is becoming an important option for waste site treatment for landfills operated as bioreactors. Various authors have examined the advantages of recycle for accelerated waste site stabilization and leachate management (Reinhart and Al-Yousfi, 1993; Delaware Solid Waste Authority, 1993; Al-Yousfi, 1992). The studies generally show that recycle is practical and successful. Leachate recycle, in the case of landfills, allows for return of volatile fatty acid products in the liquid phase, which can be consumed by acetogenic and methanogenic organisms, as the previously presented anaerobic decomposition models show. Bioreactor studies have indicated that leachate recycle improves its pH, probably by increasing carbon dioxide production and thus $CO_2$ in solution and hydrogen ion concentration. Improvement of the liquid phase favors growth of methanogens; and with lower product levels due to methanogen consumption, a likely improvement of hydrolytic capacity via nonmethanogenic bacteria will result.

In the anaerobic decomposition model presented above, incorporation of recycle as a variable necessitates the addition of a factor for the substrate contribution from liquid inflow due to gravity drainage from the segment of the waste site above the point of interest. Microorganism consumption of substrate during its flow through a waste site segment is stated in terms of the difference in concentration between inflow and outflow. The flow is stated in terms of the mobile liquid phase, which is understood, for an unsaturated system such as a waste site, to be the trickling flow, necessarily described as laminar (low velocity falling film) flow. For whole-system consideration, one approach is to let each unit depth be represented by a plate of length number $l$, across which the change in substrate (and in biomass) concentration equals: input + generation – consumption – drainage.

If the inflow is $q$ m$^3$/day, with substrate concentration $P$ kg/m$^3$, the daily substrate inflow would be $qP$ kg/day; and as this is the outflow from the layer above it, substrate inflow would be $qP_{l-1}$. Outflow would similarly be $qP_{l+1}$. If a conservative recycle rate of, say 15% to 40% is used, in recognition of likely inefficiency of leachate collection plus the need to avoid site flooding or side seepage problems, $R = 0.15$ or 0.4. For any layer (plate) $l$, the change in substrate concentration would be:

$$\frac{dP_l}{dt} = \frac{dP_{l,generation}}{dt} - \frac{dP_{l,consumption}}{dt} + (1 + R)(q_{l-1,out}(P_{l-1,out}) - q_{l+1,in}(P_{l+1,in})) \tag{7.40}$$

## The Kinetics of Aerobic Decomposition at a Waste Site

The aerobic waste system is perhaps less studied and more complex than traditional waste sites, which are most typically large landfills or dumps for which most of the system is anaerobic in nature. The upper layer of soil and wastes at these sites usually passes through an aerobic stage and may remain aerobic. Aerobic waste decomposition is nevertheless an important feature of open composting operations and shallow and incompact dumps and landfills. For aerobic decomposition, the supply of oxygen to aerobic organisms determines rate or stabilization or conversion of waste substances and organic materials. However, the waste site is inherently inhospitable to optimal oxygen supply. Oxygen used by organisms must be dissolved in the aqueous phase. Oxygen is not highly soluble; and in the unsaturated waste site, the liquid (film) phase will not have extensive sorption capacity. While the sorbed liquid film phase should offer little resistance to oxygen transport across the gas–water interface in the waste site, organism activity rates would depend upon oxygen supply and uptake. Oxygen supply is thus a limiting condition, which suggests the use of oxygen-limited kinetics for aerobic decomposition modeling.

Aerobic bio-oxidation of organic material by facultative or obligate organisms can be described in terms of aerobic digester behavior as initiated by hydrolysis of solids, acidogenic production, acid consumption and generation of final products such as carbon dioxide and water:

Organism-mediated process:
Hydrolysis (fungi, bacteria) ⇒ acetogenesis (acetogens) ⇒ methanogenesis (methanogens)

Waste conversion process:
Carbohydrates (lignocellulosics, sugars, starches), proteins, fats ⇒ sugars ⇒ acetate ⇒ $CO_2$, water + organisms (+ heat).

## AEROBIC HYDROLYSIS

As with anaerobic decomposition, a limiting step is aerobic hydrolysis, which governs growth of microbial mass, liquefaction of solid and liquid substrates in the waste site environment and generation of gas and liquid products. In most cases hydrolysis is described as a first-order biochemical reaction, represented in terms of Monod (1949) kinetics as:

$$\frac{dS}{dt} = -k_H S \tag{7.41}$$

where $S$ is the substrate converted and $k_H$ = anaerobic or aerobic rate of hydrolysis, in inverse time units, e.g., one/day. If the landfill is again considered an aerobic digester with an $j$th variety of decomposable substrates (liquid and solid) with aerobic hydrolysis rates $k_H(j)$, the above term can be written as:

$$S(j) = S_0(j)\exp[-k_H(j)t] \tag{7.42}$$

For a time step $dt$, the change in substrate $\Delta S(j)$ can be represented by $\Delta S(j) = S_0 - S(j)$, or:

$$\Delta S(j) = S_0(j)[1 - \exp(-k_H(j)\Delta t)] \tag{7.43}$$

For a landfill, mass and volume changes in total solids might be important to changes in overall landfill mass and porosity; thus summation of solid mass changes over time is useful, i.e.:

$$\sum_{j=1}^{j=n}(\Delta S(j)) = \sum_{j=1}^{j=n}(S_0(j))[1 - \exp(-k_H(j)\Delta t)] \tag{7.44}$$

where $S$ is in mass units. The volume of solid organic material converted can be estimated, as $S$ in mass units is volume = mass/density, or $V(j) = S(j)/\rho(j)$:

$$\text{Volume converted} = \sum_j \left(\frac{S(j)}{\rho_j}\right) = \sum_j \left(\frac{S(j)}{\rho_j}\right)[1 - \exp(-k_H(j)\Delta t)] \tag{7.45}$$

If the hydrolysis rates can be reliably determined for disposed biodegradable materials, total removals in mass units can thus be estimated.

## THE CHANGE FROM ANAEROBIC TO AEROBIC REGIMES

The effect of operating a waste site (model) as an anaerobic or aerobic reactor, or both, is usually to achieve stabilization of solid or liquid organic substances that may need aerobic or aerobic plus anaerobic conditions for conversion. The change from anaerobic to aerobic conditions can be induced by increasing the dissolved oxygen in the waste site through active or passive aeration. In landfills, for instance, the early decomposition stage often leads to depletion of the dissolved oxygen in the liquid phase at depths below the top soil layer. This dissolved oxygen is a combination of original pore gas oxygen and the amount dissolved in the sorbed or mobile water phase. In creating the aerobic environment, sufficient oxygen has to be supplied for the dissolved fraction to be within the range required for aerobic reactor performance.

In waste sites such as landfills, after the oxygen in the liquid phase has been depleted, it will take a period of time to replace aerobic condition as oxygen is poorly soluble in soil water. Simulation of the aerobic conditions thus requires introduction of a lag factor, as the change from anaerobic to aerobic should require sufficient time for aerobic bacteria to effectively develop to an effective level.

## LAG TIME FOR AEROBIC REACTOR DECOMPOSITION

Lag times for the aerobic organism consortia could be considered that necessary for optimum development of the most sensitive and important species members. The lag phase to overcome oxygen depletion can usually be represented in terms of the amount of time needed before the specific growth rate ($\mu$) of the microorganism group of interest would reach its maximum value. Since the lag time to maximum growth rate and thus maximum substrate consumption relates to the concentration of organisms in mass units, it is efficient to consider how organism concentration changes with time, up to the time of maximum specific growth rate ($\mu_m$), for a process with a lag rate. In this case, the portion of the specific growth vs. time ($\mu$ vs. $t$) curve incorporating the lag phase can be described with an exponential expression, for instance as proposed by Pirt (1975):

$$X = X_0 e^{\mu(t - t_{lag})} \tag{7.46}$$

This is, obviously, rearranging,

$$\ln\left(\frac{X}{X_0}\right) = \mu(t - t_{lag})$$

$$t_{lag} = t - \frac{1}{\mu_{\max}} \ln\left(\frac{X}{X_0}\right) \tag{7.47}$$

where $X$ applies only to the exponential phase of growth, i.e., to the growth phase where $\mu = \mu_m$. To examine this phase, the modeler needs to know the maximum specific growth rate of the microorganism group and initial concentration ($X_0$), where specific growth rate is constant ($\mu = 0$) or the beginning of the lag phase. The modeler should also be able to trace the change in concentration ($X$) between initial value and the end of the exponential growth stage. This would require a laboratory study or data from conditions similar to the waste site type modeled. The change in $X$ might also be deduced through a system product, for instance $CO_2$ level, once the correct relation between $CO_2$ production and the organism has been defined.

Site conditions would also make the lag phase relatively site-specific.

Bertger and Knorre (1972) examined the relation between substrate consumption and lag time to reach maximum growth and proposed a simple correction factor for the specific growth rate:

$$\mu(s, t) = \mu_{\max} \frac{s}{K_s + s}(1 - e^{-t/t_{lag}}) \tag{7.48}$$

Plots of $s$ vs. time $t$, once $K_s$ and $\mu_{\max}$, allow determination of duration of the exponential phase of substrate consumption and thus estimation of the lag time $t_{\mathrm{lag}}$.

The study by Stone et al. (1970) with field aeration of (previously anaerobic) landfill cells showed that, with active aeration, cell internal temperatures rose relatively rapidly, reaching their maximum values up to 75° F higher in 20 to 28 days. Amounts of oxygen added were 0.03 to 0.26 lbs $O_2$/lb MSW. With the higher rate of aeration of 0.61 lb/lb, for example, peak landfill temperatures were reached in 35 to 52 days with an approximately 75° F change in ambient temperature. For a second applied aeration rate of 0.28 lb/lb, peak temperatures were reached in 14 to 28 days, for a 50° F change in temperature; and for the low rate of aeration of 0.03 lb/lb, peak temperatures were reached in about 50 days, for a temperature increase of about 35° F. This indicates both that there is an optimum level for aeration above which aeration is less effective and the amount of lag is highly dependent upon level of dissolved $O_2$ for the system. The result may also suggest that close tracking of temperature increase in an organic waste decomposition environment may be one approach to determine the specific lag time for a particular site.

## AEROBIC HYDROLYSIS PRODUCT GENERATION, INCORPORATION AND USE

Product generation under aerobic conditions can be described as:

- Change in solid carbon = hydrolysis rate × solid material present
- Amount of liquid carbon = summed hydrolysis products − amount converted to acidogenic biomass
- Change in amount of acidogenic biomass present is amount present × (growth − death or decay)
- Amount of acetate carbon present = specific yield × (change in acidogenic mass) − consumption by acetate-consuming mass
- Carbon dioxide production is: yield from acidogenic processes + yield from acetotrophic processes

The generation of hydrolytic product to the leachate can be represented according to Monod (1949) kinetics by:

Change in hydrolysis product = yield × refuse fraction mass change
+ product from organism death.

### USE OF HYDROLYSIS PRODUCTS FOR GROWTH OF ACIDOGENIC BIOMASS AND ACID FORMATION

The use of hydrolysis products can be described as follows:

Change = hydrolysis product generation − incorporation into hydrolyzer/acidogen biomass + input:

$$\frac{dP_h}{dt} = -k_H(i)(S(i)) - \frac{1}{Y_a}\frac{d(X_a)}{dt} + (1 + R(j))\frac{Q(j)}{V(j)}(P_{H(j-i)} - P_{(j)}) \quad (7.49)$$

### BASIC RELATIONS FOR OXYGEN-LIMITED GROWTH

In the case of oxygen concentration dissolved in the liquid phase being a limitation, growth must be linked to oxygen concentration. One approach is the Monod (1949) kinetics specific growth rate relation, with the modification below for oxygen-limited growth:

$$\mu = \mu_m \frac{O}{K_O + O} \quad (7.50)$$

The growth rate $dX/dt$ can be stated in terms of the oxygen limitation, and if the maintenance or endogenous metabolism coefficient is included:

$$\frac{dX}{dt} = \mu X - k_{dO}X = \mu_m \frac{O}{K_O + O}X - k_{dO}X$$

$$k_{dO} = m_O Y_{X/O} \quad (7.51)$$

The general terms for substrate consumption are similarly:

$$\frac{dS}{dt} = -\frac{\mu}{Y_{X/S}}X = -\frac{\mu_m}{Y_{X/S}}\frac{O}{K_O + O}X \tag{7.52}$$

The oxygen consumption rate, also considering transfer rate, is:

$$\frac{dO}{dt} = -\frac{\mu}{Y_{X/O}}X + (\text{O}_2\ \text{Transfer}) = -\frac{\mu_m}{Y_{X/O}}\frac{O}{K_O + O}X + (k_L a)(O^* - O) \tag{7.53}$$

If the maintenance coefficient is considered and the substrate is a product $P$ of hydrolysis, the relations become:

$$\frac{dX}{dt} = -\frac{\mu_m}{Y_{X/S}}\left[\frac{S}{K_S + S}\right]\left[\frac{O}{K_O + O}\right]X + k_d X$$

$$\frac{dP}{dt} = \frac{\mu_m}{Y_{X/P}}\left[\frac{P}{K_P + P}\right]\left[\frac{O}{K_O + O}\right]X + k_d X$$

$$\frac{dO}{dt} = \frac{\mu_m}{Y_{X/O}}\left[\frac{P}{K_P + P}\right]\left[\frac{O}{K_O + O}\right]X + k_O X + k_L a[O^* - O] \tag{7.54}$$

## Oxygen as a Limiting Substrate in Aerobic Kinetics

The importance of oxygen as a limiting substrate in aerobic decomposition merits further discussion of how aerobic process kinetics can be incorporated into simplified models for waste sites. As noted in the previous section, one of the more important modifications that must be made to kinetic relationships is incorporation of oxygen availability as a growth or substrate limitation factor. When one or more dissolved substrates are growth-limiting in a medium, the specific growth rate $\mu$ is usually modified according to:

$$\mu = \mu_m \frac{S}{K_s + S}\frac{S_1}{K_{s1} + S_1}\frac{S_2}{K_{s2} + S_2}\cdots \tag{7.55}$$

In the case of oxygen plus a substrate for growth, $S_2$ = oxygen, thus:

$$\mu = \mu_m \frac{S}{K_s + S}\frac{O}{K_O + O} \tag{7.56}$$

Growth rate, with a maintenance or decay coefficient, can be stated for a segment of a reactor as:

$$\begin{aligned}\frac{dX}{dt} &= \mu X - k_d X + [q_{in}X_{in} - q_{out}X_{out}] \\ &= \mu_m \frac{S}{K_s + S}\frac{O}{K_O + O}X - k_d X + [q_{in}X_{in} - q_{out}X_{out}]\end{aligned} \tag{7.57}$$

The term $q_{in}$ represents liquid input, $x_{in}$ = inflow of organisms with liquid $- x_{out}$ = outflow of organisms, $q_{out}$ = outflow of liquid. Assuming there is no transport of organisms in and out of the landfill layer, the terms $x_{in}, x_{out} = 0$, thus:

$$\frac{dX}{dt} = \mu_m \frac{S}{K_s + S} \frac{O}{K_O + O} X - k_d X \tag{7.58}$$

If the $x_1$ organisms are hydrolyzers, the relation for solid material hydrolysis is usually presented as a first-order reaction:

$$\frac{dS}{dt} = -k_h S \tag{7.59}$$

where $S$ is the concentration in chemical oxygen demand or volatile organic matter. As there are several solid substrates present, the overall contribution to liquid substrate by hydrolyzers during the period $dt$ is given by:

$$\frac{dS_h}{dt} = \sum_{j=1}^{j=n} [S_0 \exp(-k_{h,aero} dt)] \tag{7.60}$$

Assuming input of hydrolysis products, through drainage into the reactor, of the order:

$$[1 + R][q_{l-1} S_{h,l-1} - q_{l+1} S_{h,l+1}] \tag{7.61}$$

and use by hydrolyzers during the time period $\Delta t$, the overall hydrolysis product balance can be stated as:

$$\begin{aligned}\frac{dS_h}{dt} &= \sum_{j=1}^{j=n} [-k_{h,aero} \, [S_0 \exp(-k_{h,aero} dt)]] \\ &\quad - \frac{1}{Y_h} \left[ \mu_m \frac{S_h}{K_s + S_h} \frac{O}{K_O + O} X - k_d X \right] \\ &\quad + [1 + R][q_{l-1} S_{h,l-1} - q_{l+1} S_{h,l+1}]\end{aligned} \tag{7.62}$$

where $q_{(j-i)}$ = drainage in and $q_{(j+i)}$ = drainage out. The growth of acetate utilizing organisms affects the rate of removal:

$$\frac{dX_{ac}}{dt} = \mu_{m,ac} \frac{S_a}{K_{s,a} + S_a} \frac{O}{K_{O,a} + O} X_a - k_d X_a \tag{7.63}$$

Change in acetate = yield from hydrolysis − use
for hydrogen growth − use for acetotroph growth

$$\frac{dP_{ac}}{dt} = (Y_{ac})\left[\frac{1}{Y_h}(\mu_h X_h - k_{d,h}X) - (\mu_h X_h - k_{d,h}X)\right]$$
$$- \frac{1}{Y_{ac}}(\mu_{ac}X_{ac} - k_{d,ac}X_{ac})$$
$$= (Y_{ac})\left[\frac{1-Y_h}{Y_h}\right]\left[\mu_{m,h}\frac{S_h}{K_{s,h}+S_h}\frac{O}{K_{O,h}+O}X_h\right]$$
$$- \frac{1}{Y_{ac}}\left(\mu_{ac}\frac{P_{ac}}{K_{s,ac}+P_{ac}}\frac{O}{K_{O,ac}+O} - k_{d,ac}\right)X_{ac}$$
$$+ [1+R][(q_{l-1}P_{a,l-1}) - (q_{l+1}P_{a,l+1})] \tag{7.64}$$

substituting for $\mu_h$, $\mu_a$, for the flow(s) entering and leaving the reactor segment, for $dx_h/dt$ and $dx_a/dt$: and modifying for oxygen consumption by use of the $B_1$ and $B_2$ terms, where:

$$B_1 = \frac{O}{K_{O,h}+O}$$
$$B_2 = \frac{O}{K_{O,ac}+O}$$

The oxygen concentration is given by:

$$\mathrm{d}O_2/dt = \text{Flow in} - \text{consumption}$$

The oxygen contribution of liquid drainage is:

$$O_2 \text{ Input from inflow} = (1+R)(q_{l-1}C_{O_2} - q_{l+1}C_{O_2}) \tag{7.65}$$

where $l+i \Rightarrow$ drainage from layer $j$ and $(j-i) \Rightarrow$ drainage into layer $j$ and $R =$ recycle into the layer.

Notably, liquid drainage into a waste site is not expected to have high levels of oxygen saturation, except for aerated leachate or possible water (precipitation or irrigation) entering the reactor segment. The contribution from oxygen flux to the reactor segment is:

$$\text{Influx of } O_2 = k_L a[C^* - C_{\text{inflow}}] \tag{7.66}$$

The above incorporates an oxygen supply term involving mass transfer. Thus system porosity vs. available surface area must be discussed. These are developed later in this section.

Consumption of $O_2$ by hydrolyzers can be stated as:

$$\frac{dO_{2,h}}{dt} = \mu_{m,h}\frac{O_2}{K_{O_2}+O_2}\left(\frac{1}{Y_{(X/O),h}}\right)X_h \tag{7.67}$$

Consumption of $O_2$ by acetotrophs can be similarly represented by:

$$\frac{dO_{2,ac}}{dt} = \mu_{m,ac}\frac{O_2}{K_{O_2} + O_2}\left(\frac{1}{Y_{(X/O),ac}}\right)X_{ac} \tag{7.68}$$

Thus the expression for total change in oxygen content due to the organisms present is:

$$\begin{aligned}\frac{dO_2}{dt} &= -\frac{dO_{2,h}}{dt} - \frac{dO_{2,ac}}{dt}\\ &= -\mu_{m,h}\frac{O_2}{K_{O_2} + O_2}\left(\frac{1}{Y_{(X/O),h}}\right)X_h\\ &\quad - \mu_{m,ac}\frac{O_2}{K_{O_2} + O_2}\left(\frac{1}{Y_{(X/O),ac}}\right)X_{ac}\\ &\quad + [(1 + R)(q_{l-1}C_{O_2} - q_{l+1}C_{O_2})] + k_L a[C^* - C_{\text{inflow}}]\end{aligned} \tag{7.69}$$

The above expression, which is used in a mathematical relationship for oxygen transport in the aerobic waste site reactor, assumes that the reaction is oxygen-limited but ignores the maintenance coefficient in the growth terms. According to Moser (1988), the cell mass maintenance is not affected by oxygen content.

## OXYGEN SOLUBILITY IN WATER OR LIQUID

Given that availability of oxygen is likely to be a limitation for the operation of aerobic waste sites, oxygen solubility in landfill pore water must be addressed.

The solubility of $O_2$ at 20° Celsius is quoted at 0.3 mM $O_2$ (millimoles). The conversion to mg/liter is usually made by multiplying mM by 31.3, which provides that:

$$\text{Oxygen saturation value } C^* \text{ at } 20°\,\text{C} \approx 9.39\ \text{mg/liter}$$

or

$$0.0094\ \text{kg/m}^3 \text{ of water flow}$$

Gas solubility is also subject to Henry's Law for atmospheric pressures near normal, such as would be suspected internal to landfills and to temperature variation.

According to Henry's Law the partial pressure of a gas compound in the gas phase should vary according to:

$$C^* = \frac{p}{H} \tag{7.70}$$

where $C^*$ = solubility, $p$ = partial pressure and $H$ = Henry's constant. This relation suggests that if the gas phase concentration of oxygen increases, the concentration of oxygen in landfill liquid should increase. Rehm and Reed (1981) state that, while the saturation of oxygen in the air is about 9 mg/L for an air-oxygen mixture of phases, if the gas is pure oxygen, the concentration in water can reach 43 mg/L.

The variation of saturation concentration of $O_2$ with temperature has been quoted by Crueger (1984):

$$C^* = \frac{468}{(31.6 + T(^\circ\text{C}))} \tag{7.71}$$

However, Moser (1988) noted that, while many investigators use $O_2$ solubility in water relations such as the above due to lack of reliable alternatives, other more complex relationships could be developed based upon concentrations of salts and nutrients such as glucose. These substances are said to affect oxygen solubility significantly. Given that the controlling concentrations of landfill salts in solution cannot be known in advance, water solubility of oxygen could be used as a model-starting approximation, with the expectation that solubility would be decreased by substances such as salts and glucose. Popovich's (1979) development of a solubility relationship for $C^*$ vs. glucose concentration took into account the effect of salts as:

$$\mathrm{C}^*_{\mathrm{glucose}} = \mathrm{C}^*[1 - .0012 - \mathrm{C}_{\mathrm{glucose}}]$$

Even from this relationship it can be seen that small concentrations of glucose ($C_6H_{12}O_6$, MW = 180), for example 0.001 mM, could change the solubility by 18%. This would suggest that the solubility of $O_2$ in waste site outflow should be reduced at least by some percentage from its water solubility value. There appear to be no data available for this type of correction.

It is reasonable to suggest, however, that the solubility of $O_2$ in landfill leachate under aerobic conditions should be closer to 0.8 C* (a 20% reduction due to liquid phase contaminants), where C* = water solubility:

$$\mathrm{C}^*_{leachate} = 0.8\,\mathrm{C}^*$$

Leachate-dissolved oxygen at some landfills has been indicated to be around 6.0 mg/l and can vary between 7.7 and 3 mg/l.

$$\mathrm{C}^*_{leachate} = \left(\frac{DO_{\max,liquid}}{\mathrm{C}^*_{water}}\right)\frac{(468)}{31.6 + T(^\circ\text{C})} \quad \text{mg O}_2/\text{liter} \tag{7.72}$$

Similarly, the oxygen transfer rate $N_a$ between air and leachate could be restated as:

$$N_a = K_c a[\mathrm{C}^*_{leachate} - C_{liquid}]\ \text{mg/L}$$
$$N_a = K_l a\{[374.4/(31.6 + \text{T(degrees C)}] - C_{liquid}\}$$

## OXYGEN TRANSPORT CONSIDERATIONS

If the oxygen dissolved in the soil water is insufficient, i.e., below a critical value, it would be expected that aerobic reaction rates would not be supported. Oxygen, if not naturally present in landfill soil, must be supplied by aeration. Landfill aeration is typically accomplished by pumping air into moistened soil at an appropriate rate, depth and duration to permit the growth and metabolic activity of aerobic and facultative organisms.

In terms of gas flow through a unit section or layer of the landfill reactor, the mass balance for $O_2$ in the layer can be written as:

Diffusion − uptake from moving gas − use by organisms + mass transfer into the liquid phase

$$D_{ax}\frac{\partial^2 O}{\partial z^2} - u\frac{\partial O}{\partial z} - R_{bio}(O) + k_L a[C^*_{O_2} - C_{O_2}] \tag{7.73}$$

The solution of the above form of the chemical species conservation equation for oxygen can be used for oxygen transport as well as gas transport with chemicals flux. Thus its constants are discussed to some detail. The diffusion term:

$$\text{Oxygen diffusion} = D_{ax}\frac{\partial^2 O}{\partial z^2} \tag{7.74}$$

could be assumed for a hypothetical waste site column to represent plug flow diffusion of a saturating gas such as air. The biological oxygen consumption term is represented by $R_{bio}(O)$, where $R_{bio}(O)$ = consumption by organisms. In the case of optimal delivery of oxygen, input of oxygen can be assumed to be:

$$\text{Gas-water interface oxygen transfer} = k_L a[C^*_{O_2} - C_{O_2}] \tag{7.75}$$

where $k_L a$ = oxygen transfer rate between gas and liquid.

## THE OXYGEN CONSUMPTION TERM $R(O)$

The $R_{bio}(O)$ term represents oxygen consumption rate of the organisms in the reactor sediment under steady-state conditions (supply = demand). It was defined in this section as:

$$\frac{dO_2}{dt} = -\frac{dO_{2,h}}{dt} - \frac{dO_{2,ac}}{dt}$$
$$= -\mu_{m,h}\frac{O_2}{K_{O_2} + O_2}\left(\frac{1}{Y_{(X/O),h}}\right)X_h - \mu_{m,ac}\frac{O_2}{K_{O_2} + O_2}\left(\frac{1}{Y_{(X/O),ac}}\right)X_{ac} \tag{7.76}$$

$K_{O2,h}$ = 2 saturation constant for hydrolytic organisms
$K_{O2,ac}$ = 1/2 saturation constant for acetotrophs
Steady state $\Rightarrow K_{O2,h} = K_{O2,ac} \Rightarrow B_1 = B_2 = [O_2/(K_{O2} + O_2)]$

When reaction rates are limited by oxygen transfer rate, the rate of growth becomes linear. Because water in the unsaturated waste site is likely to be in films, it is unlikely that high resistance at the gas–water interface should occur.

## Change of Oxygen Concentration with Waste Site Depth

Change of oxygen uptake rate with depth of waste site is possible, especially in large waste disposal sites; however, the transition zone is not likely to be extensive. In landfills, for instance, a (thin) transition zone exists between predominantly aerobic conditions in the upper site soil or wastes and anaerobic conditions beyond this transition zone. The effect of declining uptake would be to reduce the mass of aerobic organisms in the transition zone, which essentially means that growth level of aerobic organisms is reduced along a hypothetical transitional length. It is also likely that a declining dissolved oxygen level would favor facultative and anaerobic organisms. These conditions may be defined in terms of reduction of dissolved $O_2$, assuming that the reduction is essentially linear.

For instance if the dissolved oxygen transitional zone length is $z_T$,

$$\text{Growth rate (zone)} = f\ (\text{dissolved } O_2, z_T)$$

$$C_{O_2} = C_{O_{2,entry}}[\exp(-\beta\xi z_T)] \tag{7.77}$$

The length of the transitional zone for dissolved oxygen level (vertical direction) would be $\xi z_T$, accounting for the pore structure average tortuosity $\xi$. A declining dissolved $O_2$ concentration in this zone can be represented by:

$$\frac{dO_2}{dz_T} = -\beta C_{O_{2,entry}}[\exp(-\beta\xi z_T)] \tag{7.78}$$

where $CO_{2,\text{entry}}$ = dissolved oxygen in the liquid phase at the beginning of the transition zone, $\beta$ = rate of decline of $O_2$ over the length $z_T$ and $\xi$ = tortuosity of the waste site segment. A necessary condition is that dissolved $O_2$ is near zero at the end of the transition zone, meaning that the RHS term becomes zero. However, the rate of decline of $O_2$ in the transitional zone should be inversely proportional to the relative mass of facultative plus methanogenic organisms present. If the relative mass of active nonaerobic organisms is $X_{anaerobic}$ and active aerobic organism mass is $X_{aerobic}$:

$$\beta \propto \frac{X_{anaerobic}}{X_{aerobic}} \tag{7.79}$$

This suggests that the relation between dissolved $O_2$ concentration and relative mass of organisms may be of the form:

$$C_{O_2} = C_{O_{2,entry}}\left[\exp\left(-\left(\beta_1\frac{X_{anaerobic}}{X_{aerobic}}\right)\xi z_T\right)\right] \tag{7.80}$$

As no data are available to confirm this variation in the transitional aerobic to anaerobic zone or sheath in waste sites, no further elaboration is made for the above expression.

## Oxygen Transport and Consumption in a Column Waste Site Reactor

A more general approach is useful when the waste site is considered as a packed bed system with a column-type arrangement. Oxygen supply in this case can be visualized as associated with a connected pore system and diffusing outward from the linked gas flow pathways. Dissolved oxygen level can then be defined in terms of mass balance of diffusion, transport and mass transfer. In a previous section, this mass balance relationship was defined as:

$$D_{ax}\frac{\partial^2 O}{\partial z^2} - u\frac{\partial O}{\partial z} - R_{bio}(O) + k_L a[C^*_{O_2} - C_{O_2}] \tag{7.81}$$

This expression can then be treated as a partial differential for purposes of solution. The solution of the oxygen transport equation has a particular solution (Schugerl, 1985) similar to:

$$C_{L,O_2} = \left[\alpha_1 \exp(p_1 z) + \alpha_2 \exp(p_2 z) + \left(\frac{C^*}{G+1}\right)\right] \tag{7.82}$$

Adjusting for pore system tortuosity $\xi$:

$$C_{L,O_2} = \left[\alpha_1 \exp(p_1 \xi z) + \alpha_2 \exp(p_2 \xi z) + \left(\frac{C^*}{G+1}\right)\right] \tag{7.83}$$

In the above expression, the constants $p_1$ and $p_2$ are first and second quadratic roots of the solution of $(D_{ax})D^2 + UD + [k]z$, $z$ = depth to reactor segment and terms are as follows:

$$\alpha_1 = \frac{B_0(\frac{C^*}{G+1} - 1)(p_2 \exp(p_2))}{p_1 p_2 \exp(p_2) - p_1 p_2 \exp(p_1) - B_0[p_2 \exp(p_2) - p_1 \exp(p_1)]}$$

$$\alpha_1 = \frac{-B_0(\frac{C^*}{G+1} - 1)(p_1 \exp(p_1))}{p_1 p_2 \exp(p_2) - p_1 p_2 \exp(p_1) - B_0[p_2 \exp(p_2) - p_1 \exp(p_1)]}$$

$$p_1 = \frac{1}{2}[B_0 + \sqrt{(B_0)^2 + 4(B_0)(Da)}]$$

$$p_2 = \frac{1}{2}[B_0 - \sqrt{(B_0)^2 + 4(B_0)(Da)}]$$

$$B_0 = \frac{uL}{D_{diff}} = \text{Bodenstein number}$$

$$Da = (\Theta)\frac{u}{L}$$

$$G = \frac{Da}{St}$$

$$St = (k_L a)\frac{u}{L} \tag{7.84}$$

The parameters of this expression must be further modified to suit the particular reactor. For instance, Levenspiel (1989) states that for laminar flow through packed beds of porous particles that adsorb fluid, axial dispersion ($D$) of liquid phase through tubular pores of diameter $d$ has to take into account velocity profile, pore length distributon, molecular diffusion and adsorption coefficient:

$$D = f[\text{velocity profile, pore length distribution, molecular diffusion, adsorption coefficient}] \quad (7.85)$$

In the case where the packed bed is represented by a circular pore of average size $d$ and for low Reynolds number (10 to $10^{-4}$), Levenspiel (1989) proposed use of an expression for adsorption in smooth pores, with axial dispersion, to define dispersion $D$:

$$D = D^* + (1 + 6\delta + 11\delta^2)\frac{u^2d^2}{192D^*}$$

$$\delta = \frac{4}{k_{sorp}d}$$

$$k_{sorp} = \text{adsorption coefficient for liquid phase, 1/meter}$$

$$D^* = \text{effective diffusivity} \quad (7.86)$$

The velocity $u$ of the fluid applies to liquid flow, $d$ = average pore diameter and $\delta$ is a factor for adsorption on wall and bottom of the pore ($\delta = 4/k_{ads}d$, $k_{ads}$ = adsorption coefficient). The term $L$ must be modified, as the length the fluid 'sees' is the actual rather than Cartesian length, thus:

$$L = z\xi$$

Thus the Bodenstein number may be restated as:

$$B_0 = \frac{u(\xi z)}{[D_* + (1 + 6\delta + 11\delta^2)\frac{u^2d^2}{192D_*}]} \quad (7.87)$$

The term $C_{L,O2}$ is the dimensionless ratio of dissolved oxygen level at a length $z$ from the entry point into the reactor (liquid phase) to oxygen level at the exit from the reactor:

$$C_{L,O_2} = \frac{O_2(z)}{O_{2,out}} \quad (7.88)$$

The term $C^*$ represents the ratio of the oxygen content at liquid phase saturation to oxygen level at the exit point from the reactor.

The term $\Theta$ refers to the (oxygen consumption) reaction rate. In this section, the oxygen consumption rate was stated as:

$$\begin{aligned}\frac{dO_2}{dt} &= -\mu_{m,h}\frac{O_2}{K_{O_2} + O_2}\left(\frac{1}{Y_{(X/O),h}}\right)X_h - \mu_{m,ac}\frac{O_2}{K_{O_2} + O_2} \\ &= -\left[\mu_{m,h}\left(\frac{1}{Y_{(X/O),h}}\right)X_h + \mu_{m,ac}X_{ac}\right]\frac{O_2}{K_{O_2} + O_2}\end{aligned} \quad (7.89)$$

The Damkohler number, $Da$, is the measure of the relative importance of diffusion to reaction (Valsaraj, 2000) and can be stated in terms of the effective diffusivity $D^*$ of solute in the liquid phase, the reaction rate and the velocity of the solute as:

$$Da = D^* \frac{\Theta}{(u^*)^2} \tag{7.90}$$

The above relation assumes no sorption of $O_2$ to solid soil surface. Assuming that the diffusion velocity of the solute ($O_2$) is modified by water flux rate, water content of waste site segment and the air-water partition constant, the solute diffusion velocity $u^*$ is:

$$u_* = \frac{q_{flux,w}}{(\theta_w + \varepsilon k_{aw})} \tag{7.91}$$

where $q_{flux,w}$ = water input, $\theta_w$ = volume of water in the waste site segment and $\varepsilon$ = porosity of the waste site segment. Note that for unsaturated conditions, the porosity is a function of the matric potential $\psi$ as well as porous solid material water content $\theta_w$.

Levenspiel (1989) states that the effective diffusivity is about $10^{-6}$ to $10^{-8}$ $m^2$/second for gas in porous media and about $10^{-9}$ $m^2$/sec for liquids.

The term $\Theta$, for oxygen consumption, is now defined on the previous relation for oxygen consumption as:

$$\Theta = \left[\frac{\mu_{m,h}}{Y_{(X/O),h}} X_h + \mu_{m,ac} X_{ac}\right]\left(\frac{O_2}{K_{O_2} + O_2}\right) \tag{7.92}$$

Thus the Damkohler number $Da$ may be stated as:

$$\begin{aligned} Da &= D^* \frac{\Theta}{(u^*)^2} \\ &= D^* \left[\frac{(\theta_w + \varepsilon k_{aw})}{q_{flux,w}}\right]^2 \left[\frac{\mu_{m,h}}{Y_{(X/O),h}} X_h + \mu_{m,ac} X_{ac}\right]\left(\frac{O_2}{K_{O_2} + O_2}\right) \end{aligned} \tag{7.93}$$

The specific surface area $a$ and porosity $\varepsilon$ of the waste site were defined in Chapter 1 under Site Characterization. Velocity and porosity can also be defined in terms matric potential $\psi$ and water content $\theta_w$ of the hypothetical packed bed. The mass transfer term $k_L a$ is discussed in the following section.

For system boundary conditions, the parameters are:

$$\begin{aligned} \frac{dC_{O_2}}{dz} &= \frac{u}{D}[C_{O_{2,exit}} - C_{O_{2,entry}}], \quad \text{at } z = 0 \\ \frac{dC_{O_2}(L)}{dz} &= 0 \quad \text{at } z = L \end{aligned} \tag{7.94}$$

## DIFFUSIVITY COEFFICIENTS FOR LIQUID AND GAS SOLUTES

The expressions developed for oxygen concentration variation in a liquid downflow reactor require values of molecular diffusion of chemical species in the liquid phase, for example for the Bodenstein number. The molecular diffusion coefficient is a measure of the mean square of displacement of diffusing molecules from their original position, after a time $t$, and is:

$$D_m = \frac{(\overline{\chi})^2}{2t} \tag{7.95}$$

Units of molecular diffusivity for chemicals are typically in cm$^2$/sec.

For a waste site, considering both aerobic and anaerobic conditions, this might include a wide variety of volatile or gas species. Determinations of values are typically made for ambient temperatures, with 20°C or 25°C the most frequent experiment temperature. Because waste site temperatures can reach around 60°C or slightly above under aerobic decomposition, molecular diffusion values must be adjusted. An abbreviated list of values for molecular diffusivity of waste site chemical species is tabulated below.

Molecular diffusivity values can be corrected for temperature with the Farmer et al. (1980) convention:

$$D_{m,2} = D_{m,2}\left(\frac{T_2}{T_1}\right)^{\frac{1}{2}} \tag{7.96}$$

where $T$ = degrees Kelvin (273.16 + °C).

### PRACTICAL WASTE SITE PARAMETERS FOR DIFFUSION

The definitions above reflect the needed parameters to assess the effect of oxygen concentration in aerobic decomposition in a packed bed or soil system. Another approach to finding these parameters is from field studies, the most extensive of which are for landfills.

For landfills the vertical hydraulic conductivity has been reported to average $10^{-5}$ cm/sec, or 0.864 m/day. Thus a possible value for the liquid velocity $u$ is 0.864 m/day for a landfill. This also suggests horizontal conductivity is relatively small, as horizontal conductivity is often an order of magnitude or more smaller than vertical conductivity.

However, flow velocities in open dumps and compost piles may be more on the order of $10^{-2}$ to $10^{-4}$ cm/sec — that is, similar to flow through sand and coarse soils.

Fluid (liquid and gas) flow in landfills is known to be laminar (McBean et al., 1994) and thus will have Reynolds numbers in the range $R_e$ = 1–1000. The Schmidt numbers $S_c$ can range from about 3 to 2500 for this range of $R_e$.

The term $D^*$ is the effective diffusion coefficient of the migrating species in the soil that can be defined in terms of the hydrodynamic dispersion coefficient, for a soil

**TABLE 7.2**
**Molecular Diffusivity of Chemical Species in Air and Water**

| Chemical Species | Temperature °C | Molecular Diffusivity Air, $cm^2/sec$ | Molecular Diffusivity Water, $10^{-5}$ $cm^2/sec$ | Reference |
|---|---|---|---|---|
| Acetic acid | 20 | 0.133 | 0.88 | Thibodeaux, 1996 |
| Acetone | 20, 25 | 0.109 | 1.28 | Thibodeaux, 1996; Geankopolis, 1993 |
| Ammonia | 25, 20 | 0.28 | 1.76 | Thibodeaux, 1996 |
| Benzene | 25 | 0.088 | | Thibodeaux, 1996 |
| Butyric acid | 0 | 0.067 | | Thibodeaux, 1996 |
| Caproic acid | 0 | 0.050 | | Thibodeaux, 1996 |
| Carbon dioxide | 20 | 0.164 | 1.77 | Thibodeaux, 1996 |
| Carbon tetrachloride | 25 | 0.0828 | | Thibodeaux, 1996 |
| Chlorobenzene | 0 | 0.062 | | Thibodeaux, 1996 |
| Formic acid | 25 | 0.159 | 1.37 | Thibodeaux, 1996 |
| Hydrogen | 20, 25 | 0.410 | 4.8 | Thibodeaux, 1996; Geankopolis, 1993 |
| Hydrogen peroxide | 60 | 0.188 | | Thibodeaux, 1996 |
| Hydrogen sulfide | 0, 25 | 0.166 | 1.36 | Thibodeaux, 1996 |
| Methane | 25 | 0.277 | | McBean, 1995 |
| Nitrogen | 0, 25 | 0.13 | 2.0 | Thibodeaux, 1996 |
| Oxygen | 20, 25 | 0.206 | 2.35 | Thibodeaux, 1996 |
| Propionic acid | 0, 25 | 0.099 | 1.01 | Thibodeaux, 1996; Geankopolis, 1993 |
| Toluene | 30 | 0.088 | | Thibodeaux, 1996 |
| 1,2,4 Trichlorobenzene | 25 | 0.0676 | 0.757 | Thibodeaux, 1996 |
| n-Valeric acid | 25 | 0.067 | | Thibodeaux, 1996 |
| Water | 25 | 0.256 | | Thibodeaux, 1996 |
| Xylene | 24 | 0.071 | | Thibodeaux, 1996 |

Note: Convention in tabulation of experiment temperature: 0, 25 means 0°C for air diffusivity, 25°C, for water diffusivity.

representation, as hydrodynamic dispersion would combine both mechanical mixing of fluid and diffusion of chemical species in the fluid (McBean et al., 1995), as:

$$D^* = wD_{diff,\,fs}$$

$$D = \vartheta u + D^*$$

$$D = \vartheta u + wD_{diff,\,fs} \qquad (7.97)$$

where $D$ = liquid hydrodynamic dispersion, $D_{diff,fs}$ = diffusion coefficient of a chemical species in an open system, $\vartheta$ = dispersivity and $u$ = average bulk fluid velocity in the system. Also, $D_{diff,fs}$ is about $10^{-9}\,m^2/sec(8.64 \times 10^{-5}\,m^2/day)$ for chemical species in free solutions and $w$ has the value 0.01 to 0.5 (Freeze and Cherry, 1979).

The liquid dispersivity term $\vartheta$ represents the dispersivity tensor, or dispersivity in the flow direction. Mercer and Waddell (1992) note that longitudinal dispersion is usually larger than horizontal dispersivity (the fluid will spread out more in the direction of flow). In the case of a waste site, the direction of longitudinal flow would be predominantly in the direction of gravity drainage, with horizontal flow mainly due to preferential flow through high-permeability, horizontally oriented seams and pore pathways and capillary uptake flow. For waste sites, the longitudinal flow scale may be anywhere from zero to about 60 meters (maximum depth of landfill) and up to several hundred meters if vertical–horizontal flow to the groundwater table is considered. Mercer and Waddell (1992) plotted values of longitudinal dispersivity vs. scale of measurement, for measurement scales up to several thousand meters and showed that longitudinal dispersivity can be roughly 1/100 of the scale of measurement. This would suggest that the longitudinal dispersivity $\vartheta_L$ may be on the order of 0.01 to 0.1 for waste sites of depths up to 10 m.

## A STOICHIOMETRIC APPROACH TO DECOMPOSITION

For mass balance considerations, a stoichiometric approach provides a relatively accurate alternative to modeling the biological decomposition of organic materials. Information on empirical chemical formulation of various organic materials likely to be present in a waste site would be essential, as well as the biologically available amounts of these materials. With this information and a stoichiometric approach to decomposition process definition, mass balance relationships can be developed of inputs and products of aerobic or anaerobic decomposition.

For both aerobic and anaerobic decomposition of solid wastes, investigators have presented stoichiometric expressions that describe partial or complete decomposition. These expressions can be applied with some degree of accuracy to solid waste degradation at sites once some simplifying assumptions are made. Some of these, however, are more applicable to controlled digesters of wastes rather than to the less controlled waste site system.

For example Grady, Daigger and Lim (1999) presented a general stoichiometric approach widely applicable to the decomposition of carbohydrates. This approach sets out rules for the definition of half-reactions, for energy and for cell synthesis. These can then be mathematically summed, providing stoichiometric relations for the biological conversion process of interest. Important assumptions made include that organic substances to be decomposed can be reduced to glucose equivalents. Glucose is the basic sugar monomer; and thus carbohydrates, sugars, starches and cellulose may be described in chemical terms as polymers of glucose. This approach is useful if the chemical components of the organic substrates in a waste site can be described in terms of glucose equivalents. Inaccuracies might arise when accounting for the varieties of materials likely to be found in the MSW landfill. Some of these materials are unlikely to be fully carbohydrates, starches or cellulose, meaning that use of the Grady, Daigger and Lim (1999) *glucose equivalent* approach might require caution.

Another approach to solid waste stoichiometry is to develop expressions based upon the known chemical composition of the materials. This approach has the advantage of not being material- or material-class-specific and also lends itself to analyses suited to database methods, once chemical compositions of the degradable materials of interest have been developed. An encouraging note is that chemical formulations of the majority of materials normally disposed in waste sites can be found with extensive literature research. This makes possible the development of general stoichiometric relationships that may be applied to analyses of bioreactor performance. Though aerobic or anaerobic decomposition processes are well studied over the history of waste treatment, many stoichiometric formulations developed are not necessarily applicable to the physical and chemical heterogeneity of materials in a waste site characterized as a reactor. For this reason it is useful to develop and discuss stoichiometric relationships that may be applicable to mass balance considerations.

## The Stoichiometry of Decomposition of Wastes

McCarty (1975) has proposed an approach to development of an appropriate stoichiometric relation for an organic waste material for which only the empirical chemical formulas are (approximately) known. In this approach, the general oxidation half-reaction of an organic substrate is presented in terms of its empirical formula:

$$C_aH_bO_cN_d + (2a - c)H_2O = a(CO_2) + d(NH_4^+) + (4a + b - 2c - 4d)H^+ \tag{7.98}$$

This relation, presented in the form of the electron equivalent, assumes that the most important elements for the reaction are given by the molar concentrations of C, H, O and N, which is the case for many biologically mediated reactions. McCarty (1975) notes that the use of this generalized reaction equation ignores the phosphorous ($P$) requirement for biological synthesis. He states that the empirical formula for microbial (dry mass) composition is $C_{HON}$, but that this oversight can be addressed through considering that the stoichiometric requirement for biological (cell mass) is 14/5 ($N$). Thus, for every 14 gram moles of $N$, there would be a need to add 2.8 gram moles of $P$, which means that the molar concentration of $P$ in the empirical formula of the cell is $(0.4P)/N$. Consistent with this, it would be necessary to calculate the molar concentration of $P$ needed in the original substrate. This would be:

$$\text{grams } P = (0.4^*30.974)/(a + b + c + d)$$

While the stoichiometric relation presented by McCarty (1975) is very useful, some other considerations are useful for describing the full reaction process.

## DEVELOPMENT OF A GENERAL STOICHIOMETRIC RELATIONSHIP

A materials balance on the biological substance decomposed may be stated in terms of the sum $R$. The sum $R$ is the total of half-reactions, of the electron donor(s) $R_d$, (usually the organic substrate), the electron acceptor $R_a$ necessary for the biochemical pathway (typically oxygen dissolved in water for aerobic reactions) and cell synthesis $R_c$. McCarty (1975) and Grady and Lim (1999) have shown that consumption of the substrate has to involve a fraction $f_a$ that is used for energy, plus a fraction $f_s$ used for synthesis of cells. The summed electron-equivalent relation between $R$, $R_d$, $R_a$ and $R_c$ is given by:

Substrate energy = Substrate, initial − fraction for energy − fraction for cell synthesis

$$R = R_d - f_a R_a - f_s R_c \tag{7.99}$$

The energy use pathways may be summed similarly, as:

$$f_a + f_s = 1.0 \tag{7.100}$$

This gives that $f_a = (1 - f_s)$, or:

$$\begin{aligned} R &= R_d - (1 - f_s)R_a - f_s R_c \\ R &= R_d - R_a + f_s R_a - f_s R_c \end{aligned} \tag{7.101}$$

For aerobic decomposition, the electron acceptor $R_a$ is oxygen provided from dissolved oxygen in the reactor's liquid phase. Thus a half-reaction for oxygen can be stated as (McCarty, 1975):

$$R_a = (0.5)H_2O = (0.25)O_2 + H^+ + e^- \tag{7.102}$$

The half-reaction for biological cell synthesis is similarly stated as:

$$\begin{aligned} R_c &= 0.05C_5H_7O_2N + 0.45H_2O \\ &= 0.2CO_2 + 0.05HCO_3^- + 05NH_4 + H^+ + e^- \end{aligned} \tag{7.103}$$

The substrate donor half-reaction can be defined in terms of the empirical formula of the organic substrate, as (McCarty, 1975):

$$\begin{aligned} R_d &= C_aH_bO_cN_d + (2a - c)H_2O \\ &= aCO_2 + dNH_4^+ + (4a + b - 2c - 4d)H^+ + (4a + b - 2c - 3d)e^- \end{aligned} \tag{7.104}$$

Stating these relations according to their signs in the convenient form of:

$$(R = R_d - R_a + f_s R_a - f_s R_c)$$

$$R_d = C_aH_bO_cN_d + (2a - c)H_2O = aCO_2 + dNH_4^+$$
$$+ (4a + b - 2c - 4d)H^+ + (4a + b - 2c - 3d)e^-$$
$$-R_a = (0.5)H_2O = (0.25)O_2 + H^+ + e^-$$
$$f_s R_a = 0.5 f_s H_2O = (0.25) f_s O_2 + f_s H^+ + f_s e^-$$
$$-f_s R_c = 0.05 f_s C_5H_7O_2N + 0.45 f_s H_2O$$
$$= 0.2 f_s CO_2 + 0.05 f_s HCO_3^- + 0.5 f_s NH_4 + f_s H^+ + f_s e^- \qquad (7.105)$$

Summing these (molar equivalent) relationships to obtain R.

Assuming that mass balance considerations have been satisfied, the complete generalized stoichiometric reaction $R (= 0)$ for any organic substrate oxidized (or aerobically decomposed) can thus be stated as:

$$C_aH_bO_cN_d + 0.25(1 - f_s)O_2 + (0.05 f_s - d)NH_4^+ + (0.05 f_s)HCO_3^-$$
$$= (0.05 f_s)C_5H_7O_2N + (a - 0.2 f_s)CO_2 + (0.5 - 2a + c - 0.05 f_s)H_2O \qquad (7.106)$$

This relation allows direct estimations of molar quantities of inputs and outputs of the aerobic process if the empirical formula of the organic compound, in terms of C, H, O, N, is known.

## THE DEPENDENCE OF THE STOICHIOMETRIC RELATIONSHIP ON $f_s$ AND YIELD FACTOR $Y_{X/S}$

Notably, the relation above requires separate definition of the term $f_s$, the fraction of substrate used for cell synthesis, to solve for inputs and outputs. Since $f_s$ is a function

**TABLE 7.3**
**Molar Equivalents for Stoichiometric Relationship**

| Material | Formula | Molar Equivalent |
|---|---|---|
| Organic substrate | $C_aH_bO_cN_d$ | 1 |
| Oxygen | $O_2$ | $0.25(1–f_s)$ |
| Ammonia nitrogen | $NH4^+$ | $(0.05f_s–d)$ |
| Bicarbonate | $HCO_3^-$ | $(0.05f_s)$ |
| Biomass | $C_5H_7O_2N$ | $(0.05f_s)$ |
| Carbon dioxide | $CO_2$ | $(a–0.2f_s)$ |
| Water | $H_2O$ | $(0.5–2a+c–0.05f_s)$ |

of the true growth yield for the biomass utilizing the substance, $f_s$ must be defined in terms of the yield $Y$, or units of cell mass per unit of substrate ($Y_{X/S}$). Apart from microorganism effect upon changes in wastes, the reactor also imposes flow effects upon yields. The importance of the $f_s$ factor suggests the need to fully define this factor for use with a mixed substrate reactor such as might be envisioned for a landfill. Three approaches are possible. One of the most important would incorporate the reactor-based approach.

## REACTOR CONSIDERATIONS FOR $f_s$

Grady, Daigger and Lim (1999) stated that the fraction of the substrate used for synthesis $f_s$ is equal to 1.42 $Y$, where $Y$ = yield cell mass/mass COD. Since this yield is not grams cell/grams substrate, $f_s$ must be restated in terms of $Y_{X/S}$. In terms of the true growth yield, McCarty (1975) notes that $f_s$ is provided by $\{dX/dt\}/\{dF/dt\}$, where $dF/dt$ represents the substrate utilization rate in the reactor and $\{dX/dt\}_p$ represents the cell mass production, including from use of inactive cell materials (cell decay). The term $\{dX/dt\}_p$ was stated to be:

$$\left(\frac{dX}{dt}\right)_p = a_e\frac{dF}{dt} - f_a b X_a \tag{7.107}$$

where $a_e$ = fraction of the electron donor associated with cell synthesis at $t = 0$; also states as the growth yield coefficient or the equivalents of substrate synthesized (into biomass) per equivalent of substrate used by the microorganisms; $f_d$ = biodegradable fraction of organisms (0.85), $b$ = cell "decay" rate and $X_a$ = active concentration of organisms in the reactor. McCarty (1975) stated that $a_e$ can be defined from thermodynamic considerations as:

$$a_e = \frac{1}{1+A}$$

$$A = \frac{\left(\frac{\Delta G_p}{k^m} + 7.5\right)}{k(\Delta G_r)} \tag{7.108}$$

and $\Delta G_p$ = energy to convert cell carbon to a product = $(\Delta G^0(W)_d + 8.54)$ for organic electron donors, $k$ = efficiency of energy transfer for bacterial growth (about 0.6), ($m = 1$) if the estimated value of $\Delta G_p$ is positive and ($m = -1$) if $\Delta G_p$ is negative; and:

$$\Delta G_r = \Delta G^0(W)_d - \Delta G^0(W)_a \tag{7.109}$$

represents the difference between the free energy per electron equivalent for the donor and the acceptor.

McCarty (1975) estimated these values for a variety of electron donor and electron acceptor half-reactions. For instance, for oxygen as the electron acceptor, the value of $\Delta G^0(W)_a$ was provided as 18.675 kcals/electron equivalents. For heterotrophic reactions for a variety of organic electron donors, including domestic wastewater, protein, carbohydrates, fats and various organic aids, the value of $\Delta G^0(W)_d$ ranged between $-10$ and $-6.6$ kcals/electron equivalents.

The values estimated and tabulated by McCarty (1975) do not include a value for domestic refuse. However, by inclusion of $\Delta G^0(W)_d$ values for carbohydrates, proteins and fats, it is possible to apply the values presented if the percentages of these substances in domestic refuse are known. This is generally not possible. While the range of solid materials and substances in a waste site may not be represented, it is likely that the individual $\Delta G^0(W)_d$ values of biodegradable materials would be in the range quoted above, meaning no significant error should result from the use of a selected average of $\Delta G^0(W)_d$ values. With the average of values for domestic wastewater, proteins, carbohydrates, fats and oils, acetate, propionate and pyruvate, the approximate value for a mean $\Delta G^0(W)_d$ value would be $-7.9$ kcals:

$$\Delta G^0(W)_d = -7.9\,\text{kcals}$$

which is well within the range for all of the organic substances listed by McCarty (1975).

Use of the above value of $\Delta G^0(W)_d$ in the term $(\Delta G_p = \Delta G^0(W)_d + 8.54)$ gives a positive $\Delta G_p$ value of 0.6477 kcals, thus $m = +1$. Similarly, use of $\Delta G^0(W)_d = -7.9$ kcals and $\Delta G^0(W)_a = 18.675$ kcals in the partial expression $\Delta G_r = [\Delta G^0(W)_d - \Delta G^0(W)_a]$ provides that $\Delta G_r = -26.57$ kcals. Applying these to the McCarty (1975) expression for $A$ stated above, which is used when ammonia is available for cell synthesis, $A = -0.538$ and $a_e = 2.166$. This also indicates that as $a_e = 1/Y_{x/s}$, the growth yield in electron equivalents, $Y_{x/s}$, is given by:

$$Y_{X/S} = \frac{1}{a_e} = 0.462 \tag{7.110}$$

However, McCarty (1975) also defined the relation between $a_e$ and $f_s$ as:

$$f_s = a_e\left[1 - \frac{f_d b t_s}{1 + b t_s}\right] \tag{7.111}$$

This is based upon the fact that $f_s$ is equal to $\{dX/dt\}_p/\{dF/dt\}$, i.e., the rate of use of production of cell materials (including inert cell remains or dead material) per rate of use of the substrate. The term $\{dX/dt\}_p$ was defined as the sum:

$$\{dX/dt\}_p V = \{dX_a/dt\}_p V + \{dX_i/dt\}_p V$$

meaning that for the same reactor volume, the overall rate of cell synthesis equals rate of production of active cells plus rate of synthesis of inactive cell material. The cell synthesis rate can be defined in terms of organism growth including decay as:

$$\frac{dX_a}{dt}V = a_e\frac{dF}{dt}V - bX_a \tag{7.112}$$

This is the same as the well-known expression $-dS/dt = (1/Y_{x/s})dX/dt$, where $dX/dt = \mu X - k_d X$, $S = F$, $b = k_d =$ decay or maintenance coefficient, $a_e =$ growth yield coefficient $= 1/Y_{x/s}$ and substituting for $\mu X$. The term $\{dX_i/dt\}_p$ is defined as $\{(1 - f_d)bX_a\}$, where $f_d =$ biodegradable fraction of the organism ($f_d = 0.85$). Substitutions for $\{dX_a/dt\}$ and $\{dX_i/dt\}_p$ lead to:

$$\{dX/dt\}_p V = a_e(dF/dt)V - f_d b X_a V$$

which leads to the formerly stated expression, $\{f_s = a_e[1-(f_d bt_s)/(1+bt_s)]\}$ when $\{X_a V\}$ is substituted.

## DEFINITION OF RESIDENCE TIME $t_s$

The term $t_s$ represents the solids retention time for the reactor, which is theoretically the mass of organisms in the reactor divided by their rate of loss from the reactor. In the landfill environment, which is a fixed media, the soil system is likely to retain organisms that are filtered out of the liquid flow by the pore system or remain attached to surfaces. This suggests the only loss of organisms would be through loss of decay products through outflow. Contribution of organism mass to system liquid flows can be addressed by defining a mass balance for organisms contained in the landfill reactor.

For a layer of solid waste containing material and with typical liquid flow rates of about $1 \times 10^{-4}$ cm/sec (0.864 m/day) (Oweis et al., 1993), mass balance or organisms held in the system is given by:

> Synthesis of organism mass retained/volume = growth/volume − difference between organism in entering flow $Q_{\text{in}}$ and the exiting flow $Q_{\text{out}}$.

For active organisms concentration $X_a$ in the system, the above statement is:

$$\left(\frac{dX_a}{dt}\right)_r V = \frac{dX_a}{dt} V - (Q_{in} - Q_{out})X_a^e \qquad (7.113)$$

where $X_a^e$ represents concentration of active organisms in the flow exiting the reactor system and the subscript $r =$ concentration of organisms retained in the reactor system. If for a landfill layer, or more likely a combination thereof, active organisms are essentially filtered out of the flow (such that outflow loss per landfill segment or layer can be divided between layers), the left-hand term $[Q_{\text{in}} - Q_{\text{out}}]X_a^e$ is approximately zero and, substituting for $\{dX_a/dt\}$ in the expression above:

$$\left(\frac{dX_a}{dt}\right)_r V = \frac{dX_a}{dt} V = \left(a_e \frac{dF}{dt} - bX_a\right) V \qquad (7.114)$$

Similar mass balancing for the concentration of inactive organism material across the landfill reactor indicates:

$$\left(\frac{dX_i}{dt}\right)_r V = \frac{dX_i}{dt} V - (Q_{in} - Q_{out})X_i^e \qquad (7.115)$$

Assuming that under waste site soil filtering the same system retention conditions would apply to inactive organism material, i.e., $X_i^e = 0$, the mass balance may be restated as:

$$\left(\frac{dX_i}{dt}\right)_r V = ((1 - f_d)bX_a)V \tag{7.116}$$

Under steady-state conditions the terms $\{dX_a/dt\}_r$ and $\{dX_i/dt\}_r = 0$, which assumes no flow upsets or residence time changes. These assumptions are reasonable because for landfill reactor time-based simulations, time change units are small, usually on the order of a day or less and liquid flow changes are insufficient for reactor system upset. In this case the mass balance relations become:

$$\{dX_a/dt\}_p V = \{a_e(dF/dt) - bX_a\}V$$

$$\{dX_i/dt\}_p V = \{(1 - f_d)bX_a\}V$$

In the previous definition of the system residence time, $t_s$ = mass of organisms in the system, divided by organism loss from the system. According to McCarty (1975) this essentially applies to the *active* mass in the system. Grady and Lim's (1999) discussion of this topic indicates that some error is incurred with this assumption, as the actual degradable solids might be closer to active organisms plus other solids, i.e., the total suspended solids (TSS). However, it was shown by these authors that active organisms concentration did not change the value of $t_s$ very much. Thus $t_s$ can be defined as:

$$t_s = (X_a V)/\left(\left(\frac{dX_a}{dt}\right)_p V\right) = X_a/\left(a_e\frac{dF}{dt} - bX_a\right) \tag{7.117}$$

To solve the above, the terms $\{dF/dt\}$ and $\{X_a\}$ must be separately defined. The previous discussion indicated $\{X_a\}$ is the concentration value of organisms at the beginning of the time period of simulation, i.e., $X_0$. Additionally, the decomposition behavior of the material, $\{dF/dt\}$, can be stated in terms of the hydrolysis rate:

$$\{dF/dt\} = dS/dt = -k_h S$$

which is the same as $S = S_0 \exp(-kht)$; and as $dF = dS = \Delta S = S_0 - S$, where $dF = dS = \Delta S = [S_0(1 - \exp(-k_h \Delta t)]$, and:

$$t_s = X_0/\left(\frac{a_e}{dt}(S_0(1 - \exp(-k_h dt))) - bX_a\right) \tag{7.118}$$

This expression can be used to iteratively estimate $f_s$, as previously defined.

$$f_s = a_e[1 - (f_d b t_s)/(1 + b t_s)]$$

noting that the value of $a_e$ was indicated to be in the range of 2.166, through estimation from likely free reaction energy values for organic landfilled materials (solids and liquids) undergoing aerobic decomposition.

## THE FRACTION OF SUBSTRATE ENERGY STORED IN THE BIOMASS

Another approach to defining $f_s$ is in terms of the energy stored in the biomass synthesized from the reaction. The yield factor $Y$ can be stated in terms of a factor $\eta$, representing the substrate energy stored in (new) biomass (Luong and Volesky, 1981):

$$\eta = \frac{\gamma_b}{\gamma_s} Y \tag{7.119}$$

However, $\eta$ can also be stated in terms of the electrons available from the substrate:

$$\eta = \frac{\gamma_b}{\gamma_s} Y_{X/S} \left( \frac{\sigma_b}{\sigma_s} \right) \tag{7.120}$$

where $\gamma_b$ = reduction degree of the biomass, $\gamma_s$ = reduction degree of the substrate, $Y_{X/S}$ = grams cells/gram substrate, $\sigma_b$ = weight fraction of carbon in the biomass and $\sigma_s$ = weight fraction of carbon in the substrate. The reduction degree of the biomass $\gamma_b$ and of the organic substrate, $\gamma_s$, are defined in terms of carbon equivalents as:

$$\begin{aligned} \gamma_b &= 4 + p - n - 3q \\ \gamma_s &= 4 + b - 2c - 3d \end{aligned} \tag{7.121}$$

where $p = 7/5 = 1.4$, $n = 2/5 = 0.4$ and $q = 1/5 = 0.2$ for the biomass formula $C_5H_7O_2N$, which also indicates that $\gamma_b = 4$. The constants $b$, $c$ and $d$ are the mole percentages of the empirical substrate formula $C_aH_bO_cN_d$. The term $\sigma_b$ represents the fraction of carbon in the biomass of formula ($C_2H_5O_7N$) and is thus:

$$\sigma_b = 12(5)/[12(5) + 7 + 16(2) + 14] = 60/113 = 0.531 \text{ gm/gm},$$

The term $\sigma_s$ is similarly the carbon fraction in an organic substrate of chemical formula $C_aH_bO_cN_d$ and is thus:

$$\sigma_b = a/[a + b + c + d] \quad \text{grams carbon/gm}$$

Note that if the empirical formulas for biomass and organic substrate are in the form $C_uH_vO_wN_xS_yP_z$ and $C_aH_bO_cN_dS_eP_f$, respectively, for inclusion of moles of sulfur and phosphorous, the proper relations for $\sigma_b$, $\sigma_s$, $\sigma_b$ and $\sigma_s$ must be used.

## ACCURACY OF THE VALUE OF $\gamma_b$

It has been suggested by some investigators (Minkevich and Eroshin, 1973; Schugerl, 1987; Luong and Volesky, 1983) that other values of $\gamma_b$ are perhaps more representative of the range of conditions encountered in reactions, thus the value $\gamma_b = 4$ might be unreliable. Schugerl (1987) states, for instance, that the value of $\gamma_b$, the degree of reduction of cells, should be about 4.17, if ammonia (nitrogen) is a source. Minkevich and Eroshin (1973) state that the value of $\gamma_b$ should be about 4.2 and that the value of $\sigma_b$ is approximately 0.46. Luong and Volesky (1983) stated that for a variety of biodegradable substrates including cellulose, carbohydrates and starch, the

**TABLE 7.4**
**Reduction Degree of Organisms on Various Substrates**

| Organism Type | Name | Substrate | Value of $\gamma_b$ |
|---|---|---|---|
| Bacteria | *Bacterium* | n-Pentane | 4.607 |
| Bacteria | *Pseudomonas* | n-Alkane | 4.497 |
| Bacteria | *Pseudomonas aeruginosa* | n-Hexane | 4.32 |
| Yeast | *Candida tropicalis* | n-Alkanes | 4.385 |
| Yeast | *Candida* spp. | n-Alkanes | 4.278 |
| Yeast | *Candida* spp. | n-Alkane | 4.26 |
| Yeast | *Candida* spp. | Glucose | 4.074 |
| Yeast | *Candida* spp. | Cellulose hydrolyzate | 4.11 |
| Yeast | *Candida* spp. | Cellulose hydrolyzate | 4.122 |
| Yeast | *Candida* spp. | Cellulose hydrolyzate | 4.201 |
| Yeast | *Candida* spp. | Cellulose hydrolyzate | 4.087 |
| Yeast | *Candida* spp. | Cellulose hydrolyzate | 4.22 |
| Yeast | *Candida* spp. | Cellulose hydrolyzate | 4.214 |
| Yeast | *Candida* spp. | Cellulose hydrolyzate | 4.199 |
| Yeast | *Candida* spp. | Cellulose hydrolyzate | 4.143 |
| Yeast | *Candida* spp. | Cellulose hydrolyzate | 4.212 |
| Fungi | *Saccharomyces cerevisiae* | Glucose | 4.237 |
| Fungi | *Saccharomyces cerevisiae* | Glucose | 4.291 |
| Fungi | *Saccharomyces cerevisiae* | Ethanol | 4.469 |
| Fungi | *Saccharomyces cerevisiae* | Acetic acid | 4.416 |

value of $\gamma_b$ should vary between 4.0 and 4.4. These authors also reported several values associated with experiments determining the reducing power of biomass of different species, which are listed in the table above.

The above table suggests that the average values of $\gamma_b$ are:

- 4.168 for *Candida* spp. on cellulose hydrolyzate
- 4.264 for *Saccharomyces cerevisiae* on glucose
- 4.308 for *Candida* spp. on n-alkanes
- 4.074 for *Candida* spp. on glucose
- 4.469 for *Saccharomyces cerevisiae* on ethanol
- 4.416 for *Saccharomyces cerevisiae* on acetic acid
- 4.32 for *Pseudomonas aeruginosa* on n-hexane
- 4.32 for *Pseudomonas aeruginosa* on n-alkanes

It is thus likely that, for both aerobic and anaerobic conditions, use of the value suggested by Minkevich and Eroshin (1973), $\gamma_b = 4.2$, should be relatively accurate. These authors concluded that the value of $\gamma_b = 4.2$ is relatively constant, with a variance of only 2% and that the weight part of the carbon in the biomass, $\sigma_b = 0.46$, is also constant with a variance of 4% (reported by Luong and Volesky, 1983).

Comparison with the $\sigma_b = 0.531$ calculated using the empirical biomass formula of CHON indicates a variance of 15.4%. This variance is relatively significant. The order of magnitude of variation one, however, is the same, which suggests that if the empirical formulas of organic substances and of biomass are to be used in the process estimations, the value of $\sigma_b = 0.531$ might be more consistent with the stoichiometric expression developed earlier. With this consideration, the value of $\sigma_b = 0.531$ appears appropriate for use with the stoichiometric expression.

## THE ENERGY EXPRESSION

Reconsidering the expression stated earlier,

$$\eta = \frac{\gamma_b}{\gamma_s} Y_{X/S} \left( \frac{\sigma_b}{\sigma_s} \right) \tag{7.122}$$

Substitution in the expression for $\eta$ for $\gamma_b = 1.42$ and $\gamma_b = 0.531$ gives:

$$\eta = 2.23(Y_{X/S}) \left( \frac{1}{\sigma_s \gamma_s} \right) \tag{7.123}$$

The growth yield $Y_{X/S}$ can also be defined in terms of the number of electron equivalents available from the organic substance $Y_{ave,e}$, which can in turn be calculated from the moles of oxygen needed for combustion of one mole of the substrate by multiplying the combustion oxygen needed by four (Luong and Volesky, 1983, presented this approach in terms of one mole of a glucose substrate):

$$Y_{ave,e} = \frac{Y_{X/S}}{Y_{ave,e/s}} \tag{7.124}$$

In the above, $Y_{ave,e}$ represents the cell mass yield for the available electrons. Luong and Volesky (1983) reported averages value $Y_{ave,e} = 3.07$ for studies of 79 organisms and 69 samples of carbon sources ranging from $C_4$ to $C_6$ compounds. This suggests an average of:

$$Y_{ave,e/s} = 0.726(Y_{X/S}) = \frac{\sigma_s \gamma_s}{12} \tag{7.125}$$

from the molar definition of $(\sigma_s \gamma_s)$. The above indicates that the yield $Y_{X/S}$ can be defined for (aerobic) oxidation conditions, with ammonia nitrogen as the source, as:

$$Y_{X/S} = 0.1147(\sigma_s \gamma_s) \tag{7.126}$$

For use of this expression, $(\sigma_s \gamma_s)$ must be stated in terms of the molar fractions. From the previous definitions of $\sigma_s$ and $\gamma_s$ earlier in this section, i.e.:

$$\sigma_s = a/[a + b + c + d]$$
$$\gamma_s = [4 + b - 2c - 3d]$$

Thus an approximation for the cell mass yield from a degradable organic substance of empirical formula $C_aH_bO_cN_d$ would be:

$$Y_{X/S} = 0.1147\frac{(12a)(4+b-2c-3d)}{(12a+b+16c+14d)} \qquad (7.127)$$

The above expression permits the yields from individual substrates to be considered and can be calculated in spreadsheet fashion from determination of empirical formulation of the dry *degradable* weights of the organic substances. For instance, if the chemical (empirical) formulas of various organic materials comprising the wastes in a disposal site or a composting operation are known and the biodegradable fraction in dry weight is also known, the above expression allows calculation of the specific yield of an organism per substrate and lends itself to the convenience of spreadsheet-type calculations from information databases.

While use of the expression should be relatively accurate, considering the source of the value $Y_{ave,e} = 3.07$, from a wide range of organisms and substrates, the abbreviated empirical substrate formula $C_aH_bO_cN_d$ is understood to have small errors due to the assumption, as made by McCarty (1975), that the effect of sulfur (S) content is minor. The correct formulas must be used if the biological process products involving sulfur or phosphorous are of interest.

Notably, the above expression is based upon the available electrons in the substrate and biomass. Grady and Lim (1999) state that $Y_{X/S}$ can be converted to the cell mass yield per gram COD, by dividing by eight. According to the same authors, $f_s = 1.42\,Y$, or ($0.85 \times 1.42\,Y = 1.207\,Y$), considering that volatile solids content of microorganisms is about 0.85. This indicates that for cell mass yield per gram COD, $f_s$ can be stated as:

$$f_{s,COD} = (1.42)(0.85)\frac{Y}{8} = 1.207\frac{Y_{X/S}}{8} = 0.1284\frac{(12a)(4+b-2c-3d)}{(12a+b+16c+14d)} \qquad (7.128)$$

This equation can be considered more practical.

The previously defined term for fraction of available electrons (energy) incorporated into the microorganisms, $\eta = 2.23Y_{X/S}[1/(\sigma_s\gamma_s)]$, can thus be simply restated in terms of $Y_{X/S}$ and $(\sigma_s\gamma_s)$.

## Other Discussions of the Yield Term $Y_{ave,e}$ for the Energy Expression

The importance of this term to overall accuracy merits brief further discussion. Schugerl (1987) stated that this term could be expressed as:

$$Y_{ave,e} = (Y_{X/S}M_X)/\gamma_s \qquad (7.129)$$

From the previous definitions of $\{Y_{ave,e} = Y_{X/S}/Y_{ave,e/s}\}$ and $\{Y_{ave,e/s} = (\sigma_s\gamma_s)/12\}$, it is implied that:

$$Y_{ave,e} = (\gamma_s Y_{X/S})/(Y_{X/S}M_X) = \frac{(\sigma_s\gamma_s)}{12} \qquad (7.130)$$

which gives $\sigma_s = 0.1062$ for an organism molecular weight of 113 as is commonly accepted. Considering the previous definition of $\{\sigma_s = a/[a+b+c+d]\}$, the value $\sigma_s = 0.1062$ *fixes* determination of the carbon content of the substrate, making it inconvenient and possibly inaccurate for use in a mixed substrate bioreactor such as a waste site. Thus the former definition of $Y_{X/S}$ could be more appropriate or practical for a waste site as a reactor.

## Cell Mass Yield Factor $Y_{X/S}$ from Chemical Oxygen Demand (COD)

A third approach to definition of the substrate synthesis factor $f_s$ is through application of the formulation presented by Grady, Daigger and Lim (1999) to chemical oxygen demand (COD) values of the individual substrates. The simplicity and directness of this approach makes it very practical. For instance, Grady, Daigger and Lim (1999) have noted that the cell synthesis energy fraction $f_s$ can be defined in terms of the grams of cells formed per gram of COD removed, or $Y$,

$$f_s = 1.42Y$$

where $Y$ is in terms of the volatile solids (VS); and thus the cell mass fraction, which has 85% VS, must be converted to VS for use of $Y$:

$$f_s = 1.42(0.85)Y_{X/S}$$

where $Y_{X/S}$ = grams of cells formed per gram of COD removed ($\Delta X/\Delta(S)$. McCarty (1975) noted that the (mole) grams COD removed can be defined as:

$$\text{g mol COD} = 8[4a + b - 2c - 3d]$$

which is the same as $\{8\gamma_s\}$. The g mol of substrate (S) for the empirical substrate formula $C_aH_bO_cN_d$ is $[a+b+c+d]$, by inspection. Thus the g mole COD/g mol substrate is given by:

$$\frac{g\ \text{mol COD}}{g\ \text{mol}\ \Delta S} = \frac{8(4a + b - 2c - 3d)}{(a+b+c+d)} \tag{7.131}$$

Expressing the change in substrate $\Delta(S)$ as $\Delta\,\text{COD}(g)$:

$$g\ (\Delta\ \text{COD}) = g\ (\Delta S)\left[\frac{8(4a + b - 2c - 3d)}{(a+b+c+d)}\right] \tag{7.132}$$

The term $f_s$ may thus be restated as:

$$f_s = 1.207\left(\frac{1}{F_{COD}}\right)\left(\frac{\Delta X}{\Delta S}\right) \tag{7.133}$$

where:

$$F_{COD} = \frac{8(4a + b - 2c - 3d)}{(a+b+c+d)} \tag{7.134}$$

This value of $F_{COD}$ can be estimated using the empirical formulas of substrates, once these are known, in spreadsheet fashion for modeling. However, the value of $f_s$ is notably still dependent upon separate determination of the yield $\{\Delta X/\Delta S\}$, unless the yield can be determined by some other approach.

## Yield Estimation from Oxygen Consumption

One approach to defining yield is to use the definition of oxygen requirement from the general stoichiometric equation developed earlier, where the oxygen requirement of the reaction was stated as:

$$\text{Molar Oxygen Requirement} = 0.25(1 - f_s)O_2$$

According to Luong and Volesky (1983), in the case where the nitrogen source is ammonia and the products are carbon dioxide, cells and water, the cell mass production quantity $\Delta_X$ can be defined in terms of the grams of $O_2$ consumed. This is the same as the term $Y_{X/O}$:

$$Y_{X/O} = \frac{\Delta X}{-\Delta O_2} \tag{7.135}$$

where $-\Delta O_2$ is the $g$ of oxygen consumed. From the molar oxygen requirement as defined above, $g$ of oxygen needed would be:

$$\text{Oxygen requirement} = 32[0.25(1 - f_s)O_2] = 8(1 - f_s)$$

$$Y_{X/O} = \frac{\Delta\ X}{8(1 - f_s)}$$

$$\Delta X = 8(Y_{X/O})(1 - f_s) \tag{7.136}$$

Also considering the previous expression, where:

$$f_s = \frac{1.207(\Delta X)}{F_{COD}(\Delta S)}$$

$$\Delta X = 0.829 f_s (F_{COD})(\Delta S) \tag{7.137}$$

Equating $\{\Delta X\}$ between the two expressions and rearranging:

$$f_s = \frac{8(Y_{X/O})}{[8(Y_{X/O}) + 0.829(F_{COD})(\Delta S)]} = \frac{(Y_{X/O})}{[(Y_{X/O}) + 0.1036(F_{COD})(\Delta S)]} \tag{7.138}$$

## The Value of $Y_{X/O}$

Given that $F_{COD}$ was previously defined, determination of $f_s$ in the above expression depends only upon choice of a reasonable value of $Y_{X/O}$ and definition of $\Delta S$. If a reasonable value of $Y_{X/O}$ is used, the term $f_s$ is completely defined in terms of the organic substance. Luong and Volesky (1983) reported the values below for organisms growing aerobically in *minimal media*:

**TABLE 7.5**
**Grams Oxygen Used/Gm Cells Produced, Aerobic**

| Organism Type | Organism Name | Substrate | ($Y_{X/O}$) = Gram Biomass/ Gram Oxygen |
|---|---|---|---|
| Bacteria | *Aerobacter aerogenes* | Maltose | 1.5 |
| Bacteria | *Aerobacter aerogenes* | Fructose | 0.42 |
| Bacteria | *Aerobacter aerogenes* | Glucose | 0.4 |
| Bacteria | *Aerobacter aerogenes* | Ribose | 0.98 |
| Bacteria | *Aerobacter aerogenes* | Succinate | 0.62 |
| Bacteria | *Aerobacter aerogenes* | Lactate | 0.37 |
| Bacteria | *Aerobacter aerogenes* | Pyruvate | 0.48 |
| Bacteria | *Aerobacter aerogenes* | Acetate | 0.31 |
| Bacteria | *Pseudomonas fluorescens* | Glucose | 0.85 |
| Bacteria | *Pseudomonas fluorescens* | Acetate | 0.70 |
| Bacteria | *Pseudomonas fluorescens* | Ethanol | 0.42 |
| Bacteria | *Pseudomonas fluorescens* | Methane | 0.20 |
| Bacteria | *Pseudomonas* sp. | Methane | 0.19 |
| Bacteria | *Pseudomonas fluorescens* | Methane | 0.17 |
| Bacteria | *Rhodopseudomonas spheroides* | Glucose | 1.46 |
| Fungi (yeast) | *Candida utilis* | Glucose | 0.40 |
| Fungi (yeast) | *Candida utilis* | Acetate | 0.70 |
| Fungi (yeast) | *Candida utilis* | Ethanol | 0.61 |
| Fungi (yeast) | *Saccharomyces cerevisiae* | Glucose | 0.97 |
| Fungi (hyphal) | *Penicillium chrysogenum* | Glucose | 1.35 |

From the above table, the average gm cell yields per gram oxygen for substrates can be stated as:

| | |
|---|---|
| Bacteria, glucose substrate: | $Y_{X/O} = 0.625$ |
| Bacteria, acetate substrate: | $Y_{X/O} = 0.39$ |
| Bacteria, methane substrate: | $Y_{X/O} = 0.19$ |
| Bacteria, other sugars: | $Y_{X/O} = 0.967$ |
| Bacteria, pyruvate: | $Y_{X/O} = 0.48$ |
| Bacteria, succinate: | $Y_{X/O} = 0.62$ |
| Fungi (yeast), glucose substrate: | $Y_{X/O} = 1.15$ |
| Fungi (yeast), acetate substrate: | $Y_{X/O} = 0.70$ |
| Fungi (yeast), ethanol substrate: | $Y_{X/O} = 0.61$ |
| Fungi (hyphal), glucose substrate: | $Y_{X/O} = 1.35$ |

The applicability of these values depends upon the likely pathway for biochemical oxidation of the landfilled materials. Given that the glucose monomer is structurally common to the sugars described, glucose yields may be more convenient. However, the presence of other sugars is indicated to have significant effect upon $Y_{X/O}$; thus an overall average may be more practical:

| | |
|---|---|
| Bacteria; sugars, acids: | $Y_{X/O} = 0.62$ |
| Bacteria, methane: | $Y_{X/O} = 0.19$ |
| Fungi: | $Y_{X/O} = 1.07$ |

These values can be applied to the definition of $f_s$ for the purpose of modeling stoichiometric decomposition of the soluble organic fraction of MSW and they are representative of average values of oxygen consumption for bacteria and fungi, respectively. If more specific organisms — say of a landfill or solid waste decomposition microcosm are involved — more specific values of $Y_{X/O}$, such as listed in Table 7.5, can be used instead. Notably the value of $Y_{X/O} = 0.19$ would apply to methanotrophs oxidizing landfill gas dissolved in the liquid phase of the aerobic segment of the landfill bioreactor, if this microcosm is to be modeled.

For instance, the expression developed for hydrolysis:

$$\Delta S = [S_0(1 - \exp(-k_h \Delta t))]$$

can be applied with the $f_s$ of this section, for:

$$f_s = \frac{(Y_{X/O})}{[(Y_{X/O}) + 0.1036(F_{COD})(\Delta S)]} \tag{7.139}$$

## USE OF $f_s$ VALUES TO ESTIMATE WATER PRODUCTION FROM AEROBIC DECOMPOSITION

From the stoichiometric equation developed at the beginning of this section,

$$C_aH_bO_cN_d + 0.25(1 - f_s)O_2 + (0.05f_s - d)NH_4^+ + (0.05f_s)HCO_3^-$$
$$= (0.05f_s)C_5H_7O_2N + (a - 0.2f_s)CO_2 + (0.5 - 2a + c - 0.05f_s)H_2O \tag{7.140}$$

Moles of water produced per mole of substrate consumed is given by:

$$\text{moles} \quad H_2O = [0.5 - 2a + c - 0.05f_s]$$

The grams moles of substrate consumed for this amount is provided by:

$$\text{grams moles substrate consumed} = [a + b + c + d]$$

and the gm moles of water produced is $\{18[0.5 - 2a + c - 0.05f_s]\}$. The ratio of water produced to substrate consumed is:

$$\frac{g\ H_2O}{g\ \Delta S} = \frac{18(0.5 - 2a + c - 0.05f_s)}{(12a + b + 16c + 14d)} \tag{7.141}$$

The value $\Delta S$ in this equation refers to the amount of the specific organic substance decomposed, per time step of simulation. Solution thus depends only upon the value of $f_s$.

## $CO_2$ Produced, $O_2$ Required and Heat Produced During Aerobic Decomposition

The moles of $CO_2$ produced are similarly provided by the term $(a - 0.2f_s)$ in the stoichiometric relationship. Thus, carbon dioxide production is:

$$\frac{g(CO_2)}{g(\Delta S)} = \frac{44(a - 0.2f_s)}{(12a + b + 16c + 14d)} \tag{7.142}$$

From the stoichiometric relationship, the moles of oxygen consumed are also given by $0.25[1 - f_s]$. This would indicate that oxygen required for the decomposition reaction per unit of substance is:

$$\frac{g(O)_2}{g(\Delta S)} = \frac{32[0.25(1 - f_s)]}{(12a + b + 16c + 14d)} \tag{7.143}$$

Heat production during aerobic decomposition can be defined either in terms of amount of substance consumed or oxygen consumed (Haug, 1983; Luong and Volesky, 1983). If the chemical oxygen demand (COD) of a material is available (can be estimated from the empirical formula [McCarty, 1975]) and the quantity of its ash and moisture free content is known, the heat released is provided by multiplying the COD value by the energy released per unit COD consumed. Haug (1983) states that for aerobic oxidation, the heat release per gram COD is 3.26 kilocalories. Thus heat generation from the reaction is provided by:

Kilocals heat = (gm COD/gm substrate) × (Kilocals/gm COD consumed)

As previously noted, the molar weight (gm moles) of organic substance of empirical formula $C_aH_bO_cN_d$ is $[a + b + c + d]$. The COD value is also given by $8[4a + b - 2c - 3d]$. Heat generation is thus:

$$\frac{\text{Kilocalories}}{g(\Delta S)\ \text{consumed}} = \frac{8(4a + b - 2c - 3d)(3.26)}{(12a + b + 16c + 14d)} \tag{7.144}$$

As this value does not require the term $f_s$, it can be estimated independently for use in database simulation when several substrates are involved and when the empirical formulation of the substance decomposed is known.

Some investigators have linked heat production from aerobic digestion to the amount of oxygen consumed. For instance, Cooney et al. (1969) proposed that the amount of heat energy generated is a function of the oxygen use/heat yield factor and the amount of oxygen consumed and is independent of specific growth rate:

$$q_{heat,\ aerobic} = \left(\frac{1}{Y_{O/H}}\right) q_O \tag{7.145}$$

In this expression the term $(1/Y_{O/H})$ is indicated to match the enthalpy $(\Delta H_R^O)$ of the fermentation reaction (Moser, 1988). Luong and Volesky (1980) reported that the value of $\Delta H_R^O$ varies between 0.385 MJ/mole and 0.485 MJ/mole for filamentous fungi and between 0.385 MJ/mole and 0.565 MJ/mole for bacteria, based upon the work of Minkevich and Eroshin (1973).

The values $\Delta H_R^O$ or $1/Y_{O/H}$ above are in MJ/mole units, implying that moles of oxygen ($O_2$) consumed would be the correct unit to use. From the stoichiometric equation developed earlier in this section, for aerobic digestion the moles of $O_2$ consumed are provided by $0.25(1 - f_s)$. Thus the molar heat quantity produced from the aerobic digestion reaction would be:

$$q_{heat,\ aerobic} = (\Delta H_R^O)[0.25(1 - f_s)]$$

$$f_s = \frac{(Y_{X/O})}{[(Y_{X/O}) + 0.1036(F_{COD})(\Delta S)]}$$

$$F_{COD} = \frac{8(4a + b - 2c - 3d)}{(a + b + c + d)}$$

$$\Delta H_R^O = 0.385 - 0.485 \text{ MJ/mole (bacteria)}$$

$$\Delta H_R^O = 0.385 - 0.565 \text{ MJ/mole (filamentous fungi)} \qquad (7.146)$$

## The Stoichiometry of Anaerobic Decomposition of Solid Wastes

A similar stoichiometric reaction expression can be developed for anaerobic decomposition of solid (or liquid) organic materials, once the reaction products are known. The various relations developed for product and biomass yields can also be developed from the anaerobic reaction.

Various stoichiometric relationships have been developed for anaerobic decomposition, including for solid waste materials for which the empirical chemical compositions are known. One such relationship, practical for the present analysis and for bioreactor modeling purposes, was developed by Cardenas and Wang (1980) and presented a mass balance for the decomposition of solid wastes:

$$C_aH_bO_cN_dS_e = [a - 0.25b - 0.05c + -0.75d + 0.5e]H_2O$$

$$\Rightarrow [0.5a - 0.125b - 0.25c + 0.375d - 0.25e]CO_2$$

$$+ \ [0.5a + 0.125b - 0.25c - 0.375d - 0.25e]CH_2 + e(H_sS) \qquad (7.147)$$

Compared to the previous stoichiometric relationship developed earlier in this section, to represent aerobic decomposition, the empirical chemical formula value of sulfur content, $S$, is incorporated in the anaerobic reaction, which may be useful for estimating anaerobic decomposition products such as hydrogen sulfide — typically generated during the biodegradation of solid wastes. The COD value for an organic material of empirical formula $C_aH_bO_cN_dS_e$ is provided by $8[4a + b - 2c - 3d + 6e]$ and the gm moles of the organic are provided by $[a + b + c + d + 32.06e]$.

### Water Consumption During Anaerobic Decomposition Process

From the stoichiometric relation, it is indicated that net water uptake is involved during the anaerobic decomposition process in the molar amount of:

$$\text{moles } H_2O/\text{mole substrate use} = [a - 0.25b - 0.05c + 0.75d + 0.5e]$$

The gram moles of substrate are given by $[a + b + c + d + 32.06e]$. Thus the grams $H_2O$/gram substrate use is:

$$\frac{g\ H_2O \text{ consumed}}{g \text{ substrate consumed}} = \frac{18[a - 0.25b - 0.5c + 0.75d + 0.5e]}{[12a + b + 16c + 14d + 32.06e]} \tag{7.148}$$

### Carbon Dioxide, Methane and Hydrogen Sulfide from Anaerobic Decomposition

According to the stoichiometric relation for the anaerobic process, the moles of carbon dioxide produced are given by:

$$\text{moles}\quad CO_2 = [0.5a - 0.125b - 0.25c + 0.375d + 0.25e]$$

The grams $CO_2$ produced per gram substrate use are thus, following the previous procedure:

$$\frac{g\ CO_2}{g \text{ substrate}} = \frac{44[0.5a - 0.125b - 0.25c + 0.375d + 0.25e]}{[12a + b + 16c + 14d + 32.06e]} \tag{7.149}$$

### Methane Production from Stoichiometric Anaerobic Decomposition

Methane production on a stoichiometric basis is provided by the molar quantity:

$$\text{moles}\quad CH_4 = [0.5a + 0.125b - 0.25e - 0.375d - 0.25e]$$

The grams $CH_4$/gram substrate are thus given, following the previous procedure, by:

$$\frac{g\ CH_4}{g \text{ substrate}} = \frac{16.032[0.5a + 0.125b - 0.25c - 0.375d + 0.25e]}{[12a + b + 16c + 14d + 32.06e]} \tag{7.150}$$

### Hydrogen Sulfide Production

Hydrogen sulfide ($H_2S$) is often of some interest to the management of landfills and the aerobic management of composts as $H_2S$ is highly odorous, has health impacts if present at significant or high concentrations in air and, in the case of composting operations, signals an undesirable anaerobic condition at the bottom of the compost pile. From the general anaerobic decomposition relation given at the beginning of this section, the molar quantity of $H_2S$ produced is given by $[e]$, the molar fraction of sulfur (S) in the substrate (electron donor). Assuming this is the major or only sulfur source involved in the decomposition reaction, the grams of $H_2S$ produced is provided by:

$$\frac{g\ H_2S}{g \text{ substrate}} = \frac{34.076[e]}{[12a + b + 16c + 14d + 32.06e]} \tag{7.151}$$

## Stoichiometric Heat Production During the Anaerobic Decomposition Reaction

Heat released to the surroundings during anaerobic decomposition is generally ignored, mainly because it is significantly less than for aerobic decomposition. Although this approach is eminently practical for a waste site as a reactor, small differences in heat gradients throughout the reactor might profoundly affect overall biodegradation capacity of the system. In addition, one aspect of modeling a physical system in the environment is to incorporate temperature considerations; thus, all temperature contributions should be of interest for definition of heat balance of the system modeled.

Luong and Volesky (1983) note that whether a system is aerobic or anaerobic, the growth of organisms is accompanied by heat production. Some of the heat is conserved as energy incorporated into the biomass and the rest is released as waste heat. Assuming that heat energy wasted is more likely to affect the whole system through heat diffusion, a system heat balance must account for the fraction of the generated heat not used in biomass synthesis.

Luong and Volesky (1983) state that for an ethanol biosynthesis process, the rate of heat evolution (in the reactor) is directly proportional to the rate of substrate consumption:

$$Q_f = K_s \left( \frac{-dS}{dt} \right) \tag{7.152}$$

and that additionally, this can be related to the yield of organisms by substitution for the change in biomass concentration:

$$X = Y_{X/S}(-dS)$$

$$X = Y_{X/S} \left( \frac{1}{K_s(\Delta Q)} \right) \tag{7.153}$$

In the above expression, Luong and Volesky (1983) used term '$X$' for $dX$, the change in biomass and further noted that the value of $K_s$ could be determined and used either to estimate heat production or the concentration of product or organisms in the reactor. As ethanol biosynthesis is an anaerobic process, heat production, this provides one of the few anaerobic heat production discussions.

Notably enthalpic heat production during anaerobic decomposition is the net sum of heat of formation of the reactants. According to the first law of thermodynamics:

> During a reversible, constant pressure process, the heat supplied is equal to the increase of enthalpy

which can be interpreted as:

$$\frac{\Delta h}{\Delta T} = \frac{Q}{\Delta T} \tag{7.154}$$

In the case of an anaerobic reactor with heat generated from the stoichiometric reaction, the term $Q$ would be the heat generated, $\Delta T$ would be the change in mean temperature and $\Delta h$ would be the net enthalpic heat increase. Considering the general anaerobic stoichiometric reaction (Eq. 7.147), it would thus be useful to provide values of enthalpic heat of formation for the products and reactants involved. These are listed for convenience in Table 7.6 below.

The right-hand column indicates the signs of the values to be used and that heats of combustion should be multiplied by the number of moles from the stoichiometric relationship. For example, the enthalpic heat increase for anaerobic decomposition of a kg of substrate of empirical composition $C_aH_bO_cN_dS_e$ can be defined using the above as:

$$\begin{aligned} \Delta H = {} & (\Delta H(C_aH_bO_cN_dS_e)) + [a - 0.25c - 0.05c + 0.75d + 0.5e](-242, 830) \\ & - [0.5a - 0.125c - 0.25c + 0.375d + 0.25e](-46, 213) \\ & - [0.5a + 0.125c - 0.25c - 0.375d - 0.25e](-74, 870) \\ & - e(-20, 177) \end{aligned} \tag{7.155}$$

## VALUES OF DECOMPOSITION KINETIC CONSTANTS

Kinetic factors for MSW are typically derived from laboratory and field tests and from digester studies. Both anaerobic and aerobic values are of interest because anaerobic/aerobic cycling simulation with air flows could require that aerobic kinetic factors also be available, not only anaerobic.

**TABLE 7.6**
**Enthalpic Heat of Formation of Substances in Anaerobic Decomposition**

| Anaerobic Reaction | Component Type | Enthalpy of Formation kJ/kg mole | Multiplier |
|---|---|---|---|
| $C_aH_bO_cN_dS_e$ | Reactant (substrate) | From heat of combustion | + times moles |
| $H_2O$ | Reactant | –242,830 | + times moles |
| $NH_3$ | Product | –46,213 | – times moles |
| $CO_2$ | Product | –393,520 | – times moles |
| $CH_4$ | Product | –74,870 | – times moles |
| $H_2S$ | Product | –20,177 | – times moles |

The variables in the relationships are as described:

$t_c$ = thickness of surface tension film = $0.01\mu_m$ (Ferguson, 1993).
$k_r$ = initial surface reaction rate(= $k_H$).
$A_{h0}$ = initial area under hydrolysis attack = $N_0 \times a_h$
$X_{h0}$ = initial concentration or mass of hydrolytic organisms
$N_{h0}$ = initial number of hydrolytic organisms
$a_h$ = unit area of hydrolytic organism = $4\mu m$ □
$v_h$ = unit volume of hydrolytic organism
$\rho_h$ = unit density of hydrolytic organism
$U$ = mobile water film velocity, assumed to be refuse water velocity rates
$S, S_0$ = current and original mass of MSW fraction
$L$ = depth of refuse analyzed
$Q$ = leachate flow through refuse volume analyzed
$S(i), S_T, S_{\mathrm{TOTAL}}$ = unit and total refuse mass, in kg
$P_H, P_{H0}$ = current and original concentrations of hydrolysis products in the mobile film water flow (leachate)
$P_a$ = concentration of acetic acid in reactor (initial, leachate)
$X_H$ = concentration of hydrolytic organisms
$X_a$ = concentration of acetogens (initial)
$X_m$ = concentration of methanogens
$Y_h$ = hydrolytic organism yield on MSW fraction
$Y_a$ = yield of acetogens from hydrolysis products
$Y_{ac}$ = yield of acetic acid product/unit acetogenic mass, units = mass/mass
$Y_m$ = yield of methanogens from acetic acid
$Y_{CH_4}$ = yield of methane
$Y_{CH_4/CO_2}$ = yield of methane per unit of $CO_2$ converted by hydrogen–$CO_2$ utilizing organisms
$K_{sH}$ = □-saturation constant for acetogens in hydrolysis products
$K_{sa}$ = □-saturation constant for acetogens in acetic acid
$K_{sm}$ = 1/2-saturation constant for methanogens
$k_H$ = hydrolysis rate for MSW fraction, as reported in literature
$\mu_{mH}$ = maximum growth rate for hydrolyzers
$\mu_{ma}$ = maximum growth rate for acetogens
$\mu_{mm}$ = maximum growth rate for methanogens
$k_{dH}$ = death or decay (loss) rate for the hydrolyzer organisms
$k_{da}$ = death or decay (loss) rate for the acetogenic organisms
$k_{dM}$ = death or decay (loss) rate for the methanogenic organisms

**TABLE 7.7**
**Values of Anaerobic and Aerobic Kinetic Constants**

| Kinetic Factor | Organism | Anaerobic | Remarks | Reference |
|---|---|---|---|---|
| $Y_{PH}$ | Cellulose hydrolyzer | 47.4 gm sugar/gm cellulose | Lab. digester column, 35 C, Paper, cotton wastes | Lee and Donaldson, 1984 |
| $Y_{Pa}$ | Acetogen | 16.5 g acetic acid/g sugar | Lab. digester column, 35 C, Paper, cotton wastes | Lee and Donaldson, 1984 |
| $Y_{Pm}$ | Methanogen | 8.4 g methane/g cells | | Lee and Donaldson, 1984 |
| $Y_{PCO2(cellulose)}$ | Hydrolyzer | 1.93 gCO2/g hydrolyzer | | Lee and Donaldson, 1984 |
| $Y_{PCO2(sugar)}$ | Acetogens | 1.93 gCO2/g acetogens | | Lee and Donaldson, 1984 |
| $Y_{PCO2(acetic\ acid)}$ | Methanogen | 12.5 g product/g cells | Shredded paper cotton wastes | Lee and Donaldson, 1984 |
| $Y_H$ | Hydrolyzers | 0.837 g cells/g substrate | Shredded paper cotton wastes | Lee and Donaldson, 1984 |
| $Y_a$ | Acetogens | 0.753 g cells/g substrate | Shredded paper cotton wastes | Lee and Donaldson, 1984 |
| $Y_m$ | Methanogens | 0.753 g cells/g substrate | Shredded paper cotton wastes | Lee and Donaldson, 1984 |
| $K_{sc}$ | Cellulose | 37 kg cellulose/cu m | Shredded paper cotton wastes | Lee and Donaldson, 1984 |
| $K_{ss}$ | Sugars | 0.24 kg sugar/cu m | Shredded paper cotton wastes | Lee and Donaldson, 1984 |
| $K_{ac}$ | Acetic acid | 4.2 kg acetic acid/cu m | Shredded paper cotton wastes | Lee and Donaldson, 1984 |
| $Y_a$ | Acidogens | 0.14 | 55 kg refuse columns | Lee et al., 1993 |
| $Y_m$ | Methanogens | 0.039 | 55 kg refuse columns | Lee et al., 1993 |
| $Y_{CH4}$ | Methanogens | 0.65 | 55 kg refuse columns | Lee et al., 1993 |
| $Y_{CH4/CO2}$ | Hydrogen utilizers | 0.65 | 55 kg refuse columns | Lee et al., 1993 |

| | | | | |
|---|---|---|---|---|
| $K_a$ | 1/2 Vel. const., acidogenesis | 4.5 kg/cu m flow | 55 kg refuse columns | Lee et al., 1993 |
| $K_m$ | 1/2 Vel. const., methanogenesis | 10 kg/cu m flow | 55 kg refuse columns | Lee et al., 1993 |
| $\mu_{m,hydrolyzer}$ | Max. growth rate | 1.7/day | Lab. shredded cellulose, cotton waste column | Lee and Donaldson, 1984 |
| $\mu_{m,acetogen}$ | Max. growth rate | 0.325/day | | Lee and Donaldson, 1984 |
| $\mu_{m,methanogen}$ | Max. growth rate | 0.5/day | | Lee and Donaldson, 1984 |
| $\mu_{ma}$ | Max. growth rate, acidogen | 0.036/day | One-year MSW columns | Gonullu, 1994 (Bastuk, 1980) |
| $\mu_{mm}$ | Max. growth rate, methanogen | | | |
| $k_{da}$ | Specific death rate, acidogen | 0.10/day | One-year MSW columns | Lee et al., 1993 |
| $k_{dm}$ | Specific death rate, methanogen | 0.022/day | One-year MSW columns | Lee et al., 1993 |
| $k_{dh}$ | Hydrolyzer | 0.04 | Lab, shredded cellulosics | Lee and Donaldson, 1984 |
| $k_{da}$ | Acetogen | 0.04 | Lab, shredded cellulosics | Lee and Donaldson, 1984 |
| $k_{dm}$ | Methanogen | 0.04 | Lab, shredded cellulosics | Lee and Donaldson, 1984 |
| $P_{H(0)}$ | Hydrolysis product concentration | 28 kg/cu m flow | Lab, shredded cellulosics | Lee and Donaldson, 1984 |
| $P_{a(0)}$ | Acetic acid concentration | 1.0 kg/cu m flow | Lab, shredded cellulosics | Lee and Donaldson, 1984 |

$Q(j)$ = flow rate into $j$th reactor

$R(j)$ = leachate recycle rate into $j$th reactor

$V(j)$ = volume of reactor affected (this differs from volume of refuse. If only a portion of the landfill is wet enough to support decomposition, $V(j)$ = wetted portion of landfill reactor)

$j - 1$ = series reactor above

The values of other constants have been defined elsewhere in this section. A tabulation of some values of kinetic factors collected from literature is shown below. Hydrolysis rates are discussed separately.

# 8 Decomposition Issues

## INTRODUCTION

The discussions of Chapters 1 through 7 partially fulfill the modeling concerns that must be addressed with the simulation of waste sites, but they do not cover all of the important issues. The important issues with representation of a waste site as a granular media system are the role of surface area in conversion and mass transfer of chemicals and heat transfer, the importance of settlement in long-term waste site management, hydraulic flow and distribution of moisture and system permeability to gas flow. Initialization of the system is equally important. These topics and pertinent literature will be briefly discussed in practical terms in this section.

## WASTE SITE MODELS OF PREVIOUS WASTE SITE STUDIES

Changes in the standards for the operation of landfills over the past several decades toward greater waste isolation from the environment and change in the composition of solid waste toward greater biodegradable content pose difficulties for any short-term relief from the types of potential water and gas impacts posed by the existence of landfills in communities. Senior (1990) notes that increase in use and disposal of biodegradable packaging results in landfilling of refuse of high carbon:nitrogen (C:N) ratio, which could limit the rates of fermentation (in place) to levels of nitrogen fixation possible or require the supplementation of nitrogen. The idea of leachate recycle increases the nitrogen available in the landfill soil and added enhancement of rates has been noted when the recyclate contained sewage sludge.

The landfill, with its natural digestion of refuse resulting in generation of combustible gases, has been regarded as an anaerobic digester (Senior, 1990), with analogies attempted for study of landfills as fermenters (Rees, 1980; Rees and Grainger, 1982). Landfills as fermenters are considered complex ecosystems. Though well provided with electron donor and acceptor species compared with ordinary anaerobic waste digesters such as for sewage sludge treatment, the problems of heterogeneity of refuse, refuse location, stratification of refuse by age and organisms that are interdependent while not necessarily mixed in location create difficulties for analysis (Senior, 1990).

Williams, Pohland, McGowan and Saunders (1987) modeled leachate generation from a landfill using a mechanistic model accounting for microbially mediated landfill stabilization. The microbially mediated processes simulated were hydrolysis, acid formation and methane fermentation. The authors concluded that application to two sets of experimental data reasonably predicted gas and volatile acid generation. Limitations notably associated with the study included lack of substrate specificity, lack of a comprehensive database on landfill chemistry and uncertainty of flow distribution. McGowan et al. (1993) noted, in a subsequent paper based upon the same study, that:

> *Predictive capability of the model would have been "strengthened by more explicit data concerning substrate composition, flow distribution and circuiting and changes in microbial dynamics as substrate conversion proceeds."*

Detailed discussion of this model is of value because of its similarity to the work attempted and the recommendations made.

The authors selected a partially saturated hydraulic flow model that simulated flow through homogenous, isotropic, porous media. Important hydraulic flow relations for the flow rate and vertical flow velocity were:

$$q = k_{\text{ave}}\, i A$$

flow rate $q$, ave. hydraulic conductivity of waste layer $\text{k}_{\text{ave}}$, hydraulic gradient $i$, landfill cross-section area $A$

$$q_s = k_{\text{waste}}\, i A'$$

$A' = n_e A =$ initial cross-sectional area of the flow path; = flow through waste before reaching field capacity; $\text{k}_{\text{waste}}$ = waste hydraulic conductivity, $n_e$ = initial porosity of the waste

$$v_z = -k(\theta)(\delta\varphi/\delta z)$$

$v_z$ = vertical flow velocity, $k(\theta)$ = partially saturated hydraulic conductivity, $\varphi$ = total or hydraulic head (= pressure head or matric potential $\eta$ + elevation head (= $z$) + velocity head (neglected)), $\delta z$ = vertical distance considered.

$$\delta\theta/\delta t = -(\delta/\delta z(v_z)) = (\delta/\delta z(k(\theta).(\delta\theta/\delta z)))$$

substituting, which is the equation for partially saturated flow. Considering the volumetric water content and thus the partially saturated conductivity to depend upon depth, the Richards relation was applied:

$$(\delta\theta/\delta\omega).(\delta\theta/\delta t) = \delta/\delta z(k(\theta)\delta\theta/\delta z) - \delta k(\theta)/\delta z$$

where $(\delta\theta/\delta\omega) = C(\omega)$ = specific moisture capacity. The specific moisture capacity term $C(\omega)$ was then related to the Brooks and Corey (1984) relation for water content at which drainage occurred in soils, where $\theta_r$ = residual water content of the material, $n$ = porosity, $\omega_b$ = air entry pressure and $\lambda$ = pore size distribution coefficient. The degree of saturation for the bulk landfilled material with moisture below 92%, based on the Clapp and Hornberger (1978) relation for wetting of soil and for wetness = 92% to 100%:

$$\varphi = -m.(W - n).(W - 1)$$

$$m = \frac{\varphi_i}{(1 - W_i)^2} = \frac{b\varphi_i}{W_i.(1 - W_i)}$$

$$n = porosity = 2W_i - \frac{b\varphi_i}{W_i(1 - W_i)}$$

where $\theta_s$ = saturation water content, volume; $\omega_s$ = saturated pressure head; $b$ = empirical constant; $m$ and $n$ are constants as described; and $\omega_i$ = the pressure head at $W = 92\%$ saturation.

For the biological module of the model, the authors chose the Monod kinetic relationships for hydrolysis, acidogenesis and methanogenesis. It is worth noting that use of Monod kinetics for simulating the rate of bulk solubilization of the MSW to acids entering leachate and rates of consumption of acids and generation of methane under anaerobic conditions have been shown to be effective in the past for simulation purposes, even if not the most accurate approach.

Straub and Lynch (1982) used a first-order kinetic reaction model to simulate evolution of solubilized organics into landfill leachate as chemical oxygen demand (COD). This allowed time-varying leachate COD prediction using an average bulk solubilization rate and a maximum or saturation leachate COD value:

$$R_g = (S^m/S_0).b.(C_{\max} - C)$$

where $R_g$ = change in leachate concentration of COD with time or $(C_{i+1} - C_i)$, $C$ = COD concentration in landfill leachate, $S$ = mass of solid COD in the landfilled waste, $m$ = a derived constant = 2, $S_0$ = original or ultimate COD in solid landfilled material (12% of dry weight), $b$ = constant for rate of COD production from the waste and $C_{\max}$ = assumed maximum COD concentration in landfill leachate (40,000–50,000 mg/l). The model also simulated unsaturated moisture flow in a vertically cascaded reactor to address waste layering, as in a corresponding study of inorganic leachate strength (Straub and Lynch, 1982).

It was concluded from this analysis that convective removal of leachate COD and anaerobic activity dominate leachate organic strength and landfill stabilization and that:

> *Landfill aerobic activity is relatively insignificant unless air is applied; that landfill anaerobic activity is inhibited from ideal (laboratory) rates; that with the use of a Monod model anaerobic activity rates are inhibited by a factor of about 50%; that the model developed is useful for estimating landfill gas generation; and that retention of microorganism cells during leachate recirculation could be the main reason for the enhanced organic stabilization experienced with leachate recirculation.*

Lee et al. (1993) simulated and studied pilot-scale lysimeter production of methane from a landfill. The LEAGA-1 model was thus developed, combining a module for simulation of unsaturated landfill moisture flow with a solid waste biological decomposition module. The biological decomposition module involved three successive steps — refuse hydrolysis, acid formation and methane formation. Observed values from the laboratory scale landfill reactor were compared with simulation. It was concluded that *the hydrolysis rate constant played the most important role in the prediction of gas production and use of variable rather than constant flow rates improved gas prediction*. Important contributions of this study to the landfill study attempted include its use and modification of the refuse decomposition kinetic model first developed by Halvadakis (1983). The decomposition module used in the present simulation study is also based upon the Halvadakis (1983) refuse kinetics model, with the Lee et al. (1993) and other appropriate modifications.

The finding of the importance of the hydrolysis constant suggests predictive capacity of simulation models could be improved through use of *more specific refuse hydrolysis rates*.

Al-Yousif, Pohland and Vasuki (1992) developed the PITTLEACH landfill simulation model to enable design of landfill leachate circulation systems based upon hydraulic flow characteristics. Darcy's law and the well-known Richards equation for stimulating unsaturated one-dimensional flow in a soil column were used, among other important moisture flow relationships. The model was based upon the probabilistic hydraulics concept of optimizing the entropic tendency of unsaturated hydraulic conductivity to depend upon moisture content. The model is said to improve upon the errors associated with using best fit of hydraulic models to adjust them to experimental landfill data, such as related to effect of size distributions of buried materials on hydraulic flow. The validity of the mathematical model was tested against other well-known hydraulic expressions and found to provide greater accuracy for potential leachate recirculation designs. The need for field and performance data was noted. The importance of this model is that the importance of overcoming the effects of soil particle size distribution on hydraulic flow was recognized in attempting to reduce the statistical errors associated with hydraulic models.

One of the most widely disseminated current landfill moisture flow models is the Hydrologic Evaluation Leaching Procedure (HELP) computer model, with development sponsored by the EPA (Schroeder et al.; EPA HELP Model, 1994). Version 3, developed in 1994, replaced version 2, developed in 1988, incorporating improvements in user computer input ease, use of metric units, cover, liner and leachate collection designs using geosynthetic materials, modifications of default soil information database and water balancing calculation procedures and capacity for international use. The HELP model, aimed at landfill designers and permit writers, is stated to be a "quasi-two-dimensional hydrologic model for conducting water balance analyses of landfills, cover systems and other solid waste containment facilities." It has the capacity to use climate and precipitation data for various U.S. cities or locations; and by using a computerized finite differencing procedure, it can rapidly predict moisture inflow past landfill covers, through soil-waste layer depths and through or above liners, according to user specifications. Prediction of leachate outflow is thus made possible. The EPA HELP model (Version 3, 1994) has been used primarily in the current study to predict moisture flow past several types of final or ordinary landfill covers, including daily cover only, expected to exist at inactive MSW landfills. The relative importance of this model is discussed in other sections of this study.

Baetz and Byer (1988) note that, though the EPA HELP model (Version I) is capable of simulating moisture flow through soil covers as well as placed waste materials and can estimate the moisture occurring in landfill soil cover, it cannot assess the effectiveness of landfill operation variables on moisture controls such as during the filling period of a landfill; and that stochastic rather than average precipitation inputs should be considered. Findings of the study, from a computer simulation of a Toronto area landfill with a soil and clay cover and with climatic conditions, infiltration, evaporation, snowmelt and runoff specified, and validation by results from a similar landfill study, with records of leachate collection and treatment over two years, included that:

- Percolation estimates should use stochastic rather than average precipitation values.
- Landfill waste layering patterns significantly influence percolation.
- Landfill steepness reduces precipitation exposure time.
- Neglect of snow buildup in collection design can increase landfill internal moisture buildup.
- As cover thickness is increased, percent moisture percolation decreases, depending upon cover material, thickness and frequency of wetting (wetter conditions and thicker covers reduce percolation).

For the present study, a moisture input model sensitive to stochastic precipitation events rather than to average precipitation values could possibly increase exactness. Where long records of environmental data are available, they should improve accuracy of moisture throughflow predictions. Both Versions 2 and 3 include 5 years of default climate and precipitation data, from a data period estimated to be representative of long-time climate and precipitation trends. The option for adding to this default database is now possible, although it was assumed for this study that use of the default 5-year data, for the locations selected for analysis, posed no great error. As the major purpose of this study is to assess the biological impacts of various water inputs and to draw conclusion thereupon, use of the HELP model, with actual but average rainfall records, should be sufficient for prediction once reasonably correct initial assumptions about moisture conditions of waste and covers can be made and data improvements in water can also be made at a later date.

Noble and Arnold (1991) conducted a pilot scale landfill experiment and developed a one-dimensional finite difference landfill moisture transport model called FULFILL for simulating landfill stabilization and groundwater contamination. Darcy's law and the Richards equation for unsaturated one-dimensional flow in a soil column were also used to simulate flow. The issue of capillary rise (the adsorption and uptake and movement of water into the wastes and soil), neglected by most landfill moisture flow modeling, was also considered by this study. Differences in mathematical approaches between the FULFILL and EPA HELP models and modeling of the effect of capillarity to show moisture variation vs. landfill depth were also described in a paper by Noble and Nair (1990). The capillary rise issue is more important to the topic of predicting landfill decomposition than to prediction of leachate flow because of the critical importance of actual moisture content and availability to microbial processes. For this reason, this topic and related literature and theory are explored in greater detail in the model development section (Chapter 5) of the present work.

Knox and Gronow (1990) state that, biologically, landfills may be considered fixed film reactors. Baccini et al. (1987) state that an MSW landfill can be defined as a chemical and biological fixed bed reactor, with MSW and water as inputs and gas, leachate and residual solids as outputs. The landfill model developed by Baccini et al. (1987) introduced the concept of mean residence time of all wastes to address the fact that landfilled wastes are usually deposited over time:

$$\theta = \frac{\sum(t_i \cdot m_i)}{\sum(m_i)} \tag{8.1}$$

Over the period $t_i$ (= months or years), 0 to mean residence time and the chemical and biochemical properties of MSW incoming — where not assumed to change with time and annual waste inputs — were constant. The mean residence time concept allowed each landfill site analyzed to be characterized according to its evolution, using recorded disposal information. Using this concept, the specific area charge in mass and fill areas or specific annual flux of chemicals became functions of the mean residence time, allowing prediction. The specific area charge $z$ or annual mass input of waste into a landfill reaches a maximum before closure, then declines thereafter due to decomposition. Water content of MSW ranged from 17 to 30% by weight. The water retention capacity of the MSW, found by examining drilled cores at a Swiss landfill, was about 30% of MSW weight. The water production by biochemical reactions was estimated to be less than 30 grams/kg/year and water loss through gas flux was estimated to be under 1.5 gm/kg/year. A water balance equation was then developed and verified using 14-day sampling over a year:

$$\begin{aligned}&\text{Input (precipitation + MSW water) + Biologically produced water}\\&\quad - \text{(loss through gas, evaporation, percolation/runoff and leachate)}\\&\quad - \text{(water retention by waste)} =\end{aligned}$$

The authors noted that leachate/unit MSW mass increases as the landfill fills up, but declines to a constant value for each site once filling stops on the order of 0.4 for the leachate/precipitation ratio for the four landfills studied, with mean residence times, $\Theta$, of 5 to 10 years.

Rees (1980) studied the temperature changes in an anaerobic MSW landfill, defined as due to heat of (biochemical) reaction, specific heat of water/refuse mixtures, heat losses to air and soil and heat gains due to flow radiation aerobic metabolism. Rees (1980) concluded that even in temperate climates anaerobic landfill temperatures might reach 45°C. It was stated that landfill heat production depends upon the dry density of the MSW as deposited, water content of the MSW, specific surface of the MSW as related to particle size and landfill temperature. Heat of fermentation reaction for complex carbohydrates was characterized based on previous work by Pirt (1978) as:

$$C_6H_{12}O_6 \rightarrow CH_4 + CO_2 + \text{dry biomass} + \text{heat}$$
$$\text{1kg} \qquad 0.25\text{ kg}\ 0.69\text{ kg} \quad 0.056\text{ kg} \qquad 632\text{ kJ}$$

Gas production was shown to be linearly proportional to refuse moisture. It was stated that a rate of gas generation (65% methane, 35% $CO_2$) of 120 m/ton/year was equivalent to 172 kg of glucose decomposed to give 109,000 kJ of heat energy. In terms of the heat required to raise landfill temperatures, it stated that the drier the MSW, the longer it would take for the temperature to reach the 40°C optimum; and that for anaerobic conditions, MSW water content below 34% would make reaching this temperature difficult. Neutralization of products such as maximum ($NH_4$+), fatty acids and acetic acids could be expected to yield about 12,900 kJ/m$^2$ of MSW, based upon an assumed $NH_4$+ concentration of 16,000 gm/m$^2$ and an acetic acid concentration of 10g/dm$^2$. For heat loss to air and soil a thermal conductivity equivalent to sawdust was assumed, i.e., $58.1 \times 10^{-5}$ Joules °C/sec.cm$^2$. This allows the temperature quotient across a thickness of refuse to be assumed. For a °C gradient across 1 m of MSW, heat loss is of the order of 60 MJ/year m$^3$, reduced to 12 MJ/year for refuse 5 meters thick. Heat loss through clay is calculated at 79 MJ/year.m$^2$.

The Rees (1980) study noted the importance of aerobic metabolism in the landfill for raising temperatures and starting up in landfill reactions in winter, with the advantage of avoidance of high inhibitory concentrations of fatty acids in the landfilled wastes.

## LANDFILL SOIL SAMPLING STUDIES

Landfill soil sampling studies are very important to the initializing of wastes content, moisture and heat and microorganism content parameters. Some studies of importance are discussed below.

### ORGANICS VS. LANDFILL DEPTH

Core sampling of soils in a Spanish landfill by Gonzalez-Vila et al. (1995) showed that soils containing wastes tended to accumulate total organic carbon, carbohydrates, volatile fatty acids and humus, with levels increasing with depth and associated with age of burial and rates during the first years of burial. This *unexpected* trend was stated to be related to bacterial input. The authors also noted that the most relevant xenobiotic chemicals detected were dialkyl phthalates and traces of PCBs. Attal et al. (1992) studied the slightly different but important issue of degree of decomposition of landfilled refuse vs. age or depth. Depth profile sampling and measurement of methane content, volatile suspended solids (VSS) and methane potential was conducted experimentally.

*Important findings of the study were that biological activity type varied with depth (as could be expected), with mesophilic hydrolysis and acidification dominating near the landfill surface and thermophilic methanogenic activity deeper inside the landfill. It was also shown that VSS content of the refuse related to depth and could be used to estimate the degree of decomposition of the MSW.*

The study data include statistical information that could be used to relate waste age to degree of decomposition expected, though mathematical models of waste volatile solids content (VSS) vs. age could hardly be expected to be universally applicable.

## LANDFILL SOIL MICROORGANISM STUDIES

The levels of biological organisms in or on landfill soils depend upon the quantity and energy accessibility of the waste, climatic location of the landfill and soil distribution of the waste material. Few studies have directly sought to characterize landfill soils for organisms, with most such studies focused upon whether a particular environmental or health impact should be present depending upon the presence or absence of certain organisms. The studies discussed below represent some of the more important of those discussed under *Waste Site Ecology*.

Fedorak and Rogers (1991) have noted that microbial content of and dissemination from landfills is under-researched, compared with sewage treatment studies, though both are disposal methods. This study, which focused upon aerosols produced during waste handling, noted that paths of microbial dissemination from landfills are through groundwater movement, aerosols with microbial contamination and through higher life forms that pick up and remove microbes from the landfill environment. Prime introduction sources of organisms include disposable diapers and pet feces (Golueke, 1977); but Fedorak and Rogers (1991) state:

*MSW is relatively high in microbial content; in fact, Pahren (1987) found that coliform levels in MSW were similar to those found in sewage sludge or hospital wastes, with coliform concentrations at* 7.7 × 108, 2.8 × 109 *and* 9.0 × 108, *respectively.*

Pahren's (1987) study, according to Fedorak and Rogers (1991), showed major contributors to MSW coliforms were diapers (35.6%), paper (13.5%), plastic, rubber and leather (10.6%), food waste (8.4%), fines (42.6%), metal (10.5%) and textiles (6%), among other sources. Pahren noted that environmental temperature elevation and leachate properties are likely to reduce survival of pathogens such as coliforms, as shown by the fact that:

*Microorganism numbers in landfill leachates are "quite low," from* $1.5 \times 10^4$*/ml in new leachate to 0.3 to 1.8/ml in older leachate.*

Fedorak and Rogers (1991) also discussed the Donnelly and Scarpino (1984) study of microbial content of active and inactive landfills and solid wastes. These investigators had shown that leachates from these landfills contained 10 to 1000 fecal coliforms/gm or ml and also contained various other microbial pathogens *Clostridium, Salmonella, Klebsiella, Listeria, Proteus, Providencia, Mycobacterium, Moraxella, Acinetobacter* and *Streptococcus faeces* spp. *Clostridium* species dominated in

3-year-old (MSW) lysimeters at densities 1000 to 100,000/gm (leachates at 10/ml) and the authors stated the clostridia spores were more likely to survive landfill environments than the bacterial cells. Generally the landfilled wastes contained much higher organism counts than leachates.

MSW (in the waste lysimeters) were found to contain aerobic microorganisms around $4.3 \times 10^9$ colony forming units (cfu)/gm; and present were fecal coliforms at $4.7 \times 10^8$/gm, fecal streptococci at $2.5 \times 10^9$/gm and fungi at $2.5 \times 10^9$/gm.

The levels present in wastes were also likely to be those available to environmental dissemination by aerosols or by vectors such as gulls.

Palmisano, Maruscik and Schwab (1992) recently studied levels of fermentative bacteria in inactive landfill sites and in excavated refuse. This study is of particular interest to decomposition studies and modeling because it is one of the few providing counts and mass levels of particular decomposition bacteria classes from inside large landfills. The study was conducted at the large, well-known Fresh Kills, Staten Island, New York landfill. The authors state that:

> *Although fermentative bacteria were present at the elevated levels of* $10^5$ *to* $10^8$ *cfu/gm dry weight of refuse, only a small percentage (0 to 15%) were hydrolytic (cellulosics-solubilizing) organisms. Fat-solubilizing bacteria were 0.2 to 14% of fermentative bacteria at ambient temperature cultures. Aerobic bacteria were found in refuse sample cultures at* $10^4$ *to* $10^7$*/gm dry weight, in contrast with anaerobic bacteria at* $10^5$ *to* $10^8$ *cfu/gm dry weight. Direct bacterial counts showed total refuse bacteria levels around* $10^{10}$ *cfu/gm dry wt.*

The authors also reported that the work of Filip and Kuster (1979) with MSW incubated in a model landfill for 20 months showed microbe levels of $10^5$ to $10^6$ fungi, $10^6$ actinomycetes and $10^6$ to $10^8$ aerobic proteolytic bacteria.

> *One of the important conclusions of this study was that the more likely reason for a relatively small bacterial mass of hydrolytic organisms counted in the actual landfill (under 15% of all) was due to the use of an inappropriate growing medium for culturing cellulolytic organisms.*

While it is possible that the landfill interior — especially under anaerobic conditions — might not harbor great numbers of paper-solubilizing (hydrolytic) organisms, special care has to be taken that numbers used for hydrolytic, acidogenic and methanogenic organisms are those likely to be found under field and buried waste conditions.

Study of microbial content of refuse under aerobic, anaerobic and aged landfill conditions by Filip and Smed-Hildman (1988) showed that microbe numbers were at the following levels:

## TABLE 8.1
## Microorganism Levels in Refuse

| Microorganisms | Refuse Sample | | |
|---|---|---|---|
| | Rotted | Compacted | Old |
| Aerobic proteolytic bacteria | $1.8 \times 10^8$ | $1.2 \times 10^8$ | $4.2 \times 10^6$ |
| Anaerobic proteolytic bacteria | $2.0 \times 10^9$ | $2.5 \times 10^{10}$ | $4.4 \times 10^6$ |
| Actinomycetes | $7.3 \times 10^6$ | $2.6 \times 10^6$ | $1.9 \times 10^6$ |
| Fungi | $5.8 \times 10^6$ | $1.5 \times 10^6$ | $1.2 \times 10^5$ |

Source: Filip and Smed-Hildman, 1988

Filip and Smed-Hildmann (1988) also reported that compacted refuse had higher anaerobic bacterial counts, with other organism counts declining with refuse or landfill age.

Bogner, Miller and Spokas (1995) also recently studied MSW landfill anaerobic and aerobic biomass content in sampling cores taken at two Illinois landfills — one 17 hectares and inactive for 15 years at sampling and capped with compacted clay, and the other recently capped. The study showed stratification of biomass content with depth, through the cover downward, with highest concentrations in the top soil layers and in the mixed refuse soil layers and lowest concentrations in the compacted clay layers. The study noted natural soils favor both aerobes and facultative anaerobes, with dominance depending upon seasonal soil moisture; and measurable oxygen, at sub-atmospheric concentrations, was always present at the top of the refuse soil sequence under the cover. Some results are indicated below.

Conclusions of the Bogner (1995) landfill soil biomass study were that:

*Overall aerobic activity and anaerobic activity are codependent; i.e., as aerobic activity or aerobic biomass increases, anaerobic biomass also increases.*

The general implication is that in a functioning landfill, i.e., one with aerobic zones overlying anaerobic zones, increased upper zone aerobic activity should increase anaerobic activity — which suggests maintaining the aerobic zone might be enough to increase anaerobic activity.

*The percentage water vs. mg biomass, in g DOC/g dry soil plot, showed a stronger influence of water content than of soil volatile solids content. A quantitative relationship appears to exist between water content and biomass content, independent of soil volatile solids (available organics) content.*

In a previous study of samples taken from landfills and tested for biodegradability, Bogner (1990) had also showed biodegradability depends upon moisture and nutrients in media. The moisture dependence of the Bogner et al. (1995) microbial mass study upon soil water content is worth investigating, in the context of the

**TABLE 8.2**
**Landfill Microorganisms vs. Refuse Depth and Age**

| Landfill Layer | Depth from Surface, cms | Microbial Biomass, Dissolved Organic Carbon, mg/g dry soil | Specific Activity Ratio = measure of active biomass = activity/biomass | | % Water | % Vol. Solids | Organic Carbon | Kj. Nitrogen, ppm |
|---|---|---|---|---|---|---|---|---|
| | | | Aerobic | Anaerobic | NA | NA | NA | NA |
| Replaced topsoil | 10 | 299.8 | 1.02 | 0.44 | 24.27 | 7.65 | 3.53 | 1710 |
| Mixed topsoil | 25 | 160.0 | 1.09 | 0.44 | 20.19 | 5.57 | 2.90 | 1034 |
| Compacted clay cover | 50 | 62.00 | 0.56 | 0.10 | 13.05 | 3.16 | 2.22 | 844 |
| Compacted clay cover | 100 | 53.76 | 0.91 | 0.09 | 11.65 | 2.79 | 1.55 | 654 |
| Compacted clay cover | 150 | 37.39 | 0.72 | 0.40 | 12.45 | 2.36 | 1.71 | 545 |
| Compacted clay cover | 200 | 55.96 | 0.68 | 0.30 | 12.93 | 2.18 | 1.69 | 587 |
| Mixed refuse/compacted clay cover | 250 to 600 | 104.3 | 1.69 | 0.13 | 14.98 | 5.06 | 3.13 | 780 |

moisture vs. decomposition effects examination proposed by the current study. For practical purposes, the Bogner et al. (1995) statistical correlations were examined further. Three practically interesting statistical relationships were:

$$\% \text{ water} = 10.14 + 0.05 \text{ (mg biomass in DOC/gm soil)}$$

$$\text{anaerobic activity (mg ammonia nitrogen/gm dry soil hour)}$$
$$= -0.179 + 0.005 \text{ (mg biomass in DOC/gm soil)}$$

$$\text{aerobic activity (mg ammonia nitrogen/gm dry soil hour)}$$
$$= -0.054 + 0.011 \text{ (mg biomass in DOC/gm soil)}$$

The above indicates that:

mg biomass in DOC/gm soil = 20(10.14 – % water) = 90.91 (0.054 + aerobic activity (mg ammonia nitrogen/gm dry soil hour)) = 200 (0.179 + anaerobic activity (mg ammonia nitrogen/gm dry soil hour))

This crude statistical relationship may indicate the correlation between anaerobic or aerobic microbial mass present and water content, although water content alone cannot be a determinant of microbial mass.

Nozhevnikova et al. (1993) studied microbial distributions in the landfill profiles as well as production and removal of gas by methanotrophic organisms during gas movement through the soil. The study was done at inactive Moscow dumps, mainly at the Rameki dump of thickness (2 to 10 m deep) and the Kouchino dump (8 to 20 m deep). Pit hole samples were microbiologically and biochemically analyzed for gas generation and oxidation activity, with laboratory culturing of the two organism groups to determine rates. Temperature in the pit holes was typically 25 to 31°C. Methane:carbon dioxide gas ratios in the anaerobic zone ranged from 1.5 to 2 and averaged 1.75. The Rameki landfill was found to have three vertical zones: an aerobic 0.1 to 1.5 m depth aerobic zone, a 1 to 2 m transition zone and a 1.5 to 20 m anaerobic zone. The methane oxidizers were essentially aerobic (and thus should be in the 0 to 1.5 m zone), but in the 1 to 2 m transition zone both methane oxidizing and methane generating bacteria were present (as well as a denitrifying bacterium). In the vertical zone 10 to 100 cm from the surface, intense oxidation of methane, hydrogen and carbon monoxide occurred, with oxidizing bacteria in the range $10^{11}$ cells/gram of soil; and the zone had an aeration ratio (nitrogen:oxygen percentage ratio; air = 3.74) ranging from 3.74 to 0.0, decreasing downward. Methane oxidation rates in the oxidizing zone ranged from 0.12 to 0.6 m-moles/(day gm dry weight refuse); hydrogen oxidation rates were 0.2 to 1.75 mmole/(day gm dry weight refuse); and carbon monoxide oxidation rate was 0.02 to 0.18 mmole/(day gm dry weight refuse). Temperature decreases of 5 to 7°C decreases methane and carbon dioxide oxidation rates by 3 to 5 times and carbon monoxide oxidation rates by 10 times. Methanogenic activity was associated with the upper anaerobic zone (bottom of the aerated, methane oxidation zone).

Some important results of the aerobic oxidation zone study are tabulated below.

**TABLE 8.3**
**Characteristics of Aerobic Zone of MSW Landfill — Moscow Studies**

| Depth, m | 0.1 to 0.2 | 0.5 to 0.6 | 0.7 to 0.9 |
|---|---|---|---|
| $CH_4$% | 9.0 | 19.0 | 19.0 |
| CO% | 1.8 | 2.1 | 1.8 |
| $H_2$% | 20 | 20 | 20 |
| $O_2$% | 20 | 16 | 16 |
| Oxidation rate, $CH_4$, mmol/day gm dry wt | 0.12 to 0.6 | range | range |
| Oxidation rate, $H_2$ | 0.2 to 1.75 | range | range |
| Oxidation rate, CO | 0.02 to 0.18 range | range | range |
| $N_2$:$O_2$ ratio | 3.74 | | 0.0 |
| Methane oxidizing organism count | $10^8$–$10^{11}$ cells/gm wet ground (refuse soil sample) | $10^8$–$10^{11}$ cells/gm wet ground (refuse soil sample) | $10^8$–$10^{11}$ cells/gm wet ground (refuse soil sample) |
| Hydrogen oxidizing bacteria count | $10^7$–$10^{11}$ cells/gm wet ground (refuse soil sample) | $10^7$–$10^{11}$ cells/gm wet ground (refuse soil sample) | $10^7$–$10^{11}$ cells/gm wet ground (refuse soil sample) |
| Carbon monoxide oxidizing bacteria | $10^6$–$10^8$ cells/gm wet ground (refuse soil sample) | $10^6$–$10^8$ cells/gm wet ground (refuse soil sample) | $10^6$–$10^8$ cells/gm wet ground (refuse soil sample) |

Bacteria identified from the Russian landfills included 14 methane oxidizing groups (various *Methylococcus, Methylomonas, Methylobacter, Methylosinus* and *Methylocystis* species); domination of the methanogenic bacteria by mesophilic types, including species of *Methanosarcina* and the *Methanobacterium*; and the hydrogen oxidizing bacterial genera *Alcaligenes, Pseudomonas, Paracoccus* and *Mycobacterium*.

An important point made by the above investigators was that:

> *When the landfill's methane oxidation capacity (basically near the landfill surface) is slowed by cold weather seasons, methane emission to the atmosphere can increase.*

Other salient points are that:

> *There was a zone of maximum aeration occurring at 0.6 to 0.7 meter depth and the aeration ratio in this zone (nitrogen:oxygen ratio) may be the result of intensive (natural) out-gassing of landfill gas near this depth. This suggests gas generation and movement pressure differences between the landfill interior and the outside are sufficient to draw in air to a certain depth; and the depth where pressure forces balance can be a point of maximum (upper landfill profile aeration).*

> *The oxidation zone is generally thin; on the order of 0.6 m or 3 to 8% of landfill depth of fine soil cover.*

The transition zone between aerobic and anaerobic profiles, similarly thin, was shown by this study to contain both methanogenic and oxidative organisms. This should be expected and agrees with the Bogner et al. (1995) statement that natural soil conditions (at the top of the refuse layering sequence) should favor both aerobes and facultative organisms, because this soil profile point always contained oxygen at subatmospheric levels.

## MASS TRANSFER CONSIDERATIONS

The mass transfer coefficient is typically a dimensional correlation of media properties. For granular porous media, the Sherwood number describes this relation and correlates interface area with a characteristic length of diffusion transfer of mass. The Sherwood number has been stated as:

$$Sh = \frac{k_l d}{D} \tag{8.2}$$

where $D$ = diffusive mass transfer rate, $k_L$ = mass transfer coefficient and $d$ = layer thickness for liquid phase diffusion. The characteristic length d, as a thickness of film involved in transfer, is the average thickness of the moisture film at the interface. It thus has the value:

$$d = \frac{\theta_{ps}}{A} \tag{8.3}$$

where $\theta_{ps}$ = total volumetric moisture content (mobile + stagnant) on the surface of solid material and $A$ = area of the interface between gas (in motion) and air. Thus the Sherwood number is:

$$Sh = \frac{k_L \theta_{ps}}{DA} \tag{8.4}$$

Since $k_L$ is unknown, the Sherwood number practical for a waste site environment must be defined.

## SHERWOOD NUMBER

According to Fedikiw and Newman (1978), the Sherwood number depends only upon particle (geometry and size) and arrangement.

As discussed in Chapter 2, the waste site can be described as a granular porous medium subject to laminar or trickling flow, to which capillary flow models might apply. According to Comiti and Renaud (1989) and Comiti et al. (2000), a granular packed bed can be described in terms of its pore volume as a bundle of cylindrical but tortuous pores of average diameter $d$ and length $L$, surface area $S$ and bed height $H$, such that the tortuosity of the pore is:

$$\zeta = L/H$$

The average pore diameter $d$ can be described, as in Chapter 2, according to the Hagen Poiseuille relation for interstitial velocity $v_{int}$, where the interstitial velocity is the ratio of the superficial velocity of fluid flow through the granular to bed porosity $U_0$:

$$v_{\text{int}} = \frac{U_0}{\varepsilon} = \frac{d_{pore}^2}{32\mu}\frac{\Delta P}{L} = \frac{d_{pore}^2}{32\mu}\frac{1}{\zeta}\frac{\Delta P}{H}$$

$$d_{pore} = \left(\frac{32\mu\zeta}{\varepsilon}\left(\frac{H}{\Delta P}\right)U_0\right)^{0.5}$$

$$L = \zeta H$$

$$\frac{\Delta P}{H} = NU_0 + MU_0^2$$

$$= \left[2\gamma\zeta^2 a_{vd}^2 \frac{(1-\varepsilon)^2}{\varepsilon^3}\right]U_0 + \left[0.0968\zeta^3 \rho a_{vd} \frac{(1-\varepsilon)}{\varepsilon^3}\right]U_0^2$$

$$\zeta = \left[\frac{M^2}{N}\frac{2\gamma\eta\varepsilon^3}{(0.0968\rho)^2}\right]^{\frac{1}{4}}$$

$$a_{vd} = \left[\frac{N^3}{M^2}\frac{(0.0968\rho)^2}{(2\gamma\eta)^3}\frac{\varepsilon^3}{(1-\varepsilon)^4}\right] \tag{8.5}$$

where $U_0$ = superficial velocity or conductivity, $N$ and $M$ represent coefficients for viscous and inertial (kinetic) energy resistance, $\zeta$ = particle bed tortuosity, $\rho$ = fluid density, $\gamma$ = pore shape factor, $\varepsilon$ = packed bed void fraction or porosity and $a_{\mathrm{vd}}$ = dynamic specific surface area or specific surface area of the packed bed actually exposed to the fluid flow. Tortuosity value is a key parameter for analysis using the expressions presented above. The discussion of Chapter 6 (Tortuosity as a Function of Particle Flatness) indicates the tortuosity of disposed wastes might be in the range of $\zeta = 1.45$.

According to Seguin et al. (1996), the relation between the specific surface area of the system and that actually seen by the flow is:

$$A_d = x/(\mathrm{SSA})$$

where $x < 1$ for overlapping particles, SSA = specific surface area of the (dry) particle bed and $A_d$ = dynamic surface area = $a_{\mathrm{vd}}$. In Chapter 2, the average specific surface area is described as:

$$a_{vi} = \frac{S_p}{v_p} = \frac{6}{\varphi D_p}$$

$$V_i = \frac{M_i}{\rho_i} = \text{volume of } i\text{th type of particle}$$

$$\chi_i = \frac{V_i}{\sum_i\left(\frac{M_i}{\rho_i}\right)} = \text{volume fraction}$$

$$a_{v,\mathrm{mean}} = \sum \chi_i a_{vi}$$

$$D_m = \frac{6}{a_{v,\mathrm{mean}}} \tag{8.6}$$

In the above, $S_p$ = volume of a spherical particle with the same volume as the actual particle, $\varphi$ represents the particle sphericity, $a_{\mathrm{v,mean}}$ = mean surface area of the mix of particles, $M_i$ = mass of $i$th type of particle and $V_i$ = volume of $i$th type of particle.

For the waste site type media with the indicated hydraulic conductivity, assuming laminar flow, permeability considerations indicated mean particle size to be estimated as averaging 1.14 inches (0.029 meters). This provided that, with an indicated particle length/thickness ratio of 5 and plate-type overlap, the value of $A_d$ was on the order of 110 $m^2/m^3$ for the landfill layers containing waste.

Notably the form of $Sh$ depends upon mass transfer in a short tube of length $L$ and diameter $d$. According to Leveque this is represented by:

$$Sh = 1.615[Re^* Sc(d/L)]^{1/3} \tag{8.7}$$

if the flow is laminar and the fluid is Newtonian. The terms Re and Sc represent the Reynolds and Schmidt numbers. The Schmidt number is:

$$Sc = \frac{\mu}{\rho D_{diffusion}} \approx 1.0 \tag{8.8}$$

for gases, $10^3$ for liquids (O. Levenspiel, Jan. 1989).

The Reynolds number was defined in Chapter 2 as:

$$Re = \frac{\rho d_{pore} v_{interstitial}}{\mu}$$

$$d_{pore} = 4\frac{\varepsilon}{a_v(1-\varepsilon)}$$

$$v_{interstitial} = \frac{U_0}{\varepsilon}$$

$$U_0 = k\mu\frac{\Delta P}{L} = k\mu\frac{\Delta P}{\zeta H} \text{ (in terms of packed bed resistance to flow)}$$

$$\frac{\Delta P}{H} = NU_0 + MU_0^2 = \left[2\gamma\zeta^2 a_{vd}^2 \frac{(1-\varepsilon)^2}{\varepsilon^3}\right] U_0 + \left[0.0968\zeta^3 \rho a_{vd} \frac{(1-\varepsilon)}{\varepsilon^3}\right] U_0^2$$

$$k = \text{intrinsic permeability} = (\mu U_0 \zeta)\frac{H}{\Delta P}$$

$$= \frac{\mu}{[2\gamma\zeta a_{vd}^2 \frac{(1-\varepsilon)}{\varepsilon^3}^2] + [0.0968\zeta^2 \rho a_{vd} \frac{(1-\varepsilon)}{\varepsilon^3}]U_0}$$

$$Re = \frac{4\rho\mu}{a_{vd}(1-\varepsilon)([2\gamma\zeta a_{vd}^2 \frac{(1-\varepsilon)^2}{\varepsilon^3}] + [0.0968\zeta^2 \rho a_{vd} \frac{(1-\varepsilon)}{\varepsilon^3}]U_0)}$$

$$U_0 \approx K = \text{conductivity}$$

$$k = \frac{K\mu}{\rho g}$$

$$k_{gas} = k\, k_{rel,gas}$$

$$K_{gas} = \frac{\rho_{gas} g}{\mu_{gas}} k = \frac{\rho_{gas} g}{\mu_{gas}} \frac{K\mu}{\rho g} = \left(\frac{\rho_{gas}\mu_{liquid}}{\rho_{liquid}\mu_{gas}}\right) K_{liquid} \tag{8.9}$$

When the fluid is gas, the area available changes according to degree of saturation because flow of gas only through the open, connected pores is possible. Thus $K_{\text{liquid}}$ and $K_{gas}$ must be stated in terms of relative degree of saturation.

$$K_{liquid} = K(\theta) = K(\psi)$$

The discussions of unsaturated hydraulic conductivity provide that the conductivity can be expressed in terms of water content or matric potential. The Sherwood number and mass transfer coefficient can thus be restated as:

$$Sh = 1.615\left[Re\left(4\frac{\varepsilon}{a_{vd}(1-\varepsilon)}\right)\frac{\mu}{\rho D_{diffusion}}\frac{1}{\zeta H}\right]^{\frac{1}{3}}$$

$$= 4.07\left[\frac{1}{\zeta H}\frac{\rho\mu\varepsilon}{a_{vd}^3(1-\varepsilon)^3([2\gamma\zeta a_{vd}^2\frac{(1-\varepsilon)^2}{\varepsilon^3}] + [0.0968\zeta^2\rho a_{vd}\frac{(1-\varepsilon)}{\varepsilon^3}]U_0)}\right.$$

$$\left.\frac{\mu}{\rho D_{diffusion}}\right]^{\frac{1}{3}}$$

$$k_L = \frac{D_{diffusivity}a_{vd}Sh}{\theta_{ps}} \tag{8.10}$$

Length of transfer considerations indicate change between full saturation and dryness changes the pore volume available to gas, so only the pore diameter term $d$ need be modified. The only modification needed would be for gas volume in the medium pore space.

Values of moisture saturation developed from water mass balance for the system using a precipitation input model such as the EPA HELP model or other reliable water inflow models allow the unsaturated conductivity to be stated in terms of saturated conductivity and moisture content or matric potential.

The units for $k_L$ must be in $M/L^2T$ and is the overall transfer coefficient.

## APPLICATION OF TRANSPORT MODEL TO GAS FLUX

Landfill gas carries with it a wide variety of chemicals. Appendix 2 lists these chemicals in representative listing for both landfill gas and for leachate. Because leachate concentrations could be expected to be in equilibrium with concentrations, partitioning between the two would be expected. From a treatment perspective, the stripping of these from the gas would be of treatment interest. One approach to this is to assume that the organisms capable of removing or converting the substances of interest are present in the soil; and if the right conditions are applied, removal of the chemical could be effected.

## GAS–LIQUID TRANSFERS

Flux of chemicals between liquid leachate and landfill gas is expected to sweep out volatile chemicals from liquid in the pores. Treatment for removal of landfill gas substances would involve transfer of chemicals from gas to liquid. Description of gas to liquid transfer is thus important to conversion or removal of contaminants.

As decomposition occurs in the landfill reactor, volumes of solid, liquid and air or gas fractions change. These dynamic changes in volumes affect the rates of transfer of chemicals into and from landfill gas.

## MASS FLUX

The mass transfer rate or flux can be described in terms of the driving force/resistance — i.e., pressure difference, across the gas film (water vapor) between the gas–liquid interface and the gas itself — or concentration difference between the gas–liquid interface and the interior of the liquid film:

$$\varphi = \Delta P/(1/k_{gas}) = \Delta C/(1/k_{liquid}), \text{ or} \tag{8.11}$$

$$\varphi = k_{gas}.(p_{gas} - p_{int}) = k_{liquid}.(C_{int} - C_{liq}) \tag{8.12}$$

where $\varphi$ = flux; $\Delta P$, $p_{gas}$ and $p_{int}$ are pressure difference, pressure in the gas and pressure at the interface, respectively; $C_{int}$, $C_{liq}$ = concentration of the chemical or substance of interest at the interface and inside the liquid film, respectively; and $k_{gas}$, $k_{liquid}$ are mass transfer coefficients for the gas and liquid films, respectively, including the gas–liquid interface area involved in transfer.

For a particular substance, the partial pressure exerted in the gas can be related to its gas concentration by Henry's Law, which is:

$$p_i = H.C_i$$

where $i$ denotes the chemical of interest and $H$ = Henry's constant. In the same way, if Henry's Law holds (Sinclair and Mavituna, 1983):

$$p_i = H.C_i$$

$$p_{gas} = H.C_{liq(eq)}$$

$$p_{gas(eq)} = H.C_{liq}$$

Then:

$$\varphi = k_{gas}.H.(C_{liq(eq)} - C_i) = k_{liq}.(C_{int} - C_{liq})$$

extracting $C_{int}$:

$$C_{int} = (k_{gas}.H.C_{liq(eq)} + k_{liq}.C_{liq})/(k_{gas}.H + k_{liq}),$$

which, substituted into the expression for flux per unit area–time, gives:

$$\varphi = [k_{liq}.k_{gas}.H.(C_{liq(eq)} - C_{liq})]/(k_{gas}.H + k_{liq}), \text{ and}$$

$$\begin{aligned}\varphi &= k_{gas}.(p_{gas} - p_{int}) = k_{liq}.C_{int} - C_{liq})\\ &= k_{liq}.k_{gas}.(p_{gas} - p_{gas(eq)})/(k_{gas}.H + k_{liq})\\ &= [k_{liq}.k_{gas}.H.(C_{liq(eq)} - C_{liq})]/(k_{gas}.H + k_{liq})\end{aligned}$$

where $k_L, k_G$ are the overall mass transfer coefficients for the liquid and gas films, respectively. These values are:

$$1/k_L = 1/k_{liq} + 1/(k_{gas}.H), \text{ and}$$
$$1/k_G = 1/k_{gas} + H/(k_{liq})$$

Incorporating the surface area of film involved Ap, square meters and letting the time unit be one/day:

$$\varphi/\text{day} = A_p.[[k_{liq}.k_{gas}.H.(C_{liq(eq)} - C_{liq})]/(k_{gas}.H + k_{liq})] \tag{8.13}$$

This is also, by substituting the overall mass transfer coefficient $k_L$ for liquid, i.e., for transfer of the chemical or substance from the gas to the liquid film:

$$\varphi/\text{day} = A_p.k_L(C_{liq(eq)} - C_{liq}) \tag{8.14}$$

and for overall mass transfer from the liquid film to the gas, using $k_G$:

$$\varphi/\text{day} = A_p.k_G.(C_{liq(eq)} - C_{liq}) \tag{8.15}$$

To solve the above mass transfer coefficient, concentrations of chemical in the gas or liquid must be assumed or known. These are to be provided from landfill gas and leachate quality studies. Henry's constants for the compounds of interest must also be used. The gas and liquid film transfer coefficients are taken to be those for a water film and for a water vapor film. The wetted film surface area is to be assumed based upon overall surface area of the solids in the reactor times the fraction of solids wetted or the open pore area of the wetted fraction of the reactor. Surface area considerations for the present study provide gas–liquid interface area in terms related to moisture content and landfill bioreactor porosity.

## REMOVAL OF CHEMICAL IN LIQUID FILM

Assuming a microorganism-mediated removal of chemical in the liquid film forced by metabolic or cometabolic reaction, the net removal during a unit time period is:

rate of removal = rate of conversion × concentration in film × area of film, or:

$$r_i = k_i.C_i.A_p$$

As $C_i$ at any time is previous concentration + flux in (from gas) to flux out (from gas), or:

$$C_i = C_L + (k_G - k_L).(C_{liq(eq)} - C_{liq})$$

Thus the removal rate $r_i$ for the chemical is:

$$r_i = k_i.A_p.[C_L + (k_G - k_L).(C_{liq(eq)} - C_{liq})] \tag{8.16}$$

To solve the relationship, the value of $k_i$, the removal rate for the organisms of interest must be known, in gm/(gm organisms.m$^3$.day) × film thickness, $C_i$ in g/cc or m$^3$, $A_p$ = film area in $m^2$ for consistency of units.

## APPLICATION OF TRANSPORT MODEL TO GAS CHEMICALS FLUX

Landfill gas carries with it a wide variety of chemicals. Chapter 2 listed these chemicals in representative listing for both landfill gas and for leachate. Because leachate concentrations could be expected to be in equilibrium with concentrations, partitioning between the two would be expected. From a treatment perspective, the stripping of these from the gas would be of treatment interest. One approach to this is to assume that the organisms capable of removing or converting the substances of interest are present in the soil; and if the right conditions are applied, removal of the chemical can be effected.

Flux of chemicals between liquid leachate and landfill gas is expected to sweep out volatile chemicals from liquid in the pores. Treatment for removal of landfill gas substances would involve transfer of chemicals from gas to liquid. Description of gas to liquid transfer is thus important to conversion or removal of contaminants.

As decomposition occurs in the landfill reactor, volumes of solid, liquid and air or gas fractions change. These dynamic changes in volumes affect the rates of transfer of chemicals into and from landfill gas. For example, volumes of solids in the (unit) reactor can be described as:

Volume (solid) = Initial volume − Δ Volume(solid) + Volume (biomass), or:

$$V_s = V_{s0} - \Delta V_s + V_b$$

where $V_s$ = total solids content, $V_{s0}$ = initial solids content, $\Delta V_s$ = change due to decomposition or conversion and $V_b$ = volume of biomass. Where densities $\varrho$ of $t$ are applied, the volumes are of the form $V = \varrho.M$, where $M$ = mass of solid. In this case $\varrho_s$ = overall density of the solids in the landfill unit and $\varrho_b$ = average density of the biomass in the reactor. Thus $V_s$ becomes:

$$V_s = \varrho_s.M_{s0} - \varrho_s.(\Delta M_s) + \varrho_b.M_b = [\varrho_s.(M_{s0} - \Delta M_s) + \varrho_b.M_b]$$

Similarly, the volume of liquid can be described by mass balance as:

Volume (liquid) held = (Vol. input + Vol. from decomposition + Vol. previously present) to (Vol. drained + Vol. consumed (organisms) + Vol. evap.)

$$V_L = (V_{in} + V_{decomp} + V_0) - (V_{leach} + V_{bc} + V_{evap})$$

and where density and mass/unit volume are considered and assuming approximately the same liquid density for all volumes:

$$V_L/m^3 = \varrho_L.[(M_{in} + M_{decomp} + M_0) - (M_{leach} + M_{bc} + M_{evap})]$$

The unit volume of gas is then the total unit volume to vol. solid to vol. gas, or:

$$V_G/m^3 = 1 - [\varrho_s.(M_{s0} - \Delta M_s) + \varrho_b.M_b] - \varrho_L. [(M_{in} + M_{decomp} + M_0) - (M_{leach} + M_{bc} + M_{evap})] \quad (8.17)$$

For a unit volume of landfill, with solid, liquid and gas volumes $V_L$, $V_L$ and $V_G$ present, fractional volumes of liquid, solid and gas are also given by $f_L$, $f_S$ and $f_G$, respectively; and the ultimate porosity of the reactor, i.e., if water and gas is removed, is given by:

$$\begin{aligned}\varepsilon_{ult} &= \text{ultimate porosity} = f_L + f_G, \text{ where } f_L = V_L/V_{Lnf},\\ f_S &= V_S/V_{Lnf} \text{ and } f_G = V_G/V_{Lnf}\end{aligned}$$

and $V_{nf}$ = unit volume of landfill (= 1.0 m$^3$). The value $\varepsilon_{ult}$ is impractical for gas–liquid transfer considerations. At the start of the simulation period, a value of $\varepsilon$, the porosity of the reactor, may be known or assumed. During the simulation period the porosity may increase or decrease due to gain or loss. Thus at the end of the simulation period:

$$\varepsilon = \varepsilon_0 + \Delta_\varepsilon$$

$$\Delta\,\varepsilon = \Delta(V_G/\text{m}^3) + \eta(V_L/\text{m}^3)$$

$$\Delta(V_G/\text{m}^3) = (V_{G0} - \Delta\,V_s - \Delta\,V_L)/\text{m}^3$$

Settlement or compression/expansion of the reactor can be considered to describe change in pore volume and thus gas volume, if a suitable expression is used.

## BIODEGRADATION RATES FOR WASTE SITE ORGANIC CHEMICALS

The organic substances of reduction interest in landfill gas and thus in leachate were summarized in Chapter 3. The listed chemicals include many of moderate to high volatility. In the case of use of the system for biofiltration, the potential for methane removal in aerobic soil is also of interest. Solubility, Henry's constant, molecular weight, diffusion coefficients and other needed properties of the listed compounds were compiled for development and exercise of the gas model. Important among these properties are known biodegradation rates in the liquid phase, under both aerobic and anaerobic conditions.

Howard (1991) compiled a comprehensive list for biodegradation of many of the work site effluent chemicals listed in Chapter 3 as present in landfill gas or leachate. His discussion of sources of this information noted that aerobic biodegradation rates for soils were assumed to be the same as in natural surface waters and when only soil rates were available, the half-life for water biodegradation was assumed to be equal to the half-life in soil. It was also noted that anaerobic rates were expected to be generally much slower than for aerobic biodegradation and, where experimental data could not be located, was assumed to be four times longer than for aerobic biodegradation.

## PARTITIONING BETWEEN GAS AND LIQUID

Many investigators have postulated that in soil film–vapor partitioning models, gas-to-water partitioning of diffusing and slowly moving vapors approximate Henry's Law. Smith et al. (1983) state that if the concentration of the dissolved gas or vapor is small and the temperature and pressure for the gases or vapors are far removed from their critical values, solubility according to Henry's Law can be assumed. Batterman et al. (1995) state that Henry's Law would also apply in the case of partitioning of dilute solutions. Low concentration in the landfill water phase would be the case for methane or oxygen in a landfill environment. Apart from the Henry's Law relationships previously discussed for partitioning of organic chemicals such as hydrocarbons between gas and water, — i.e., $C_{gas} = HC_{liq}$ — linear partitioning between the dissolved substance and the solid phase is often assumed, with the sorption to the solid phase specific to the chemical involved but dependent upon the organic content ($f_{\text{oc}}$) of the solid:

$$C_{solid} = K_d C_{liq}$$
$$K_d = K_{oc}\, f_{oc}$$

where $K_{\text{oc}}$ = octanol–water partitioning coefficient of the chemical. Mass balance for partitioning of the compound provides that amount in the liquid phase, including sorption to solids, is also proportional. The mass sorbed in the solids and liquid phase is thus:

$$C_{solid} + C_{liq} = [K_{oc} f_{oc} + 1]C_{liq} \tag{8.18}$$

Removal of $C$ from the sorbed (liquid) phase can be stated as:

$$dC_{liq}/dt = -k_b C_{liq}$$

where $k_b$ is the anaerobic or aerobic rate of removal of the substance. For aerobic biodegradation, the removal term is modified for oxygen limitation. The sorption/removal rate for the substance can thus be stated, including the mass transfer term, as:

$$[1 + K_{oc} f_{oc}]dC_{liq}/dt = k_L A(C_g/H - C_{liq}) - k_b C_{liq} \tag{8.19}$$

Substitution in the general expression (for dispersion + advection), where $k_L$ = mass transfer coefficient and $A$ = landfill moisture dependent interface area for mass transfer permits the removal rate to be solved.

## WASTE SITE SETTLEMENT

With a thorough model representation of biological degradation processes in a waste site decomposition, it is possible to translate the conversion of waste materials into changes in site soil and waste mass and volume. Biological conversion of solid materials is also time-dependent over long periods, suggesting that change in solid mass and volume of organic materials is cumulative and chemically irreversible. Simulation of the solid materials conversion process thus offers an opportunity to define system changes.

The important phenomenon for settlement prediction is the change in volume. Overall volume change due to decomposition is the sum of conversions of individual materials. For a range of materials of type $i = 1 \ldots$ n and grain size variation $j = 1 \ldots$ m, as outlined in Chapter 2, the mass $S_i$ is given in kilograms or grams dry matter by:

$$S_i(t) = \sum_{j=1}^{j=m} S_{i,j}(t) \tag{8.20}$$

The change in volume of a solid material $S_i$ due to biological degradation is thus $\Delta\, S_I$, or:

$$\text{Change in volume} = \Delta\, V_i = \Delta\left[\frac{S_i(t)}{\rho_i}\right] = \sum_{j=1}^{j=m} \Delta\left(\frac{S_{i,j}(t)}{\rho_i}\right) \tag{8.21}$$

The solid material density $\rho_i$ would be known and its initial mass $S_{0,i}$ in the mixture of granular materials in the waste site is assumed to be known from standard waste composition analysis or from site sampling and screening. Considering *hydrolysis* of solid biodegradable organic materials as the primary essential process for conversion to liquids (organic acids) or gases, the change in mass of an organic waste for a small time period $\Delta t$ can be stated, as in Chapter 7, according to first-order kinetics of hydrolysis:

$$\Delta S_i(t) = S_{0,i} \exp(-k_h(i)\Delta t)$$

$$\Delta V_i = \frac{\Delta S_i(t)}{\rho_i} = \sum_{j=1}^{j=m} \left(\left(\frac{S_{0,i}}{\rho_i}\right) \exp(-k_h(i)\Delta t)\right) \tag{8.22}$$

In the above, the *specific hydrolysis rate* $k_h(i)$ of the $i$th biodegradable solid waste material must be known — for instance, the anaerobic or aerobic decomposition rate for paper, wood, vegetables, meat, grass or leaves (available from literature searches or experimental studies). The total change in volume due to biological degradation, for a short period of time $\Delta t$ is the sum of change in specific volumes for waste materials of type $i$:

$$\begin{aligned} \Delta V_{Total,Bio} &= \sum_i \Delta V_i = \sum_i \left(\frac{\Delta S_i(t)}{\rho_i}\right) \\ &= \sum_i \left[\sum_{j=1}^{j=m} \left(\left(\frac{S_{0,i}}{\rho_i}\right) \exp(-k_h(i)\Delta t)\right)\right] \end{aligned} \tag{8.23}$$

The above expression only requires the specific *density, mass* and *hydrolysis or conversion rate* of the biodegradable materials in the site to estimate the change in volume over a time period $\Delta t$. For a full time period of interest, for instance for a site operated and/or inactive for several years, the cumulative change in volume would be the sum of volume changes during time period $\Delta t$.

The sum of volume changes is also a sketch of the decomposition history of every waste type present, including the influence of environmental factors. This means that the waste decomposition pattern is not intrinsically monotonic but will show the effect of aerobic conditions, anaerobic conditions, moisture limitation and temperature variation. For a historic period for the waste site, environmental factors must be incorporated. The hydrolysis or conversion rate should be targeted as an indicator of environmentally induced changes in decomposition rate, as it is the kinetic variable most influenced by climate and other environmental factors. Consider wastes buried in soil or exposed in the environment for an initial aerobic phase of decomposition $t_{\text{aerobic}}$, followed by an anaerobic decomposition phase $t_{\text{anaerobic}}$. Also, the final anaerobic phase includes a period of leachate recycle $T_{LR}$, as for a landfill bioreactor or a period of aerobic treatment (induced air diffusion) $T_{+\text{air}}$. The change in site volume due to decomposition of waste materials can be stated as:

$$dV_{(total)} = dV_{anaerobic} + dV_{aerobic} + dV_{leachate\ recycle,\ anaerobic} + dV_{aerobic\ treatment} \quad (8.24)$$

which is:

$$\begin{aligned}
\Delta V_{Site,Bio} &= \Delta V_{Total,Bio}(t) = \sum_i \Delta V_{i,bio}(t) = \sum_i \left(\frac{\Delta S_i(t)}{\rho_i}\right) \\
&= \sum_i \left(\frac{\Delta S_i(t)}{\rho_i}\right)_{aerobic} + \sum_i \left(\frac{\Delta S_i(t)}{\rho_i}\right)_{anaerobic} \\
&\quad + \sum_i \left(\frac{\Delta S_i(t)}{\rho_i}\right)_{anaerobic,recycle} \\
&\quad + \sum_i \left(\frac{\Delta S_i(t)}{\rho_i}\right)_{aerobic\ treatment} \\
&= \sum_i \left[\sum_{j=1}^{j=m} \left(\left(\frac{S_{0,i}}{\rho_i}\right) \exp(-k_{h,aerobic}(i) t_{aerobic})\right)\right] \\
&\quad + \sum_i \left[\sum_{j=1}^{j=m} \left(\left(\frac{S_{0,i}}{\rho_i}\right) \exp(-k_{h,anaerobic}(i) t_{anaerobic})\right)\right] \\
&\quad + \sum_i \left[\sum_{j=1}^{j=m} \left(\left(\frac{S_{0,i}}{\rho_i}\right) \exp(-k_{h,anaerobic}(i) t_{recycle})\right)\right] \\
&\quad + \sum_i \left[\sum_{j=1}^{j=m} \left(\left(\frac{S_{0,i}}{\rho_i}\right) \exp(-k_{h,aerobic}(i) t_{aerobic\ treatment})\right)\right]
\end{aligned} \quad (8.25)$$

If there is no recycle or no aerobic treatment, the site is only subject to an aerobic phase followed by an anaerobic phase. A shallow waste site could also be essentially aerobic, without an anaerobic phase of waste decomposition.

The above expression is also explicit for grain size. For instance, volume reduction for grains of size of $j = 13\,\text{mm}$, for the $i =$ paper waste, has been incorporated. Sensitivity of volume change to grain size is potentially useful in simulations where the age for wastes can be statistically related to waste grain size or for a particular location or climatic region.

For grain size sensitivity, the theoretical assumption can be made that if all grains of a solid waste material having the same average size are exposed to the same environmental agents, reduction in volume of each grain of the same size for a certain waste type should be similar. This assumption does not apply to all situations, as conditions (anaerobic or aerobic, local water content, local temperature) are likely to vary across and within the site. The importance of establishing initial conditions of wastes (and soils) for a site before a model is used for simulation is highlighted. Defining initial conditions could reduce errors related to spatial heterogeneity in site conditions.

If the history of a type of waste rather than its grain size distribution is of interest, the overall volume reduction can be stated in terms of the waste type only:

$$\begin{aligned}
\Delta V_{Site,Bio} &= \Delta V_{Total,Bio}(t) = \sum_i \left(\frac{\Delta S_i(t)}{\rho_i}\right) \\
&= \sum_i \left[\left(\frac{S_{0,i}}{\rho_i}\right) \exp[-k_{h,aerobic}(i)t_{aerobic}]\right] \\
&\quad + \sum_i \left[\left(\frac{S_{0,i}}{\rho_i}\right) \exp[-k_{h,anaerobic}(i)t_{anaerobic}]\right] \\
&\quad + \sum_i \left[\left(\frac{S_{0,i}}{\rho_i}\right) \exp[-k_{h,anaerobic}(i)t_{recycle}]\right] \\
&\quad + \sum_i \left[\left(\frac{S_{0,i}}{\rho_i}\right) \exp[-k_{h,aerobic}(i)t_{aerobic\ treatment}]\right] \qquad (8.26)
\end{aligned}$$

The above expression allows the volume reduction of individual waste types to be followed as well as the overall volume reduction.

Note that the term $S$ refers to the dry, biodegradable mass fraction of the waste undergoing decomposition. If 50 kg of cardboard buried in a waste site has 14% moisture by weight, the dry matter would be 43 kg. If only 60% of the dry matter is biodegradable, the value of $S$ for buried cardboard would be 25.8 kg.

As settlement is of considerable importance to waste site stabilization and construction on former disposal sites, the topic has drawn the interest of a number of investigators over the years.

Wall and Zeiss (1995) state that long-term compression or settlement in a waste site can be classified into three successive periods: initial compression, primary compression and secondary compression. *Initial compression* or settlement is associated with the nearly instantaneous compaction of (waste site) void space and grains when a load is applied from above (Tuma and Abel-Hady, 1973). *Primary compression* is said to be due to the expression of soil gas and pore water due to an applied load and it also occurs within a relatively short period (30 days) after a load is applied (Morris and Woods, 1990; Sowers, 1973). Wall and Zeiss (1995) and El-Fadel and Khoury (2000) indicate that for unsaturated waste sites, primary compression may not be due to water or gas expression, as there would be no pore pressure and liquid and gas is free to escape. It is possible that primary compression might represent a period of deformation and rearrangement of particles that begins after the applied load reaches a critical distribution; and the process of rearrangement itself might release water and gas trapped in the fine pore structure.

It has generally been concluded that much of the ground settlement at landfills, considered *secondary compression*, is due to decomposition (Park et al., 2002; El-Fadel and Khoury, 2000; Wall and Zeiss, 1995; Coduto and Huitric, 1990; Rao, 1977; Sowers, 1973, 1974) and is long term; but it can be represented in terms of vertical strain that decreases logarithmically with time. Coduto and Huitric (1990) postulated that 18 to 24% of vertical compression of municipal solid waste in landfills could be due to decomposition. Ling et al. (1998) indicate that overall settlement in a landfill can be 30 to 30% of initial height. However, El-Fadel and Khoury (2000) state that long-term settlement in landfills due to decomposition can be up to 40% of the original height, but on average 15% of long-term settlement is due to decomposition of wastes.

Lee and Park (1999) state that the effect of waste decomposition should be considered for proper prediction of settlement in landfills. For modeling the effect of decomposition on settlement, the secondary compression phase would be of interest. Wall and Zeiss (1995) describe secondary compression or settlement as due to the creep (or strain) of the refuse skeleton and biological decay. El-Fadel and Khoury (2000) describe secondary, long-term compression in waste sites as due to physicochemical changes such as corrosion, oxidation and combustion, as well as biological decay. Because of the inherent difficulty, very few settlement models attempt to separate out the decomposition effect. The recently presented models by El-Fadel and Khoury (2000); El-Fadel et al. (1999); Ling et al. (1998); Bleicker et al. (1995); and Simpson and Zimmie (2001) adopt time-dependent soil mechanics approaches for which mathematical models with coefficients represent the various phases of consolidation. However, Simpson and Zimmie (2000) adapted the power creep model developed by Kumar (1998) and Edil (1990) for soil consolidation, for settlement $S(t)$ at any time as a function of material compressibility (strain/applied compressive stress). The strain ($\Delta H$ = change in height $H$) due to decomposition (secondary compression) could be positioned at the end of the initial consolidation period ($t_r$) and the coefficients for strain-stress ratio relationship could be determined from biodegradation studies. Simpson (2000) notes that the power creep law overestimated settlement and that more data was needed for accuracy of the coefficients $m$ and $n$ (see below):

$$S(t) = H\Delta\sigma m\left(\frac{t}{t_r}\right)^n = \text{settlement}$$

$$m = \frac{S(t)}{H}\frac{1}{\Delta\sigma} = \text{compressibility at } t = t_r$$

$$n = 0.256\log(m) - 0.662 \tag{8.27}$$

The above expression implicitly includes the effect of decomposition in the coefficient $m$ and the applied compressive stress (which will vary as material is lost during degradation). The examination of biodegradation effect on waste site settlement by Park et al. (2002) indicated that, except for the power-creep law as mentioned above, long-term settlement can be successfully predicted using several methods — but only if *accelerated logarithmic compression* is included to represent decomposition. In the Edil et al. (1990) expression for secondary compression:

$$\frac{S(t)}{H} = e(t) = \Delta\sigma\left[a + b\left(1 - \exp\left(-\left(\frac{\lambda}{b}\right)t\right)\right)\right] \tag{8.28}$$

$e(t)$ represents vertical strain or change in height and $\lambda/b$ is the secondary compression rate, where $b$ = compressibility of the material. However, the rate $\lambda/b$ also applies to primary compression, implying the same rate variation throughout settlement. Park et al. (2002) found this to be incorrect, as the rate increases during decomposition. Accelerated logarithmic compression represents a faster settlement rate for the waste site — occurring just after the end of the primary consolidation phase — and thus involves a higher compression index (strain/stress ratio variation coefficient). This coefficient could be found from experimental data from which the compressibility and biodegradation rate for mixed municipal wastes can be determined.

The model presented by Gabr et al. (2000) incorporates compression as a function of change in volume of the waste site, where one portion of the volume change is due to decomposition and the other to change under weight external load. While this is similar to the approach followed by Sowers (1973), incorporation of volume change permits analysis of actual waste conversion as a volume change. For instance, the overall volume change is presented as:

$$\Delta V_s(t) = V_{init}\left[C_m(t)\left(\Delta\sigma_{oct} - \Delta u(t)\right)\right] = \text{interparticle volume change with time}$$

$$\Delta V_v(t) = V_{init}[D_m(t)\Delta\tau_{oct}] = \text{intraparticle volume change with time}$$

$$\Delta V(t) = \Delta V_s(t) + \Delta V_v(t) \tag{8.29}$$

In the above expression $\Delta V_s(t)$ represents the change in pore space due to compression forces, while $\Delta V_v(t)$ represents change in particles, $C_m$ = compressibility coefficient, $D_m$ = time-dependent coefficient, $s$ = compressive octahedral, stress and $\tau$ = octahedral shear stress. The intraparticle change is obviously important for modeling decomposition effect, though as stated, it is defined as a gradient of interparticle shear or of change in bulk compressibility.

Gabr et al. (2000) state that settlement can be related to waste biodegradation and landfill gas production and that a comprehensive model should take into account the increase in void ratio due to solid- to-gas conversion, saturation variation and changes in particle size and distribution. Solid waste conversion is said to cause the void ratio to increase, the saturation degree (volume water/volume solids) to increase and the hydraulic conductivity to increase, which increases compressibility (strain/applied stress). Modeling of the effect of biological decomposition on settlement must thus evolve from development of void ratio vs. waste conversion (hydrolysis) in granular media, saturation degree vs. void ratio and hydraulic conductivity vs. shrinking volume and grain size.

According to Sowers (1973), the decomposition varies according to void ratio. At any time $t$, the void ratio of the system, $\varepsilon$, is given by:

$$\varepsilon = 1 - \theta_w - V_{solids}$$

$$V_{solids} = V_{inert} + V_{organic,residual} + V_{organic,rottable}$$

$$V_{solids} = V(f_{bio,org} + f_{nonbio,org} + f_{inert})$$

$$\frac{dV_{solids}}{dt} = \frac{dV(f_{bio,org})}{dt} = \text{change in solid volume of particles due to decomposition} \tag{8.30}$$

where $\theta_w$ represents the volumetric moisture content of a layer of waste–soil material and $V_s$ represents the volume of solids in that layer. The negative change in solids volume due to biodegradation would obviously increase change in voids and thus the effect of compression and shear. However, the change in void ratio due to change in solids volume is part of the overall change in voids. This points out the value of extending models such as presented by Gabr et al. (2000), with explicit incorporation of the change in void ratio due to solids volume reduction.

The relative independence of the waste site settlement models discussed from biological degradation models illustrates a disadvantage for modeling the highly important effect of decomposition on settlement. The review of settlement models incorporating biodegradation by El-Fadel and Khoury (2000) indicates that very few of these models are explicit for biodegradation effect. The use of kinetic models can allow estimation of the waste volume converted to liquid or gas as a result of biodegradation.

The need is thus for waste site compression models that explicitly incorporate waste conversion as a contributor to long-term settlement. The majority of models mentioned do not have this refinement and thus are unlikely to have great sensitivity to variation in the biodegradation process.

# 9 Sensitivity Analysis and Conclusions

## INTRODUCTION

For modeling, the waste site environment — albeit complex — must be represented in terms of the most important variables. These variables, including heat, moisture, substrate heterogeneity and climate effects, have been defined in a new approach to waste site analysis of this type that incorporates the effects of moisture at macroscopic and microscopic levels. However, for the model to be useful, it must be representative as much as possible to the real world of waste sites. This means that, while the effect of all of the variables involved cannot be perfectly known, for accuracy of simulation the parameters chosen for a model must be both appropriate and representative. *Appropriateness* means that parameters chosen for a process are valid according to theory and experience and that expressions developed to represent waste site processes are within the stable range. A representative model must also be appropriate to both specific and general cases.

For all models variables must be defined as discussed. Their relative effect upon outputs should be examined by varying parameters to determine sensitivities of values.

For the landfill as a bioreactor, the simulation of long-term decomposition, effects of final covers or soil caps, recycle and gas and leachate management, testing of the effect of the major variables upon each of these options would verify how the model behaves when subject to likely ranges of inputs and which inputs have the greatest effects. Most waste site studies involving assessment of decomposition have concluded that moisture content is a major factor controlling biodegradation. Anex (1996) stated that it is generally accepted that moisture addition accelerated landfill waste decomposition and gas production even though there is no consensus on optimal moisture content. Jones and Lee (1993) concluded that use of highly impermeable caps can result in landfills similar to "dry tombs," e.g., waste sites in which many wastes are preserved until cover failure because the moisture content is kept too low to allow appreciable degradation. Thus, waste decomposition should be modeled as sensitive to moisture input. In landfill leachate generation and flow models, sensitivity to moisture dynamics is implicit in that the actual effect of moisture on biodegradation rates is not usually included. An attempt is made in this work to illustrate how the effect of moisture can be explicitly included in kinetic expressions.

It is also likely that heat is another variable of major importance, the dynamics of which affect biological reaction rates — as is well known from waste treatment processes. Heat or moisture extremes have profound effects on waste decomposition, chemicals generation and transport. Thus, sensitivity of the waste model to heat and moisture should be established as a first line of establishing appropriateness.

For the waste site model, important inputs, in order of precedence as building blocks for the model, are:

- A database for substrates (wastes and soil physical and chemical characteristics)
- Soil analysis methods to establish moisture content variation, grain size distribution and bulk flow properties
- Constants for aerobic and anaerobic decomposition
- Development of kinetic rates and expressions
- Waste site ecology
- Moisture inflow, uptake, distribution, outflow and effect upon biodegradation
- Heat capacity of the system and heat flow dynamics
- Effect of temperature on biodegradation processes
- Mass transfer and conversion of chemicals
- Effect of liquid (leachate) recirculation
- Effect of aerobic and anaerobic regimes
- Biofiltration effect (e.g., reduction of methane in the upper aerobic zone of landfills by methanotrophic bacteria)
- Site settlement due to biodegradation

For processes, submodels of the above would be developed and incorporated into a comprehensive model. The variety of processes involved suggests the complexity of a waste site and the importance of examining individual process parameters, inputs and outputs to validate and determine which inputs have the greatest effect on model outputs.

Sensitivity testing should provide insight into which model inputs have greatest influence upon results. It should also provide insight into the degree of response that would be expected if a variable is deliberately or otherwise changed.

## INFORMATION IN DATABASE FOR MSW FRACTIONS AS SUBSTRATE

The physical and chemical characterization of MSW fractions developed for the waste properties database provides a very specific compilation of property information for MSW fractions as bioreactor substrates. The accuracy of some values included must be determined through future experiments. In addition, in some categories broad assumptions must be made with regard to specific properties such as kinetic (hydrolysis and maximum growth) rates for decomposition under aerobic or anaerobic soil conditions, where data was poor or lacking, and of waste properties such as moisture content at the point of simulation. Similar assumptions had to be made regarding the surface likely to be involved in mass transfer.

## RANGE OF ANAEROBIC AND HYDROLYSIS RATES

Aerobic and anaerobic hydrolysis rates for the materials are perhaps some of the most important inputs provided by a wastes database. Examination of the database indicates that the range of kinetic rates vary widely, with the highest values associated with cellulosic and green food or yard wastes fraction. Many of the rates included are based upon laboratory experiments conducted in specific and likely more controlled and hospitable environments for microorganism growth than the landfill. Use of the values in simulation thus requires careful consideration of whether such rates could exist in the field.

Wherever available, experimental data should be used in statistical programs for direct estimation of long-term rates for materials deterioration. This directly provides mass disappearance rates. It also addresses the important issue of lignocellulosic materials' increasing resistance to biological deterioration with exposure age. Application of first-order hydrolysis of solid material at rate $-k_h$, for instance of the form:

$$dS/dt = (-k_h S) + r$$

where $r$ = entirely resistant fraction, supposes either that $r$ is lignin — which has great resistance to biodeterioration — or $r$ is a term including material ash or inorganic matter. First-order curve slopes of deterioration data independently provide indication of confirmation of how material resistance will affect rate in the long term. However, long-term rather than short-term rates are of issue to landfills — even for the least resistant waste fractions, poor water availability and nutrient availability can extend degradation by an order of magnitude as is evidenced in digs for the recent landfill study at the Edinburgh landfills indicating preservation of normally biodegradable material.

Hydrolysis rates varied widely. Within the waste paper category, hydrolysis rates were indicated to vary between 0.001 and 0.02/day, which is a range of two orders of magnitude. In the food wastes category, the range of hydrolysis rate from laboratory experiment and from composting studies was up to three orders of magnitude. The effect of hydrolysis rate range upon model accuracy is thus large. The errors associated with use of a single value of hydrolysis rate are also apparent. Use of a single rate for the whole waste system should add to the difficulty of — rather than improve — decomposition model reliability. It is thus recommended that future models establish as comprehensive a range of hydrolysis rates as possible so that wastes properties are better reflected. While this might add to the tasks required of modeling, there would be less compromise of accuracy.

## CHEMICAL CHARACTERIZATION OF WASTE FRACTIONS

The development of stoichiometric relationships and expressions for chemical oxygen demand of Chapter 3 indicates that model predictability is improved when chemical information on wastes fractions exists. Molar values for C, H, O, N or S are unlikely to vary much and so can provide a solid and systematic basis for reaction mass balance considerations. However, little current or applicable data of this kind is available; and values obtained from literature, except for those provided by the ASME study

(1987), show no fine discrimination between waste fractions. Even when available, prediction with molar expressions must include effects of environmental or reactor conditions, which would decrease or increase model outputs. The model included reactor conditions such as oxygen availability in solution and temperatures.

Values tabulated for microorganism yield factors, based upon COD or volatile solids content considerations, are 0.25 to 0.4 average for the range of paper fractions analyzed and 0.5 to 0.9 for food wastes fractions. These values are well within the range of microorganism cell to substrate yields reported in literature for bioconsumption of food or paper materials as substrates; thus, the values can be said to be reliable.

However, more resistant materials such as bottles, cans and wood, while listed as having significant COD value, should produce little by way of yield of organisms in the short term, though indications are otherwise. The kinetic rates suggest this outcome. This suggests that only volatile solids content (VS) values rather than chemical COD values be used for substrate values with stoichiometric expressions for databases of the type developed for this study, for consistency, if short-term biodegradation analyses are being made. It also suggests that these types of predictions only include those wastes known to have significant short-term response (5 to 20 years).

Since the COD factors expressions could obviously still be used, this should present little difficulty to use of a database.

The database also provides for and allows estimation of the value of volatile solids (VS). The convenience of stoichiometric expressions is that they provide a quick and direct method of including the effect of wastes heterogeneity in modeling landfill biodegradation while providing reliability.

The recommendation is thus made that with use of the database, consideration of values of biological VS content should be of first priority if short-term predictions are considered — before development or use of kinetic expressions such as for the present study — to address or overcome the hurdle of analysis of landfilled wastes of considerable heterogeneity in the type of reactor described.

The importance of the above should not be underestimated. Many waste fractions, even though they have physical importance in a landfill as a reactor system, have little short-term bioreactive capacity when expressed in terms of substrate value. Plastics, if they are indicated to biodegrade under very long-term scenarios as indicated by the discussions of Chapter 3, should also be very slowly bioreactive. For long-term predictions, however, most materials are indicated to undergo biodegradative decay. The hydrolysis rates provided by the database allow long-term prediction without loss of sensitivity to heterogeneity of material to biodegradation. If the goal of such an analysis is for long-term prediction, all wastes can be included with their proper characterizations without any loss of consistency.

## MOISTURE SORPTION FACTORS FOR MUNICIPAL WASTE MATERIALS

The development and use of moisture sorption factors for MSW materials as done for the present study provides a first step in the full characterization of MSW materials according to their likely response to moisture input. The moisture screening model allowed a choice of material for biodegradation based upon its water activity that can be directly or indirectly related to microorganism response models, which typically include consideration of the water activity of material.

Review and testing of likely models for response to water availability of this kind indicate poor reliability when both temperature and moisture are considered. For instance, the Davey (1989) model, which combines growth with water activity in the polynomial approximative form:

$$k = C_0 + C_1/T + C_2/T + C_3 a_w + C_4 a_w^2$$

was premised upon the finding that, while the shape for the curve (quite similar to that for growth vs. substrate), it did not vary much with temperature (degree Kelvin); *apparent curvature* with water activity $a_w$ was observed for the normal 0.0 to 1.0 (dry to saturation) range of this parameter. Testing the range of high saturation levels had to be reached before growth could begin, and thus the model showed no substrate biodegradation until the material was indicated to have reached a relatively high saturation. Additionally, the response was erratic and unreasonable, i.e., beyond possible expectations. Testing of other models also indicated that response over the range of moisture content was unreliable; thus the moisture response models available could not be adapted.

This suggests that, while the waste wetness screening model could be used, the next research step would be to develop a reliable new moisture response model for this type of reactor. For the present type of model, accuracy of determination of the moisture activity is essential. Approximations were set up for testing according to the discussion below.

## MOISTURE RESPONSE OF MATERIALS TO THE ENVIRONMENT

Though the thermodynamics of moisture loss to the atmosphere are reasonably well established, relatively few approaches were relevant to the model developed. This meant that testing of any models developed solely for the purpose of water sensitivity of materials could not be fully tested. However, some discussion of testing issues and relevance to database information is useful.

Under steady state conditions, the moisture loss from a material could be expected to be equivalent to the heat energy required for vaporization:

$$(IT_{gas} - T_{solid}) = k_{solid}(dT_s/dy) = h_{gf}(IT_{gas} - T_{solid}) - \lambda(p_s - p)K_{mass}$$

where $k_{solid}$ = thermal conductivity of the solid, $dT_s$ = temperature variation across the solid's thickness $dx$, $h_{gf}$ = heat transfer coefficient for the gas liquid film at the surface of the material, $T_{gas}$ = temperature of the gas surrounding the solid, $T_{solid}$ = temperature at the surface of the solid, $\lambda$ = latent heat of vaporization, $p_s$ = saturated vapor pressure for moisture at the surface of the solid, $p$ = partial vapor pressure or moisture in the gas and $K_{mass}$ = mass transfer coefficient for the material. The term $p_s$ directly relates to the water activity of the material as material water activity governs or controls moisture available to the surface:

$$p_s/p_{s,atmospheric} = a_{w,material}$$

At steady state at the material surface:

$$k_{solid}(dT_s/dy) = 0$$
$$h_{gf}(T_{gas} - T_{solid}) = \lambda(p_s - p)K_{mass}$$

Thus, with substitution for $p_s$, the moisture of the mass transfer term $K_{mass}$ could be approximated as:

$$K_{mass} = [h_{gf}(T_{gas} - T_{solid})]/[\lambda(p_{s,atm}\, a_{w,m} - p_{wtr})]$$

In a landfill atmosphere the bulk air can be assumed to be saturated by all accounts so that $p_{s,atm}$ = (rel. humidity = 1) $\times$ saturation vapor pressure (PVS) according to Antonie's equation:

$$\text{PVS} = a/(T\ \text{deg}\ .K) + b$$

where $a$ and $b$ have the values −2238 and 8.896 for water, indicating:

$$K_{mass} = [h_{gf}(T_{gas} - T_{solid})]/[\lambda(\{a/(T\ \text{deg}\ .K) + b\}a_{w,m} - p_{wtr})]$$

The value of $p_{wtr}$ for water vapor in the bulk gas is similarly:

$$p_{wtr} = (\text{PVS} \times \text{relative humidity of gas}) = a/(T\ \text{deg}\ .K) + b, \text{ implying:}$$

$$K_{mass} = [h_{gf}(T_{gas} - T_{solid})]/[\{a/(T\ \text{deg}\ .K) + b\}\ \lambda[a_{w,m} - 1)]$$

Examination of the above expression leads to the conclusion that, if an equivalence is made between $T_{solid}$ and $T_{gas}$, and if the original value of $a_{w,m}$ is known, an expression for mass loss per unit time could be developed. An equivalence was made by considering that the temperature at the solid surface for a gas water vapor film model could be approximated from the perfect gas law according to:

$$(T_{gas} - T_{solid})/(p_s - p)\lambda\, 0.5\, \text{mmH}_g/\text{deg C}$$
$$\text{for } T_{solid} = T_{gas} - 0.5(p_s - p)$$

Substitution for $T_{solid}$ in the expression for $K_{mass}$, with proper adjustment for degree temperature units, leads to an expression of the form:

$$K_{mass} = [h_{gf}T_{solid}(0.5(p_s - p))]/[\{a/(T\ \text{deg}\ .K) + b\}\ \lambda\ [a_{w,m} - 1)]$$

However, examination also indicates that unless the temperature and pressure values vary, the transfer rate is constant, indicating a controlled and unlikely situation for a landfill (bioreactor) subject to gas throughflow.

The object of moisture adjustment calculation is to determine loss naturally occurring in the landfilled material between two temperatures or saturation pressures. The assumption was made that both initial and final temperature of the landfill layer would be known, thus any moisture loss due to small temperature differences between the landfilled bulk gas initial moisture transfer rate of constant value was made to simulate the initial temperature condition. A change in temperature for the solids could be incorporated.

The change in mass transfer of moisture could be estimated by presuming that a change in temperature of the solid material between a time period *dt* is known. This permits the new water activity of the material to be calculated because the starting moisture content and thus water activity are presumably known according to the relations discussed in Chapter 2:

$$a_{w,1}/w_1 = A + Ba_{w,1} + Ca_{w,1}^2$$

The values of $A$, $B$ and $C$ had been estimated for the materials landfilled for the present study. It would be expected that the values of $A$, $B$ and $C$ should vary with temperature and according to whether moisture loss or gain is considered. However, for the relatively limited range of temperatures likely to be encountered inside a landfill, it would be unlikely the average values of $A$, $B$ and $C$ from moisture sorption data at ambient temperature would be in great error because desorption/sorption curves are indicated to be practically the same for most of the materials involved. Also, temperature effects on fitting constants are indicated to be minimal for small ranges of temperature.

The latent heat of evaporation is also temperature dependent. According to Piver and Lindstrom (1991), it varies according to temperature as:

$$\lambda(T) = 598.88 - 0.547(T - 273.16)$$

## TESTING APPROACH

It is likely that for most cases of waste site reactor simulation the change in temperature over a time period *dt* would be known, thus estimation of the new saturation vapor pressures for a current temperature would be known. Also, variation of the latent heat of evaporation permits the variation in mass transfer to be accounted for according to $dK_{mass}(T)$.

If, then, one of the landfilled materials is indicated to have an initial temperature of 25°C and the system undergoes an average temperature change $dT = 10°C$ in 4 days, neglecting daily variation for the purpose of long-term simulation with the assumption that moisture exchanges would adjust according to long-term temperature trends in the soil, and the values of constants are:

$$h_{gf} = 2.456 \text{ K/kg water}$$

$$T_{solid} = 10°C$$

$$T_{solid,0} = 25°\text{C}$$

$$a = -2238$$

$$b = 8.896$$

$$d(\lambda) = 5.47 \times 1000 \text{ kcal/kg water}$$

$$a_{w,m} = 0.94 \text{ (moist material)}$$

$$h_{gf}(T_{gas} - T_{solid}) = \{a/(T \deg .K) + b\} [\{(a_{w,m,2} - 1)\}]$$

requiring solution for the new water activity, moisture changes are indicated to be in the range 0.0017 kg per unit of exposed surface for this change of average temperature.

Discussion of potential moisture loss modeling is presented here for the purpose of inclusion, and because it is a topic of importance to fine-scale tuning of moisture parameters of the model. For example, if the moisture availability in terms of material water activity is an important growth factor for microorganisms attached to the material, it provides a means of direct estimation of moisture content change with change in temperature. Provision of moisture sorption factors A, B and C as general physical characteristics of the material in the database provides the information development of a relevant model of the present type.

A worthy area of research would thus be further refinement of the model for water sensitivity by development of a simplified but generally applicable model suitable to estimating water loss as specific exchanges between the materials included in the landfill interior and the bulk gas under flow.

## OTHER PROPERTIES ESTIMATED FOR THE DATABASE

In cases where values of hydrolysis rates could not be obtained, default values were used based upon similarity to other materials of known rate. For instance, no values could be reliably developed for items such as leather or glass. It is believed that the decay rate for glass may be similar to that for stone, i.e., on the order of hundreds or even thousands of years depending upon source. Leather, which is indicated to be attacked by fungi, should decay at rates slightly above those for wood. Other recalcitrant materials such as plastics, metals and rubber were discussed in detail in Chapter 7.

It is thus recommended that, because the correct values of some materials are imperfectly known for the purposes of this study, database values included are not necessarily definitive, and could be used and updated as information becomes available.

## CONSTANTS FOR AEROBIC AND ANAEROBIC DECOMPOSITION

The discussions of Chapters 2 and 3 and under stoichiometry indicate that, while rates available in literature could be used, they should not be used unless they appear to be reliable averages for the materials considered. For instance, cell yield per unit of material is one of the most substrate-specific model inputs.

The values of $Y_{X/S}$, based upon stoichiometric expressions developed, ranged from:

0.2 to 0.5 for paper materials
0.4 to 0.7 for yard wastes
0.5 to 0.7 for wood
0.7 to 0.9 for food wastes

This range is relatively small. Simplified weighted averages could be used to test model sensitivity to cell yield. It is obvious that this factor controls the overall biodegradation rate and thus should be examined for sensitivity.

## SOIL MOISTURE CONTENT

Model sensitivity to moisture is one of the most important parameters of the present model to be validated. The expressions developed for water activity of materials, moisture variation between dryness and saturation vs. flow and for surface area cover many of the more complicated issues involved in such an analysis.

## MOISTURE INFLOW EFFECT OF COVER

While predictions with the models are routine, response of biodegradation to moisture content or availability requires careful incorporation of leachate flow terms in general kinetic expressions.

If, for instance, the rate of recycle is set at 0.4 for the anaerobic and aerobic routines, as representative of older landfills without good liners and possibly landfill bioreactors, at which leachate control is implemented through pumping and reinjection or spraying, leachate inflows and outflows can be integrated to show the effect of inflow of liquid COD or of organic acids could. Slow variation of the recycle value — and comparison with nonrecycle options ($R = 0$) could allow prediction of leachate recycle effects upon gas generation and organism growth. Development of the aerobic and anaerobic decomposition, heat, water input and other model sections can thus provide the major testing of leachate and gas recycle.

## TEMPERATURE AS A DECOMPOSITION FACTOR

The kinetic rates were developed to include temperature variation of reaction according to Arrhenius law. This approach has been found to be widely applicable to biological processes, and various investigators have developed correlative forms of reaction equations for growth response to temperature. As temperature was in turn dependent upon the moisture input model and thus prediction instability, the biokinetic models were used with set values of temperature for the layers. Thus temperature effects were incorporated and indicate some effect upon biodegradation vs. landfill depth. Moisture storage predictions must be refined before sensitivity testing; temperature initialization is a first priority of future refinement of a model.

## BIOFILTRATION EFFECT

Development of the general gas flow model necessitates the development of a database similar to that for MSW characteristics, for landfill gas properties and for leachate quality and chemical constituents. This provides a basis for analysis of model performance for removal of gas phase consitituents. One purpose of gas generation and flow modeling would be to allow landfill or waste site systems to be used as bioftration systems, as it could be assumed that microorganisms could be present that might act as sinks for destruction of trace chemicals in solution.

Similarity of gas flow, heat and oxygen flow models means that analytical solutions could be developed as for the heat equation of Chapter 6, or approximate solutions, once the proper conditions are specified and boundary conditions set.

The approach for biofiltration simulation, limited in favor of the present phase, should consider the *aerobic layers or segments of the landfill as containing methanotrophs that could act as sinks for methane produced from anaerobic layers.*

## SETTLEMENT EFFECT

A landfill layer and column settlement testing module can be developed to give results that are dimensionally correct on the order of thousandths of a meter for short periods of simulation, and strain decreased with time but reliability cannot be guaranteed by this stage of model development.

However, plotting behavior indicates sensitivity to volumetric water content over the time periods of study, Thus, in soil settlement models, settlement effect can be incorporated in the compression index term, which is the slope of the strain vs. time curve.

## DISCUSSION

The waste site biodegradation model approach for the present work provides information for simulation of layer and material-specific biodegradation under moisture input variation as the result of cover conditions in the field, geographic temperatures to incorporate temperature variation with depth, heat generation by biological reaction, gas generation and transport, moisture and chemical flux and chemicals biodegradation via the gas transport model flux and gas recirculation for the purpose of biofiltration or reduction of methane. The scope of effort requires development of input values, mathematical models and fully integrated moisture sensitivity both on a macroscopic and microscopic scale limited by the quantity of model outputs or results.

## MOISTURE INPUT

Data for moisture input were reliably passed to the model bioreactor segment via a cover moisture input MATLAB routine developed for the study but not included at this stage.

However, one of the difficulties for estimation of moisture in the layer — which also affects values of long-term prediction with the heat generation and flow model — was that the program routines for solution of the nonlinear relation for moisture content of layers of the water balance module results in unstable solutions when flow in the layer is large, or, more likely, small, and for the narrow range of numerical values typically associated with landfill layer saturation (0.4 to 0.6 for MSW, 0.2 to 0.44 for sand) used to simulate drainage between layers. In many cases, these types of moisture relations are solved using transformations into ordinary differential equations that are then solved using partial differencing. Resolution of the issue includes use of a smaller value for the initial guess.

Simulations with the heat generation and transport model indicates that relatively stable moisture inputs are needed to guarantee that temperature predictions are useful. The temperature model is similarly sensitive to moisture flows, as it incorporates effect of leachate flows and water content. The system temperature makes use of a stable heat generation model important, as temperature values generated therefrom are to be integrated with biokinetic relationships for biodegradation predictions. A priority is thus to include a stable moisture behavior predictor in the landfill bioreactor model.

## CONCLUSIONS

The modeling approaches presented with databases provide a first look at development of a bioreactor system model applicable to existing water inflow models for landfills or as a stand-alone tool for mixed, heterogenous solid substrate bioreactor analyses which could be used to examine several gas or liquid effluent treatment options.

Simple test modules indicate the biological component performs as expected and that material and landfill, long-term biodegradation and gas effluent generation, with sensitivity to moisture and heat, can be predicted by the model.

The approaches fully incorporate for the first time the moisture balances of importance to biological reactions such as substrate moisture content and sorption properties. This may overcome to a large extent the difficulty of addressing waste heterogeneity as a bioreactor constraint in waste site systems. It also allows representation of the waste site as a segmented, packed bed reactor system subject to trickling and poorly distributed flows of moisture, saturated by gas and subject to internal heat generation.

Complete sensitivity analyses and predictive results for models would validate the assumptions of this work and fulfill its various objectives of developing a new predictive tool.

## RECOMMENDATIONS

It is clearly indicated that validation of heat and moisture behavior of this type of landfill model, and of system gas treatment capacity, would be most promising and would provide valuable areas of new research. These would add to the reliability of predictions.

New research priorities for the topic of this study should include:

1. Determination of limitations upon aerobic or anaerobic biodegradation in this type of relatively harsh environment. While some assumptions had to be made with regard to the applicability of current theory to the study, there is a need for theory to support predictive links between biodegradation in anaerobic vs. aerobic environments.
2. Development of more reliable water availability vs. growth substrate consumption rate prediction models. While it is unlikely that landfills would be too dry even under the most arid capped conditions to curtail biodegradation at some level, the large variation in moisture availability to organisms when the surfaces of materials become too dry could have potentially serious limiting effects upon landfill biodegradative capacity. This suggests the importance of models capable of predicting the availability of water — rather than only the water content — of the variety of materials indicated to be potentially biodegradable in the long term.
3. Validation of predictive capacity of the model by forward and backward exercise.
4. Validation of the effect of heat generation upon system flows and reactions.
5. Validation of mass transfer effects important to use of reactors as potential systems with capacity for gas quality improvement should be studied. Surface area considerations indicate that the system's high porosity could decrease the availability of moisture to soil organisms. Hydrolysis is a surface phenomenon, and gas outflow could essentially dry the surfaces of materials contacted. This could increase the difficulty of enhanced biodegradation attempts when the landfill has had poor moisture supply. High porosity similarly affects the capacity for optimal dissolved oxygen supply, nutrient supply or methane oxidation and decreases the effectiveness of treatment of aerobic treatment or of using the uncontrolled system for biofiltration.

# Appendix 1
## Waste Properties

| MSW Component | Material | Percent Mass, kg/kg | Percent Water, as Fresh | % Dry Weight Ash | Volatile Matter, % Dry Weight | Maximum Water Sorption, g/100 g dry matter | Specific Gravity | Bulk Density, kg/cu.m | Percent Oxidizab. Dry Weight | Aerobic Hydrolysis Constant, per day |
|---|---|---|---|---|---|---|---|---|---|---|
| Newspapers (newsprint) | Paper | 4.6 | 5.97 | 1.43 | 98 | 11.32 | 0.92 | 443.8 | 92.6 | 0.02 |
| Books (bond) | Paper | 0.6 | 10.24 | 5.38 | 75.94 | 12.4 | 0.92 | 560.6 | 84.38 | 0.0116 |
| Magazines (bond) | Paper | 0.9 | 4.11 | 22.47 | 66.39 | 9 | 0.92 | 642.5 | 73.42 | 0.0116 |
| Office Papers (bond) | Paper | 2.4 | 10.24 | 5.38 | 75.94 | 12.4 | 0.92 | 361.8 | 84.38 | 0.0136 |
| Telephone Directories (light kraft) | Paper | 0.3 | 10.24 | 5.38 | 75.94 | 29.85 | 0.92 | 560.6 | 84.38 | 0.0175 |
| Third Class Mail (bleached kraft) | Paper | 2.4 | 4.56 | 13.09 | 73.32 | 12.78 | 0.92 | 361.8 | 82.35 | 0.0175 |
| Other Commercial Printing (bond) | Paper | 3.5 | 10.24 | 5.38 | 75.94 | 12.4 | 0.92 | 361.8 | 84.38 | 0.0136 |
| Tissue Paper & Towels (absorbent) | Paper | 1.8 | 6 | 6.7 | 87.3 | 37.2 | 0.92 | 89 | 87.3 | 0.015 |
| Paper Plates & Cups (foodboard) | Paper | 0.5 | 4.71 | 6.5 | 84.2 | 17.3 | 0.92 | 361.8 | 88.79 | 0.0054 |
| Plastic Plates & Cups (coated board) | Paper | 0.3 | 4.71 | 2.64 | 84.2 | 22.5 | 0.92 | 361.8 | 92.65 | 0.005 |
| Disposable Diapers (absorbent) | Paper | 0.6 | 60.2 | 3.6 | 3.7 | 20 | 0.92 | 119 | 36.2 | 0.0008 |
| Other NonPackaging Paper (unbleached sulfite cellulose) | Paper | 2.8 | 5.83 | 6.7 | 79.7 | 11.02 | 0.92 | 361.8 | 87.47 | 0.0175 |
| Clothing | Clothes | 2.2 | 10 | 2.2 | 84.34 | 27.31 | 1.55 | 307.7 | 87.8 | 0.0003 |
| Towels, Sheets & Pillowcases (cotton linen) | Textile | 0.4 | 14.7 | 2.4 | 84.34 | 12.5 | 1.55 | 307.7 | 82.9 | 0.00026 |
| Corrugated boxes (kraft pulp) | Paper | 8 | 5.2 | 5.06 | 77.47 | 11.66 | 0.92 | 422.8 | 89.74 | 0.0688 |
| Milk Cartons (waxboard) | Paper | 0.3 | 3.45 | 1.17 | 90.92 | 17.3 | 0.92 | 361.8 | 95.38 | 0.0027 |
| Folding Cartons (paperboard) | Paper | 2.6 | 6.11 | 6.5 | 75.59 | 22.5 | 0.92 | 361.8 | 87.39 | 0.00186 |
| Other Paperboard | Paper | 0.2 | 6.11 | 6.5 | 75.59 | 23.4 | 0.92 | 361.8 | 87.39 | 0.0215 |
| Bags & Sacks (kraft) | Paper | 1.1 | 5.83 | 1.01 | 83.92 | 10 | .92 | 1265 | 93.16 | 0.0175 |
| Wrapping Papers (kraft) | Paper | 0.1 | 5.83 | 1.01 | 83.92 | 13.8 | 0.92 | 293.7 | 93.16 | 0.0175 |
| Other Paper Pkg. (kraft) | Paper | 0.7 | 5.83 | 1.01 | 83.92 | 13.8 | 0.92 | 148.3 | 93.16 | 0.0175 |
| Soft Drink & Beer | Glass | 2.3 | 1.67 | 96.49 | 0.93 | 0.2 | 2.47 | 806.2 | 1.84 | 0.000006 |
| Wine & Liquor | Glass | 0.8 | 1.67 | 96.49 | 0.93 | 0.2 | 2.47 | 806.2 | 1.84 | 0.000006 |
| Food & Other | Glass | 2.5 | 1.67 | 96.49 | 0.93 | 0.2 | 2.47 | 806.2 | 1.84 | 0.000006 |
| Beer & Soft Drink | Ferrous | 0 | 12 | 118.9 | 0.0002 | 2 | 7.86 | 1192.5 | 0.005 | 0.000055 |
| Food & Other | Ferrous | 0.9 | 12 | 118.9 | 0.0002 | 2 | 7.86 | 1192.5 | 0.005 | 0.000055 |
| Other | Ferrous | 0.1 | 12 | 118.9 | 0.0002 | 2 | 7.86 | 1192.5 | 0.005 | 0.000055 |
| Beer & Soft Drink | Aluminum | 0.4 | 2 | 93.03 | 0.0002 | 2 | 2.73 | 227.3 | 4.97 | 0.000068 |
| Other Cans | Aluminum | 0 | 2 | 93.03 | 0.0002 | 2 | 2.73 | 227.3 | 4.97 | 0.000068 |
| Foil & Closures | Aluminum | 0.2 | 2 | 93.03 | 0.0002 | 2 | 2.73 | 107.1 | 4.97 | 0.000068 |
| Soft Drink Bottles (PET) | Plastics | 0.2 | 0.2 | 5 | 79.7 | 2.5 | 0.95 | 1330 | 94.8 | 0.0002 |

| Anaerobic Hydrolysis Constant, per day | Water Sorption Factor, A | Water Sorption Factor, B | Water Sorption Factor, C | Maximum Water Content, Fraction | Percent Dry Weight Carbon, C | Percent Dry Weight Hydrogen, H | Percent Dry Weight Oxygen, O | Percent Dry Weight Nitrogen, N | Percent Dry Weight Sulfur, S | Specific Heat, kJ/kg K |
|---|---|---|---|---|---|---|---|---|---|---|
| 0.000325 | 3.64 | 22.46 | −17.27 | 0.113223471 | 49.14 | 6.1 | 43.03 | 0.05 | 0.16 | 1.34 |
| 0.0029 | 5.23 | 10.28 | −3.07 | 0.080380489 | 43.4 | 5.82 | 44.32 | 0.25 | 0.2 | 1.34 |
| 0.0029 | 5.23 | 10.28 | −3.07 | 0.080380489 | 32.91 | 4.95 | 38.55 | 0.07 | 0.09 | 1.34 |
| 0.0034 | 5.23 | 10.28 | −3.07 | 0.080380489 | 30.7 | 5 | 44 | 0.09 | 0.12 | 1.34 |
| 0.004375 | 1.565 | 16.96 | −14.85 | 0.271987355 | 43.4 | 5.82 | 44.32 | 0.25 | 0.2 | 1.34 |
| 0.004375 | 3.68 | 21.5 | −17.38 | 0.128170522 | 37.87 | 5.41 | 42.74 | 0.17 | 0.09 | 1.34 |
| 0.0034 | 5.23 | 10.28 | −3.07 | 0.080380489 | 43.4 | 5.82 | 44.32 | 0.25 | 0.2 | 1.34 |
| 0.00375 | 1.96 | 22.04 | −20.43 | 0.279936475 | 43.5 | 6 | 44 | 0.3 | 0.2 | 1.34 |
| 0.00135 | 1.136 | 20.29 | −15.66 | 0.17337995 | 45.3 | 6.17 | 45.5 | 0.18 | 0.08 | 1.34 |
| 0.00125 | 1.136 | 20.29 | −15.66 | 0.17337995 | 45.3 | 6.17 | 45.5 | 0.18 | 0.08 | 1.34 |
| 0.0002 | 1.96 | 22.04 | −20.43 | 0.279936475 | 16.2 | 10 | 71.4 | 0.28 | 0.04 | 1.34 |
| 0.004375 | 2.26 | 14.94 | −8.23 | 0.111469684 | 43.4 | 5.82 | 44.32 | 0.25 | 0.2 | 1.34 |
| 0.000075 | 1.308 | 13.12 | −10.89 | 0.282548147 | 55 | 6.6 | 31.2 | 4.12 | 0.13 | 1.34 |
| 0.000065 | 3.77 | 19.68 | −15.64 | 0.128009159 | 28.6 | 7.7 | 54 | 0.63 | 0.14 | 1.335 |
| 0.0172 | 2.36 | 16.32 | −10.122 | 0.116832691 | 43.73 | 5.7 | 44.93 | 0.09 | 0.21 | 1.34 |
| 0.000675 | 1.136 | 20.29 | −15.66 | 0.17337995 | 59.18 | 9.25 | 30.13 | 0.12 | 0.1 | 1.34 |
| 0.000465 | 1.733 | 17.23 | −14.51 | 0.224485822 | 44.74 | 6.1 | 41.92 | 0.15 | 0.16 | 1.34 |
| 0.005375 | 1.733 | 17.23 | −14.51 | 0.224485822 | 35.5 | 6.8 | 49.6 | 1 | 0.11 | 1.34 |
| 0.004375 | 4.68 | 12.84 | −7.62 | 0.100997552 | 44.9 | 6.08 | 47.34 | 0 | 0.11 | 1.34 |
| 0.004375 | 0.95 | 20.28 | −13.98 | 0.137902636 | 44.9 | 6.08 | 47.34 | 0 | 0.11 | 1.34 |
| 0.004375 | 0.95 | 20.28 | −13.98 | 0.137902636 | 44.74 | 6.1 | 41.92 | 0.15 | 0.16 | 1.34 |
| 1.50E-06 | 30 | 30 | −15 | 0.02222 | 1.37 | 0.14 | 0.22 | 0.04 | 0.03 | 0.662 |
| 1.50E-06 | 30 | 30 | −15 | 0.02222 | 1.37 | 0.14 | 0.22 | 0.04 | 0.03 | 0.662 |
| 1.50E-06 | 30 | 30 | −15 | 0.02222 | 1.37 | 0.14 | 0.22 | 0.04 | 0.03 | 0.662 |
| 1.38E-05 | 28 | 8 | −3 | 0.030300184 | 1.9 | 0.3 | 0 | 0.05 | 0.01 | 0.63221 |
| 1.38E-05 | 28 | 8 | −3 | 0.030300184 | 1.9 | 0.3 | 0 | 0.05 | 0.01 | 0.63221 |
| 1.38E-05 | 28 | 8 | −3 | 0.030300184 | 1.9 | 0.3 | 0 | 0.05 | 0.01 | 0.63221 |
| 0.000017 | 30 | 5 | −3 | 0.031246777 | 3.05 | 0.78 | 0 | 0.09 | 0.01 | 0.934 |
| 0.000017 | 30 | 5 | −3 | 0.031246777 | 3.05 | 0.78 | 0 | 0.09 | 0.01 | 0.934 |
| 0.000017 | 30 | 5 | −3 | 0.031246777 | 3.05 | 0.78 | 0 | 0.09 | 0.01 | 0.934 |
| 0.0002 | 121 | 100 | −19 | 0.004950152 | 84.38 | 14.18 | 0 | 0.06 | 0.03 | 1.57 |

| MSW Component | Material | Percent Mass, kg/kg | Percent Water, as Fresh | % Dry Weight Ash | Volatile Matter, % Dry Weight | Maximum Water Sorption, g/100 g dry matter | Specific Gravity | Bulk Density, kg/cu.m | Percent Oxidizab. Dry Weight | Aerobic Hydrolysis Constant, per day |
|---|---|---|---|---|---|---|---|---|---|---|
| Milk Bottles (HDPE) | Plastics | 0.3 | 0.2 | 1.19 | 79.7 | 2.5 | 0.95 | 950 | 98.61 | 8.20E-07 |
| Other Containers (PPE) | Plastics | 1.2 | 0.2 | 5 | 79.7 | 0.1 | 0.9 | 905 | 94.8 | 2.42E-06 |
| Bags & Sacks (LDPE) | Plastics | 1 | 0.2 | 1.19 | 79.7 | 2.5 | 0.92 | 915 | 98.61 | 1.37E-06 |
| Wraps (PVC) | Plastics | 1.3 | 0.2 | 2.06 | 79.7 | 0.07 | 0.92 | 1350 | 97.74 | 0.000685 |
| Other Plastic Pkg. (PS, etc) | Plastics | 1.6 | 0.2 | 0.45 | 79.7 | 0.06 | 0.92 | 160 | 99.35 | 1.80E-06 |
| Wood Pkg. (pine) | Wood | 5.5 | 12 | 0.75 | 77.62 | 27.94 | 0.43 | 187.8 | 87.25 | 0.00011 |
| Other Wood | Wood | 4.1 | 6 | 1.34 | 80.92 | 27.94 | 0.45 | 187.8 | 92.66 | 0.00011 |
| Fish (cooked trout) | Food | 0.035 | 63.41 | 1.67 | 52.7 | 45.9 | | 288.3 | 34.92 | 0.00055 |
| Beef (cooked beef) | Food | 0.54 | 50 | 3.11 | 56.34 | 63.8 | | 288.3 | 46.89 | 0.00055 |
| Pork (cooked) | Food | 0.635 | 42.9 | 3.11 | 56.34 | 28.1 | | 288.3 | 53.99 | 0.00055 |
| Lamb (cooked, Refuse, = pork sorp.) | Food | 0.013 | 40.3 | 3.2 | 34.1 | 40.91 | | 288.3 | 56.5 | 0.00055 |
| Chicken (cooked) | Food | 2.165 | 57.3 | 1.05 | 56.34 | 30.77 | | 288.3 | 41.65 | 0.00055 |
| Eggshells | Food | 0.106 | 3 | 90 | 4.4 | 11 | | 288.3 | 7 | 0.00055 |
| Coffee Grounds | Food | 0.194 | 62.9 | 0.5 | 67 | 53.7 | | 288.3 | 36.6 | 0.00055 |
| Tea Leaves | Food | 0.0211 | 86 | 3.2 | 74 | 57.6 | | 288.3 | 10.8 | 0.00055 |
| Bread (white) | Food | 0.4 | 35.3 | 2.6 | 85 | 17.9 | | 288.3 | 62.1 | 0.00055 |
| Fats, Oils | Food | 0.185 | 0 | 0 | 97.64 | 0 | | 288.3 | 100 | 0.00055 |
| Orange (Peel and Seeds 26%, sorp = citrus pulp) | Food | 0.273 | 64.5 | 0.8 | 26.7 | 103.9 | | 288.3 | 34.7 | 0.028 |
| Tangerine (sorp = citrus pulp) | Food | 0.044 | 72.5 | 3.18 | 71.46 | 103.9 | | 288.3 | 24.32 | 0.028 |
| Grapefruit (Peel, Seeds, Core, Membrane, 50%, sorp = citrus pulp) | Food | 0.233 | 44.53 | 0.31 | 71.46 | 103.9 | | 644.4 | 55.16 | 0.028 |
| Apple (Core and Stems, 8%) | Food | 0.33 | 77.2 | 0.24 | 15.24 | 90.6 | | 644.4 | 22.56 | 0.028 |
| Avocado (Skin and Seed, 24%) | Food | 0.07 | 55.12 | 0.83 | 21.62 | 75.5 | | 644.4 | 44.05 | 0.00055 |
| Banana (Skin or Peel, 35%, sorp = edible fruit) | Food | 0.46 | 48.25 | 0.52 | 16.53 | 45.7 | | 644.4 | 51.23 | 0.00055 |
| Cherry, Grape (Sweet, Raw, Pits and Stems, 10%) | Food | 0.134 | 72.65 | 0.4 | 17.2 | 145.8 | | 644.4 | 26.95 | 0.00055 |
| Peach (Pits and Skins, 24%) | Food | 0.07 | 66.6 | 1.4 | 9.54 | 54 | 1.32 | 1530 | 32 | 0.00055 |
| Pear (Core and Stems, 8%) | Food | 0.018 | 77.1 | 3.25 | 15.92 | 145.8 | | 644.4 | 19.65 | 0.028 |
| Pineapple (Core 6%, Crown 10%, Parings 32%, sorp = fruit) | Food | 0.051 | 93.7 | 0.46 | 15.53 | 66.82 | | 644.4 | 5.84 | 0.028 |

| Anaerobic Hydrolysis Constant, per day | Water Sorption Factor, A | Water Sorption Factor, B | Water Sorption Factor, C | Maximum Water Content, Fraction | Percent Dry Weight Carbon, C | Percent Dry Weight Hydrogen, H | Percent Dry Weight Oxygen, O | Percent Dry Weight Nitrogen, N | Percent Dry Weight Sulfur, S | Specific Heat, kJ/kg K |
|---|---|---|---|---|---|---|---|---|---|---|
| 2.05E-07 | 121 | 100 | −19 | 0.004950152 | 84.54 | 14.18 | 0 | 0.06 | 0.03 | 2.30274 |
| 6.05E-07 | 121 | 100 | −19 | 0.004950152 | 45.04 | 5.6 | 1.56 | 0.08 | 0.14 | 1.57 |
| 1.37E-06 | 121 | 100 | −19 | 0.004950152 | 84.54 | 14.18 | 0 | 0.06 | 0.03 | 2.30274 |
| 0.000171 | 121 | 100 | −19 | 0.004950152 | 45.14 | 5.61 | 1.56 | 0.08 | 0.02 | 1.57 |
| 4.50E-07 | −0.31 | 34.83 | −14.14 | 0.049064384 | 86.91 | 8.42 | 3.96 | 0.21 | 0.02 | 1.382 |
| 2.75E-05 | 0.838 | 14.89 | −12.28 | 0.289912903 | 51 | 6.2 | 41.8 | 0.1 | 0.08 | 2.09 |
| 2.75E-05 | 2.2 | 9.67 | −8.3 | 0.280029681 | 49.7 | 6.1 | 42.6 | 0.1 | 0.08 | 2.303 |
| 0.000138 | 0.817 | 18.7 | −17.83 | 0.592113741 | 59.59 | 9.47 | 24.65 | 1.02 | 0.19 | 3.63 |
| 0.000138 | 1.06 | 15.12 | −14.61 | 0.636307579 | 59.59 | 9.47 | 24.65 | 1.02 | 0.19 | 3.08 |
| 0.000138 | 0.705 | 23.8 | −20.94 | 0.280334702 | 59.59 | 9.47 | 24.65 | 1.02 | 0.19 | 2.901 |
| 0.000138 | 4.73 | 14.88 | −15.12 | 0.222618729 | 59.59 | 9.47 | 24.65 | 1.02 | 0.19 | 3.203 |
| 0.000138 | 2.207 | 7.29 | −6.504 | 0.334015714 | 59.59 | 9.47 | 24.65 | 1.02 | 0.19 | 3.53 |
| 0.000138 | −0.5 | 30 | −16 | 0.07406557 | 0 | 0 | 0 | 0 | 0 | 3.31 |
| 0.0018 | 0.531 | 29.77 | −28.44 | 0.536510303 | 48 | 6.4 | 37.6 | 2.6 | 0.4 | 3.8 |
| 0.000138 | 0.974 | 21.133 | −20.37 | 0.574998686 | 48 | 6.4 | 35.8 | 4.41 | 0.4 | 3.8 |
| 0.000138 | 18.32 | −29.45 | 16.74 | 0.178248093 | 48 | 6.4 | 37.6 | 2.6 | 0.4 | 2.6 |
| 0.000138 | 1400 | 1200 | −500 | 0.000476147 | 73.14 | 11.54 | 14.82 | 0.43 | 0.07 | 1.968 |
| 0.0047 | 8.25 | −9.8 | 2.51 | 1.04104412 | 44.11 | 5.11 | 38.34 | 1.02 | 0.11 | 3.85 |
| 0.0047 | 8.25 | −9.8 | 2.51 | 1.04104412 | 44.11 | 5.11 | 38.34 | 1.02 | 0.11 | 3.85 |
| 0.0047 | 8.25 | −9.8 | 2.51 | 1.04104412 | 44.11 | 5.11 | 38.34 | 1.02 | 0.11 | 3.85 |
| 0.0047 | 2.22 | 9.61 | −10.73 | 0.908021901 | 48 | 6.4 | 37.6 | 2.6 | 0.4 | 3.78 |
| 0.000138 | 0.64 | 25.31 | −24.625 | 0.753280628 | 48 | 6.4 | 37.6 | 2.6 | 0.4 | 3.31 |
| 0.000138 | 5.61 | 5.3 | −13.1 | −0.457011458 | 48 | 6.4 | 37.6 | 2.6 | 0.4 | 3.56 |
| 0.000138 | 1.85 | 6.94 | −5.33 | 0.288957377 | 53.9 | 7.1 | 38.4 | 0.26 | 0.34 | 1.34 |
| 0.000138 | 2.3 | 13.52 | −13.96 | 0.537164816 | 51.45 | 6.34 | 41.92 | 0.2 | 0.09 | 1.34 |
| 0.007 | 5.1 | −3.21 | −1.21 | 1.469224771 | 48 | 6.4 | 37.6 | 2.6 | 0.4 | 3.751 |
| 0.007 | 8.392 | −13.18 | 6.744 | 0.511204351 | 48 | 6.4 | 37.6 | 2.6 | 0.4 | 3.802 |

| MSW Component | Material | Percent Mass, kg/kg | Percent Water, as Fresh | % Dry Weight Ash | Volatile Matter, % Dry Weight | Maximum Water Sorption, g/100 g dry matter | Specific Gravity | Bulk Density, kg/cu.m | Percent Oxidizab. Dry Weight | Aerobic Hydrolysis Constant, per day |
|---|---|---|---|---|---|---|---|---|---|---|
| Plum, Prune (Pits, 6%) | Food | 0.006 | 80.1 | 0.7 | 13.37 | 25 | 1.334 | 1526 | 19.2 | 0.028 |
| Strawberry (Caps and Stems, 6%) | Food | 0.011 | 86.06 | 0.4 | 7.5 | 80 | | 644.4 | 13.54 | 0.028 |
| Cantaloupe, Honeydew (Cavity Contents, 5%; Rind, 49%) | Food | 0.31 | 75.67 | 0.53 | 9.1 | 185.83 | | 644.4 | 23.8 | 0.028 |
| Watermelons (Casaba, Cavity Contents, 11%, Rind, 29%) | Food | 0.45 | 55.2 | 0.8 | 11.55 | 185.83 | | 644.4 | 44 | 0.028 |
| Potato (Baked, Flesh and Skin) | Food | 0.82 | 71.2 | 1.16 | 27.64 | 42.64 | | 539.3 | 27.64 | 0.00055 |
| Broccoli (Raw) | Food | 0.106 | 90.7 | 0.92 | 9.23 | 65.4 | | 539.3 | 8.38 | 0.00055 |
| Cabbage (Raw) | Food | 0.106 | 92.52 | 0.72 | 7.65 | 313.8 | | 539.3 | 6.76 | 0.00055 |
| Carrots (Raw) | Food | 0.088 | 78.8 | 0.77 | 12.4 | 105.32 | | 539.3 | 20.43 | 0.00055 |
| Cauliflower (Raw, sorp = cabbage) | Food | 0.11 | 92.3 | 0.66 | 8 | 80.6 | | 539.3 | 7.04 | 0.00055 |
| Corn (Cob 45%) | Food | 0.25 | 38.3 | 1.7 | 80.1 | 10 | | 539.3 | 60 | 0.00042 |
| Celery (Raw) | Food | 0.053 | 94.7 | 0.89 | 4.41 | 143.6 | | 539.3 | 4.41 | 0.00055 |
| Cucumber, Eggplant (Raw, sorp = marrow) | Food | 0.014 | 96.05 | 0.38 | 1.35 | 185.83 | | 539.3 | 3.57 | 0.00055 |
| Garlic, Onions (raw) | Food | 0.123 | 92.24 | 0.42 | 8.34 | 89.22 | | 539.3 | 7.34 | 0.00055 |
| Green Beans (Cooked) | Food | 0.011 | 89.22 | 0.73 | 11.4 | 31.8 | | 539.3 | 10.05 | 0.00055 |
| Green Pepper (Raw, Sweet) | Food | 0.041 | 92.77 | 0.62 | 6.61 | 331.4 | | 539.3 | 6.61 | 0.0243 |
| Lettuce, Spinach (Raw) | Food | 0.466 | 95.9 | 0.48 | 3.62 | 313.8 | | 539.3 | 3.62 | 0.00055 |
| Tomato (Red, Ripe, Raw) | Food | 0.095 | 93.95 | 0.61 | 5.44 | 182.8 | | 539.3 | 5.44 | 0.0162 |
| Cooked Rice | Food | 0.185 | 75 | 0.3 | 24.8 | 31.45 | | 288.3 | 24.7 | 0.112 |
| Leaves | Yard | 2.7 | 9.97 | 3.82 | 66.93 | 85.3 | | 196 | 86.21 | 0.0304 |
| Grass | Yard | 2.95 | 75.24 | 1.62 | 94.9 | 76.7 | | 159.7 | 23.14 | 0.0256 |
| Brush | Yard | 8.85 | 53.94 | 2.34 | 35.64 | 66.4 | | 104 | 43.72 | 0.0058 |
| Leather | Leather | 0.3 | 10 | 9.1 | 68.46 | 33.8 | | 293.4 | 80.9 | 0.0006 |
| Shoe, Heel, Sole (Leather, Styrenics, ABS) | L, P, | 0.3 | 1.15 | 29.74 | 67.03 | | | 293.4 | 69.11 | 0.0004 |
| Major Appliances (Metal) | Ferrous | 0.9 | 0.2 | 0 | 3 | | | 296.7 | 99.8 | 0.000068 |
| Small Appliances (Metal) | Ferrous | 0.5 | 0.2 | 0 | 3 | | | 296.7 | 99.8 | 0.000068 |

| Anaerobic Hydrolysis Constant, per day | Water Sorption Factor, A | Water Sorption Factor, B | Water Sorption Factor, C | Maximum Water Content, Fraction | Percent Dry Weight Carbon, C | Percent Dry Weight Hydrogen, H | Percent Dry Weight Oxygen, O | Percent Dry Weight Nitrogen, N | Percent Dry Weight Sulfur, S | Specific Heat, kJ/kg K |
|---|---|---|---|---|---|---|---|---|---|---|
| 0.007 | 1.85 | 6.94 | −5.33 | 0.288957377 | 47.73 | 5.9 | 43.57 | 0.32 | | 1.34 |
| 0.007 | 5.031 | 2.415 | −7.7623 | −3.174395381 | 48 | 6.4 | 37.6 | 2.6 | 0.4 | 3.93 |
| 0.007 | 3.86 | 3.994 | −7.43 | 2.352226993 | 48 | 6.4 | 37.6 | 2.6 | 0.4 | 4 |
| 0.007 | 3.86 | 3.994 | −7.43 | 2.352226993 | 48 | 6.4 | 37.6 | 2.6 | 0.4 | 4 |
| 0.000138 | 0.687 | 9.916 | −8.266 | 0.4277315 | 48 | 6.4 | 37.6 | 2.6 | 0.4 | 3.63 |
| 0.000138 | 3.304 | 9.5 | −13.32 | −1.944248416 | 49.06 | 6.62 | 37.55 | 1.68 | 0.2 | 3.93 |
| 0.000138 | 2.026 | 12.23 | −15.42 | −0.860396104 | 49.06 | 6.62 | 37.55 | 1.68 | 0.2 | 3.98 |
| 0.000138 | 5.92 | 3.09 | −9.96 | −1.054394148 | 49.06 | 6.62 | 37.55 | 1.68 | 0.2 | 3.88 |
| 0.000138 | 2.024 | 13.77 | −18.73 | −0.340840391 | 49.06 | 6.62 | 37.55 | 1.68 | 0.2 | 3.98 |
| 0.000105 | 3.5 | 12 | −12.5 | 0.333155646 | 48.7 | 5.69 | 44.85 | 0.58 | 0.15 | 4.9 |
| 0.000138 | 0.223 | 5.411 | −8.62 | −0.334995391 | 49.06 | 6.62 | 37.55 | 1.68 | 0.2 | 3.88 |
| 0.000138 | 3.86 | 3.994 | −7.43 | 2.352226993 | 49.06 | 6.62 | 37.55 | 1.68 | 0.2 | 4.078 |
| 0.000138 | 0.22 | 5.204 | −8.73 | −0.302562249 | 49.06 | 6.62 | 37.55 | 1.68 | 0.2 | 3.88 |
| 0.000138 | 1.53 | 19.4 | −20.76 | 5.806222594 | 49.06 | 6.62 | 37.55 | 1.68 | 0.2 | 3.902 |
| 0.006075 | 6.88 | 3.85 | −10.45 | 3.549459083 | 49.06 | 6.62 | 37.55 | 1.68 | 0.2 | 3.98 |
| 0.000138 | 0.837 | 15.32 | −15.25 | 1.100583773 | 49.06 | 6.62 | 37.55 | 1.68 | 0.2 | 4.053 |
| 0.00405 | 1.511 | 9.3 | −11.67 | −1.165933451 | 49.06 | 6.62 | 37.55 | 1.68 | 0.2 | 4.028 |
| 0.028 | 0.825 | 7.69 | −5.37 | 0.317902403 | 48 | 6.4 | 37.6 | 2.6 | 0.4 | 4.95 |
| 0.0076 | 0.699 | 14.82 | −13.15 | 0.421872413 | 52.15 | 6.11 | 30.34 | 6.99 | 0.16 | 2.5 |
| 0.0064 | 0.799 | 20.79 | −20.495 | 0.912301041 | 48.18 | 5.96 | 36.43 | 4.46 | 0.42 | 2.5 |
| 0.001667 | −9.57 | 41.8 | −31.12 | 0.899155321 | 46.65 | 6.61 | 40.18 | 1.21 | 0.26 | 2 |
| 0.0003 | 2.26 | 3 | −2.16 | 0.322534656 | 60 | 8 | 11.6 | 10 | 0.4 | 1.507 |
| 0.00009 | 2.26 | 3 | −2.16 | 0.322534656 | 60 | 8 | 11.6 | 10 | 0.4 | 1.507 |
| 0.000017 | 28 | 8 | −3 | 0.030300184 | 1.9 | 0.3 | 0 | 0.05 | 0.01 | 0.63221 |
| 0.000017 | 28 | 8 | −3 | 0.030300184 | 1.9 | 0.3 | 0 | 0.05 | 0.01 | 0.63221 |

| MSW Component | Material | Percent Mass, kg/kg | Percent Water, as Fresh | %Dry Weight Ash | Volatile Matter, %Dry Weight | Maximum Water Sorption, g/100 g dry matter | Specific Gravity | Bulk Density, kg/cu.m | Percent Oxidizab. Dry Weight | Aerobic Hydrolysis Constant, per day |
|---|---|---|---|---|---|---|---|---|---|---|
| Furniture & Furnishings (Wood, Padding) | Wood | 4.7 | 8 | 1.5 | 80.92 | | | 187.6 | 90.5 | 0.00011 |
| Carpets & Rugs (PP, PVC) | Polymer | 1.4 | 10 | 2.06 | 84.34 | 1.5 | | 118.7 | 87.94 | 0.00025 |
| Rubber Tires (Vulc. Rubber) | Rubber | 2 | 2.9 | 36 | 56.2 | 1.1 | 1.04 | 131.5 | 61.1 | 0.00184 |
| Batteries (Lead Acid) (PP) | Polymer | 0.1 | 0 | 0 | 80.9 | 0.1 | | 1796 | 100 | 0.00011 |
| Clear Glass Containers | Glass | 1.45 | 1.67 | 96.49 | 0.93 | | 2.47 | 1063 | 1.84 | 0.000006 |
| Green Glass Containers | Glass | 0.47 | 1.67 | 96.49 | 0.93 | | 2.47 | 1063 | 1.84 | 0.000006 |
| Brown Glass Containers | Glass | 0.38 | 1.67 | 96.49 | 0.93 | | 2.47 | 1063 | 1.84 | 0.000006 |
| Miscellaneous Glass | Glass | 0.11 | 1.67 | 96.49 | 0.93 | | 2.47 | 1063 | 1.84 | 0.000006 |
| Aluminum: | Aluminum | 0.91 | 2 | 93.03 | 0 | | | 251.6 | 4.97 | 0.000068 |
| Beverage Containers | Aluminum | 0.28 | 2 | 93.03 | 0 | | | 251.6 | 4.97 | 0.000068 |
| Other Aluminum Containers | Aluminum | 0.5 | 2 | 93.03 | 0 | | | 251.6 | 4.97 | 0.000068 |
| Miscellaneous Aluminum | Aluminum | 0.06 | 2 | 93.03 | 0 | | | 251.6 | 4.97 | 0.000068 |
| Food Containers (Canned Goods) | Ferrous | 0.94 | 0.2 | 118.9 | 0 | | 7.8 | 296.7 | −19.1 | 0.000055 |
| Other Ferrous Metals | Ferrous | 1.06 | 0.2 | 118.9 | 0 | | 7.8 | 296.7 | −19.1 | 0.000055 |
| Inorganic Non-Haz. Waste: | Other | 2.28 | 0.02 | 0 | 0 | | | 480.6 | 99.98 | 0.00003 |
| Bi-Metal Cans | Other | 0.01 | 0.02 | 0 | 0 | | | 480.6 | 99.98 | 0.00006 |
| Non-Bulk Ceramics | Other | 0.08 | 0.02 | 0 | 1.56 | | | 480.6 | 99.98 | 0.00003 |
| Miscellaneous Inorganics (Fines, Vacuum Catch) | Other | 1.19 | 20 | 20 | 54 | | | 480.6 | 60 | 0.00003 |
| Hazardous Waste: | Other | 0.37 | | | 80 | | | 326.3 | 100 | 0.0007 |
| Pesticides | Other | 0.01 | | | 80 | | | 326.3 | 100 | 0.0007 |
| Non-Pesticide Poisons | Other | 0.02 | | | 80 | | | 326.3 | 100 | 0.0007 |
| Paints/Solvents/Fuel | Other | 0.04 | | | 80 | 31.4 | | 980.7 | 100 | 0.0007 |
| Dry Cell Batteries | Other | 0.02 | 14.3 | 100 | 3 | 2 | | 296.7 | −14.3 | 0.0003 |
| Car Batteries (PPE, Metals) | Other | 0.02 | 0.2 | 5 | 79.7 | 0.1 | | 1796 | 94.8 | 0.0003 |
| Upholstered (PS, wood) | Other | 0.468 | | | 60 | | | 65.6 | 100 | 0.00011 |
| Steel | Other | 0.159 | 12 | 118.9 | 0 | | 7.8 | 296.7 | −30.9 | 0.000055 |
| Aluminum | Other | 0.028 | 2 | 93.03 | 0 | | 2.73 | 251.6 | 4.97 | 0.000055 |
| Wood | Other | 0.214 | 35.1 | 1.5 | 80.92 | 26.4 | 0.45 | 187.6 | 63.4 | 0.00011 |
| Mixed | Other | 0.106 | 0 | 0 | 0 | 0 | 0 | 0 | 100 | 0.000055 |
| Stoves (Ferrous) | Other | 0.064 | 12 | 118.9 | 0 | 0.2 | | 296.7 | −30.9 | 0.000055 |

| Anaerobic Hydrolysis Constant, per day | Water Sorption Factor, A | Water Sorption Factor, B | Water Sorption Factor, C | Maximum Water Content, Fraction | Percent Dry Weight Carbon, C | Percent Dry Weight Hydrogen, H | Percent Dry Weight Oxygen, O | Percent Dry Weight Nitrogen, N | Percent Dry Weight Sulfur, S | Specific Heat, kJ/kg K |
|---|---|---|---|---|---|---|---|---|---|---|
| 2.75E-05 | −0.31 | 34.83 | −14.14 | 0.049064384 | 88.91 | 8.42 | 3.96 | 0.21 | 0.02 | 2.303 |
| 6.25E-05 | 252.17 | 28.8 | −217.9 | 0.01584359 | 34.9 | 7.12 | 49.89 | 1.25 | 0.15 | 1.926 |
| 0.00046 | 94 | 4.9 | −11.48 | 0.01143765 | 49.5 | 5.2 | 14.8 | 1.61 | 0.67 | 1.758 |
| 2.75E-05 | 121 | 100 | −19 | 0.004950152 | 0 | 0 | 0 | 0 | 0 | 1.926 |
| 1.50E-06 | 30 | 30 | −15 | 0.02222 | 1.37 | 0.14 | 0.22 | 0.04 | 0.03 | 0.662 |
| 1.50E-06 | 30 | 30 | −15 | 0.02222 | 1.37 | 0.14 | 0.22 | 0.04 | 0.03 | 0.662 |
| 1.50E-06 | 30 | 30 | −15 | 0.02222 | 1.37 | 0.14 | 0.22 | 0.04 | 0.03 | 0.662 |
| 1.50E-06 | 30 | 30 | −15 | 0.02222 | 1.37 | 0.14 | 0.22 | 0.04 | 0.03 | 0.662 |
| 0.000017 | 30 | 5 | −3 | 0.031246777 | 3.05 | 0.78 | 0 | 0.09 | 0.01 | 0.934 |
| 0.000017 | 30 | 5 | −3 | 0.031246777 | 3.05 | 0.78 | 0 | 0.09 | 0.01 | 0.934 |
| 0.000017 | 30 | 5 | −3 | 0.031246777 | 3.05 | 0.78 | 0 | 0.09 | 0.01 | 0.934 |
| 0.000017 | 30 | 5 | −3 | 0.031246777 | 3.05 | 0.78 | 0 | 0.09 | 0.01 | 0.934 |
| 1.38E-05 | 28 | 8 | −3 | 0.030300184 | 1.9 | 0.3 | 0 | 0.05 | 0.01 | 0.63221 |
| 1.38E-05 | 28 | 8 | −3 | 0.030300184 | 1.9 | 0.3 | 0 | 0.05 | 0.01 | 0.63221 |
| 7.50E-06 | 28 | 8 | −3 | 0.030300184 | 0 | 0 | 0 | 0 | 0 | 1.842 |
| 0.000015 | 28 | 8 | −3 | 0.030300184 | 0 | 0 | 0 | 0 | 0 | 0.5 |
| 7.50E-06 | 39.04 | 53.64 | −29.2 | 0.0157513 | 6.6 | 1.7 | | 0.2 | 0.05 | 0.754 |
| 7.50E-06 | 5.14 | 262.7 | −238 | 0.03348478 | 35.7 | 4.73 | 20.08 | 6.26 | 1.15 | 1.842 |
| 0.000175 | | | | | | | | | | 2 |
| 0.000175 | | | | | | | | | | 2 |
| 0.000175 | | | | | | | | | | 2 |
| 0.000175 | 1.532 | 142.9 | −126.9 | 0.056996805 | | | | | | 2.09 |
| 0.000075 | 0 | 0 | 0 | | 0 | 0 | 0 | 0 | 0 | 1.842 |
| 0.000075 | 121 | 100 | −19 | 0.004950152 | 45.04 | 5.6 | 1.56 | 0.08 | 0.02 | 1.926 |
| 2.75E-05 | −0.31 | 34.83 | −14.14 | 0.049064384 | 86.91 | 8.42 | 3.96 | 0.21 | 0.02 | 1.382 |
| 1.38E-05 | 28 | 8 | −3 | 0.030300184 | 1.9 | 0.3 | 0 | 0.05 | 0.01 | 0.63221 |
| 1.38E-05 | 28 | 8 | −3 | 0.030300184 | 3.05 | 0.78 | 0 | 0.09 | 0.01 | 0.934 |
| 2.75E-05 | 5.94 | 3.22 | −5.38 | 0.264471059 | 42.7 | 6.1 | 48.8 | 0.5 | 0.06 | 2.303 |
| 1.38E-05 | | 0 | 0 | | 0 | 0 | 0 | 0 | 0 | 0 |
| 1.38E-05 | 28 | 8 | −3 | 0.030300184 | 4.54 | 0.63 | 4.28 | 0.05 | 0.01 | 0.63221 |

| MSW Component | Material | Percent Mass, kg/kg | Percent Water, as Fresh | % Dry Weight Ash | Volatile Matter, % Dry Weight | Maximum Water Sorption, g/100 g dry matter | Specific Gravity | Bulk Density, kg/cu.m | Percent Oxidizab. Dry Weight | Aerobic Hydrolysis Constant, per day |
|---|---|---|---|---|---|---|---|---|---|---|
| Refrigerators (Ferrous) | Other | 0.238 | 12 | 118.9 | 3 | 0.2 | | 296.7 | −30.9 | 0.000055 |
| Dishwashers (Ferrous) | Other | 0.0106 | 12 | 118.9 | 3 | 0.2 | | 296.7 | −30.9 | 0.000055 |
| Others | Other | 0.104 | | | 16 | 0.2 | | 296.7 | 100 | 0.000055 |
| Ferrous | Other | 0.17 | 12 | 118.9 | 0 | 0.2 | | 296.7 | −30.9 | 0.000055 |
| Non-Ferrous | Other | 0.194 | 12 | 95 | 0 | 0.2 | | 178 | −7 | 0.000055 |
| Misc. Wood | Other | 0.399 | 35.1 | 1.5 | 80.92 | 26.4 | | 187.6 | 63.4 | 0.00011 |
| Rugs/Carpet/Textiles (PS, PP, mainly PVC) | Other | 0.369 | 0.2 | 2.06 | 84.34 | 0.32 | 1200 | 1350 | 97.74 | 0.00015 |
| Tires (Vulc. rubber) | Other | 0.232 | 2.9 | 36 | 56.2 | | 0 | 131.5 | 61.1 | 0.000184 |
| Miscellaneous | Other | 0.483 | 0 | 0 | 0 | 0 | 0 | 0 | 100 | 0.00015 |
| Metal | C&D | 0.65 | 13.6 | 118.9 | 0 | | | 178 | −32.5 | 0.000055 |
| Concrete | C&D | 0.024 | 6.8 | 0 | 23.7 | 12.6 | 2300 | 1056 | 93.2 | 0.000005 |
| Masonry | C&D | 0.001 | 13.6 | 0 | 23.7 | 0.11 | 670 | 400 | 86.4 | 0.000005 |
| Rock (limestone) | C&D | 0.02 | 13.6 | 13.6 | 23.7 | 1.5 | 2700 | 1465 | 72.8 | 0.00005 |
| Green Wood | C&D | 3.304 | 12.1 | 2.6 | 72 | 66.2 | | 178 | 85.3 | 0.00011 |
| Asbestos | C&D | 0.001 | 0 | 0 | 0 | 0.1 | 1510 | 119 | 100 | 0.00004 |
| Drywall (gypsum wallboard) | C&D | 0.008 | 10.6 | 3.9 | 76.8 | 2.93 | 1340 | 163.2 | 85.5 | 0.00043 |
| Plastics (PVC) | C&D | 0.64 | 14.3 | 5.3 | 80 | | | 1350 | 80.4 | 2.42E-06 |
| Other (Glass, Asphalt, etc., composition based upon asphalt) | C&D | 5.2 | 0 | 0 | 0 | | | 237.3 | 100 | 0.000005 |
| Dirt | C&D | 0.154 | 1.2 | 0 | 5 | | | 593.3 | 98.8 | 0.00005 |
| Glass | C&D | 0.09 | 1.2 | 0 | 0 | | | 386 | 98.8 | 0.00005 |
| Urban Wood (Lumber, Pallets) | C&D | 0 | 35.1 | 33 | 19.4 | 27.27 | | 178 | 31.9 | 0.00005 |
| Paper (brown sulfite cellulose) | C&D | 1.4 | 6.8 | 16.1 | 69 | 11.66 | 0.92 | 119 | 77.1 | 0.0008 |
| Raw Sludge | Sludge | 0.736 | 65 | 36.88 | 51.91 | | | 740 | −1.88 | 0.00014 |
| Screenings | Sludge | 0.02 | 65 | 36.88 | 51.91 | | 0.8 | 810.4 | −1.88 | 0.0037 |
| Grit | Sludge | 0.132 | 97.3 | 36.88 | 51.91 | | | 1351.5 | −34.18 | 0.0037 |
| Scum | Sludge | 0.054 | 97.3 | 36.88 | 51.91 | | | 961.1 | −34.18 | 0.0037 |
| I. V. Bag (HDPE) | Medical | 0.0173 | 14.3 | 5.33 | 80 | | | 98.6 | 80.37 | 0.000342 |
| Sharps (Steel, Plastic) | Medical | 0.0146 | 12 | 118.9 | 60 | | | 178 | −30.9 | 0.0009 |
| Sharps Containers (Ferrous) | Medical | 0.0034 | 13.6 | 118.9 | 80 | | | 119 | −32.5 | 8.20E-07 |
| Bloody Items | Medical | 0.0179 | 10.6 | 3.9 | 76.8 | | | 178 | 85.5 | 0.0005 |
| Apparatus (Plastic) | Medical | 0.039 | 13.6 | 118.9 | 0 | | | 98.6 | −32.5 | 0.00007 |
| Drug Containers (PET) | Medical | 0.0012 | 14.3 | 5.3 | 80 | | | 385.6 | 80.4 | 0.00034 |
| Solution Containers (Glass) | Medical | 0.013 | 1.2 | 0 | 0 | | 2.54 | 119 | 98.8 | 0.00034 |

| Anaerobic Hydrolysis Constant, per day | Water Sorption Factor, A | Water Sorption Factor, B | Water Sorption Factor, C | Maximum Water Content, Fraction | Percent Dry Weight Carbon, C | Percent Dry Weight Hydrogen, H | Percent Dry Weight Oxygen, O | Percent Dry Weight Nitrogen, N | Percent Dry Weight Sulfur, S | Specific Heat, kJ/kg K |
|---|---|---|---|---|---|---|---|---|---|---|
| 1.38E-05 | 28 | 8 | −3 | 0.030300184 | 4.54 | 0.63 | 4.28 | 0.05 | 0.01 | 0.63221 |
| 1.38E-05 | 28 | 8 | −3 | 0.030300184 | 4.54 | 0.63 | 4.28 | 0.05 | 0.01 | 0.63221 |
| 1.38E-05 | 0 | 0 | 0 | | 4.54 | 0.63 | 4.28 | 0.05 | 0.01 | 0.6322 |
| 1.38E-05 | 28 | 8 | −3 | 0.030300184 | 1.9 | 0.3 | 0 | 0.05 | 0.01 | 0.63221 |
| 1.38E-05 | 28 | 8 | −3 | 0.030300184 | 4.54 | 0.63 | 4.28 | 0.05 | 0.01 | 0.934 |
| 2.75E-05 | 2.2 | 9.67 | −8.3 | 0.280029681 | 41.4 | 6.4 | 51.1 | 0.02 | 0.06 | 2.303 |
| 3.75E-05 | 314.6 | −151 | −97.26 | 0.015064509 | 45.14 | 5.61 | 1.56 | 0.08 | 0.14 | 1.926 |
| 0.00046 | 94 | 4.9 | −11.48 | 0.01143765 | 49.5 | 5.2 | 14.8 | 1.61 | 0.67 | 1.758 |
| 3.75E-05 | 0 | 0 | 0 | | | | | | | 1.842 |
| 1.38E-05 | 28 | 8 | −3 | 0.030300184 | 4.54 | 0.63 | 4.28 | 0.05 | 0.01 | 0.63221 |
| 0.00005 | 8.98 | 102.6 | −90 | 0.046317958 | 4 | 1.2 | 19.4 | 0.01 | 2.03 | 0.88 |
| 0.00005 | −3.8 | 66.76 | −52.9 | 0.099355087 | 0 | 0 | 0 | 0 | 0 | 1.0844 |
| 0.00005 | 117.17 | 460.14 | −469.9 | 0.009305034 | 0 | 0 | 0 | 0 | 0 | 0.88 |
| 2.75E-05 | −15.2 | 52.6 | −35.9 | 0.665748002 | 50.12 | 6.4 | 42.26 | 0.14 | 0.08 | 1.34 |
| 0.00006 | 7.07 | 305.14 | −216.04 | 0.010395841 | 0 | 0 | 0 | 0 | 0 | 1.047 |
| 0.000108 | 49.5 | 226.223 | −244.58 | 0.032079651 | 55.1 | 5.7 | 31.6 | 2.67 | 0.2 | 1.089 |
| 6.90E-07 | 121 | 100 | −19 | 0.004950152 | 71.87 | 11.84 | 3.26 | 0.03 | 0.17 | 2.30274 |
| 0.0000005 | 0 | 0 | 0 | | 82.33 | 10.69 | 0 | 0.81 | 6.16 | 0.92 |
| 0.00005 | 5.138 | 262.7 | −238 | 0.033487023 | 34.7 | 4.76 | 35.3 | 0.14 | 0.2 | 1.842 |
| 0.00005 | 30 | 30 | −15 | 0.022222 | 1.37 | 0.14 | 0.22 | 0.04 | 0.03 | 0.837 |
| 2.75E-05 | 2.2 | 9.67 | −8.3 | 0.280029681 | 48.3 | 5.97 | 42.44 | 0.29 | 0.11 | 2.093 |
| 0.0002 | 2.36 | 16.32 | −10.122 | 0.116832691 | 29.84 | 4.5 | 31.03 | 0.06 | 0.02 | 1.34 |
| 0.000035 | 5.138 | 262.7 | −238 | 0.033487023 | 28.55 | 4.09 | 16.03 | 3.62 | 1.36 | 0 |
| 0.000925 | 5.138 | 262.7 | −238 | 0.033487023 | 28.55 | 4.09 | 16.03 | 3.62 | 1.36 | |
| 0.000925 | 5.138 | 262.7 | −238 | 0.033487023 | 28.55 | 4.09 | 16.03 | 3.62 | 1.36 | |
| 0.000925 | 5.138 | 262.7 | −238 | 0.033487023 | 28.55 | 4.09 | 16.03 | 3.62 | 1.36 | |
| 8.55E-05 | 121 | 100 | −19 | 0.004950152 | 51.5 | 5.9 | 36.5 | 0.06 | 0.1 | 2.30274 |
| 0.000225 | 28 | 8 | −3 | 0.030300184 | 1.9 | 0.3 | 0 | 0.05 | 0.01 | 0.63221 |
| 2.05E-07 | 28 | 8 | −3 | 0.030300184 | 45.04 | 5.6 | 1.56 | 0.08 | 0.02 | 1.57 |
| 0.000125 | 1.39 | 22.74 | −19.16 | 0.201124082 | 55.1 | 5.7 | 31.6 | 2.67 | 0.2 | 1.34 |
| 1.75E-05 | 121 | 100 | −19 | 0.004950152 | 60 | 7.2 | 22.6 | 0 | 0 | 1.57 |
| 0.000085 | 121 | 100 | −19 | 0.004950152 | 60 | 7.2 | 22.6 | 0 | 0 | 1.57 |
| 0.000085 | 30 | 30 | −15 | 0.02222 | 1.37 | 0.14 | 0.22 | 0.04 | 0.03 | 0.662 |

| MSW Component | Material | Percent Mass, kg/kg | Percent Water, as Fresh | % Dry Weight Ash | Volatile Matter, % Dry Weight | Maximum Water Sorption, g/100 g dry matter | Specific Gravity | Bulk Density, kg/cu.m | Percent Oxidizab. Dry Weight | Aerobic Hydrolysis Constant, per day |
|---|---|---|---|---|---|---|---|---|---|---|
| Gauze (Cellulosic) | Medical | 0.0008 | 10.6 | 3.9 | 76.8 | | | 98.6 | 85.5 | 0.0003 |
| Medicine Cups (PS) | Medical | 0.0003 | 14.3 | 5.3 | 80 | 0.06 | | 98.6 | 80.4 | 0.0002 |
| Plastic Bags (HDPE) | Medical | 0.043 | 14.3 | 5.3 | 80 | 2.5 | | 119 | 80.4 | 1.10E-06 |
| Disposable Linens (cotton) | Medical | 0.1833 | 10.6 | 3.9 | 76.8 | 20.12 | 1.55 | 184.9 | 85.5 | 0.0003 |
| Food Service (PS) | Medical | 0.124 | 14.5 | 5.3 | 80 | 0.06 | | 119 | 80.2 | 0.0002 |
| Packaging (PVC) | Medical | 0.053 | 14.3 | 5.3 | 80 | 0.07 | 0.92 | 98.6 | 80.4 | 0.0008 |
| Paper | Medical | 0.008 | 24 | 7.9 | 63.6 | | 0.92 | 361.8 | 68.1 | 0.0008 |
| Paper Towels | Medical | 0.04 | 24 | 7.9 | 63.6 | | 0.92 | 89 | 68.1 | 0.015 |
| Cloth | Medical | 0.003 | 10.6 | 3.9 | 76.8 | | 1.55 | 307.7 | 85.5 | 0.0003 |
| Other | Medical | 0.173 | 0 | 0 | 0 | | | 0 | 100 | 0.0002 |
| Drainage Sets (HDPE, Rubber?) | Medical | 0.0003 | 1.2 | 0 | 0 | | 1.5 | 119 | 98.8 | 2.42E-06 |
| Animal Bedding | Medical | 0.0036 | 48.9 | 10.6 | 34.9 | | | 443.8 | 40.5 | 0.0008 |
| RMW Containers (PP) | Medical | 0.021 | 0 | 14.3 | 80 | 0.1 | | 905 | 85.7 | 0.000001 |
| Pathological Waste | Medical | 0.0036 | 48.9 | 10.6 | 34.9 | | | 288.3 | 40.5 | 0.0004 |
| On-Site Incineration (Ashes) | Medical | 0.1144 | 10 | 63.2 | 2.7 | | | 1470 | 26.8 | 0.000005 |
| Mixed Paper | Medical | 0.061 | 24 | 7.9 | 63.6 | | 0.92 | 361.8 | 68.1 | 0.0008 |
| Office Paper | Medical | 0.038 | 24 | 7.9 | 75.94 | | 0.92 | 361.8 | 68.1 | 0.0008 |
| Computer Printout | Medical | 0.009 | 24 | 7.9 | 63.6 | | 0.92 | 295.9 | 68.1 | 0.0008 |
| Batteries (PP, Metals) | Medical | 0.0003 | 14.3 | 100 | 16 | 0.1 | | 1796 | −14.3 | 0.0002 |
| Decubitus Pads (absorbent cotton) | Medical | 0.007 | 24 | 7.9 | 63.6 | 27.9 | | 119 | 68.1 | 0.0003 |
| Kitchen Waste | Medical | 0.128 | 14.3 | 5.5 | 80 | | | 288.3 | 80.2 | 0.0008 |
| Corrugated Cardboard | Medical | 0.077 | 63.6 | 7.9 | 77.47 | | 0.92 | 422.8 | 28.5 | 0.0688 |
| Newspapers/Magazines | Medical | 0.044 | 63.6 | 7.9 | 81.12 | 11.32 | 0.92 | 361.8 | 28.5 | 0.0456 |
| Animal Remains | Maint | 0.05 | 26.14 | 0.68 | 70 | | | 356 | 73.18 | 0.0145 |
| Sand and Gravel | Cover, L | 20 | 16 | 25 | 10 | 7 | | 2391 | 59 | 0.0001 |
| Silt and Clay | Cover, L | 15 | 16 | 25 | 15 | | | 2500 | 59 | 0.0001 |
| Air | Contents | 0.06 | 100 | 0 | 0 | | 0.0012 | 1.29 | 0 | 0 |
| Landfill Gas | Contents | 0.06 | 0 | 0 | 0 | | 0.00135 | 1.35 | 100 | 0 |

| Anaerobic Hydrolysis Constant, per day | Water Sorption Factor, A | Water Sorption Factor, B | Water Sorption Factor, C | Maximum Water Content, Fraction | Percent Dry Weight Carbon, C | Percent Dry Weight Hydrogen, H | Percent Dry Weight Oxygen, O | Percent Dry Weight Nitrogen, N | Percent Dry Weight Sulfur, S | Specific Heat, kJ/kg K |
|---|---|---|---|---|---|---|---|---|---|---|
| 0.000075 | 1.39 | 22.74 | −19.16 | 0.201124082 | 55.1 | 5.7 | 31.6 | 2.67 | 0.02 | 1.335 |
| 0.00005 | 121 | 100 | −19 | 0.004950152 | 87.1 | 8.45 | 3.96 | 0.21 | 0.02 | 1.382 |
| 2.75E-07 | 121 | 100 | −19 | 0.004950152 | 84.54 | 14.18 | 0 | 0.06 | 0.03 | 2.30274 |
| | | | | | | | | | | |
| 0.000075 | 3.77 | 19.68 | −15.64 | 0.128009159 | 55.1 | 5.7 | 31.6 | 2.67 | 0.2 | 1.335 |
| 0.00005 | 121 | 100 | −19 | 0.004950152 | 51.5 | 5.9 | 36.5 | 0.06 | 0.1 | 1.382 |
| 0.0002 | 121 | 100 | −19 | 0.004950152 | 29.84 | 4.5 | 31.03 | 0.06 | 0.02 | 1.57 |
| 0.0002 | 5.23 | 10.28 | −3.07 | 0.080380489 | 29.84 | 4.5 | 31.03 | 0.06 | 0.02 | 1.34 |
| 0.00375 | 1.96 | 22.04 | −20.43 | 0.279936475 | 41.7 | 5.8 | 44 | 0.27 | 0.09 | 1.34 |
| 0.000075 | 1.308 | 13.12 | −10.9 | 0.283348663 | 55 | 6 | 31.2 | 4.12 | 0.13 | 1.335 |
| 0.00005 | 0 | 0 | 0 | | 0 | 0 | 0 | 0 | 0 | 1.34 |
| | | | | | | | | | | |
| 6.05E-07 | 121 | 100 | −19 | 0.004950152 | 84.54 | 14.18 | 0 | 0.06 | 0.03 | 2.3074 |
| 0.0002 | 3.64 | 22.46 | −17.27 | 0.113223471 | 49.14 | 6.1 | 43.03 | 0.05 | 0.16 | 1.34 |
| 2.50E-07 | 121 | 100 | −19 | 0.004950152 | 45.04 | 5.6 | 1.56 | 0.08 | 0.02 | 1.57 |
| 0.0001 | 1.06 | 15.12 | −14.61 | 0.636307579 | 19.6 | 7.6 | 61.3 | 0.63 | 0.17 | 2.6 |
| | | | | | | | | | | |
| 0.000005 | 5.14 | 262.7 | −238 | 0.03348478 | 28 | 0.5 | 0.8 | 0 | 0.5 | 0.795 |
| 0.0002 | 5.23 | 10.28 | −3.07 | 0.080380489 | 29.84 | 4.5 | 31.03 | 0.06 | 0.02 | 1.34 |
| 0.0002 | 5.23 | 10.28 | −3.07 | 0.080380489 | 43.41 | 5.82 | 44.32 | 0.25 | 0.2 | 1.34 |
| 0.0002 | 0.62 | 23.2 | −20.2 | 0.276084307 | 41.7 | 5.8 | 44 | 0.27 | 0.09 | 1.34 |
| 0.00005 | 0 | 0 | 0 | | 45.04 | 5.6 | 1.56 | 0.08 | 0.02 | 1.57 |
| | | | | | | | | | | |
| 0.000075 | 1.533 | 2.864 | −0.812 | 0.278921782 | 41.7 | 5.8 | 44 | 0.22 | 0.09 | 1.335 |
| 0.0002 | 0.223 | 5.41 | −8.62 | −0.334883207 | 51.5 | 5.9 | 36.5 | 0.27 | 0.1 | 3.8 |
| 0.0172 | 2.36 | 16.32 | −10.122 | 0.116832691 | 32.91 | 4.95 | 38.55 | 0.07 | 0.9 | 1.34 |
| 0.0114 | 3.64 | 22.46 | −17.27 | 0.113223471 | 49.14 | 6.1 | 43.03 | 0.05 | 0.16 | 1.34 |
| 0.003625 | 2.57 | 6.56 | −6.621 | 0.398419212 | 59.6 | 9.5 | 24.7 | 1.02 | 0.19 | 2.6 |
| 0.000025 | 5.14 | 262.6 | −238 | 0.03359728 | | | | | | 0.795 |
| 0.000025 | 5.14 | 260 | −200 | 0.015346717 | | | | | | 1.842 |
| 0 | 0 | 0 | 0 | | | | | | | 1 |
| 0 | 0 | 0 | 0 | | | | | | | 1 |

# Appendix 2
## Landfill Gas Properties

| CHEMICAL | Formula | Molecular Weight | CAS # | Average Leachate Concentration mg/l (Harris and Gaspar, 1989) | RANGE, Leachate, mg/l (Venkataramani and Ahlert, 1984) | Average Landfill Gas Concentration, ppm | Solubility, mg/l | Density, gm/cc | Boiling Point, deg. C |
|---|---|---|---|---|---|---|---|---|---|
| Chlorobenzene | | | 108-90-7 | 6.2 | | 0.24 | 500 | 1.106 | 132 |
| Chloroethene | | | 75-01-4 | 54.7 | | 1.28 | 2763 | 0.864 | 138.5 |
| Chloroform | | | 67-66-2 | 2.7 | | 0.06 | 7220 | 1.485 | 62 |
| 1,1-Dichloroethane | | | 75-34-3 | 66.8 | 180 | 2.52 | 550 | 1.175 | 57.3 |
| 1,2-Dichloroethane | | | 107-06-2 | 3.19 | 4.7-490 | 1.05 | 8690 | 1.25 | 83.5 |
| 1,4-Dichlorobenzene | | | 541-73-1 | 3.24 | | 0 | 49 | 1.458 | 174 |
| trans-1, 2-Dichloroethene | | | 156-60-5 | 63.87 | | 5.09 | 6300 | 1.257 | 60.3 |
| Ethylbenzene | | | 100-41-4 | 63.5 | | 14.64 | 152 | 0.867 | 136.3 |
| 2-Hexanone | | | 108-10-1 | 61.25 | | 0.78 | 20000 | 0.83 | 128 |
| Methylene chloride | | | 75-09-2 | 552 | | 19.7 | 20000 | 1.33 | 40 |
| 4-Methyl-2-Pentanone | | | 108-10-1 | 67.8 | | 1.93 | 17000 | 0.798 | 116.8 |
| Tetrachloroethene | | | 127-18-4 | 8.8 | 0-590 | 7.15 | 2200 | 1.623 | 121.2 |
| Toluene | | | 108-88-3 | 327 | 0-16200 | 51.6 | 534.8 | 0.867 | 110.6 |
| Trichloroethene | | | 79-01-6 | 28.1 | 0-490 | 3.8 | 1070 | 1.465 | 86.7 |
| Trichloroethylene | | | 79-01-6 | 28.1 | 0-7700 | 3.8 | 1070 | 1.465 | 86.7 |
| Trichlorofluoromethane | | | 75-69-4 | 3.8 | | 0.99 | 1100 | 1.476 | 26.63 |
| 1,1-Trichloroethane | | | 71-55-6 | 59.7 | | 0.69 | 150 | 1.34 | 74.1 |
| Vinyl chloride | | | 75-01-4 | 14.87 | 0.014-0.023 | 7.04 | 2763 | 0.911 | −13.4 |
| Xylenes | | | 95-47-6 | 392 | 0-3300 | 14.52 | 130 | 0.864 | 138.5 |
| SEMIVOLATILES in $\mu g$/l: | | | | | | | | | |
| Benzoic acid | | | 65-85-0 | 289 | | | 2900 | 1.27 | 249 |
| bis (2-Ethylhexylphthalate) | | | 117-81-7 | 17.54 | | | 0.285 | 0.99 | 385 |
| Diethyl phthalate | | | 84-66-2 | 59.58 | | | 1080 (25) | 1.12 | 298 |
| 2,4-Dimethylphenol | | | 106-67-9 | 4.52 | | | 4200 | 0.965 | 211.5 |
| Napthalene | | | 91-20-3 | 4.28 | | | 31.7 | 1.145 | 218 |
| Phenanthrene | | | 85-01-8 | 129 | | | 2.67 | 1.174 | 339 |
| Phenol | | | 108-95-2 | 107 | | | 84000 | 1.07 | 182 |
| OTHER LEACHATE CHEMICALS: | | | | | | | | | |
| Acetic acid | C2H4O6 | 60.05 | 64-19-7 | | 120-2750(Ole) | | INF | 1.049 | 117.9 |
| Acetaldehyde | C2H4O | 44.05 | 75-07-0 | ? | ? | | INF | 0.783 | 20.8 |
| Acrolein | C3H4O | 56.06 | 107-02-8 | | | | 400000 | 0.841 | 52.7 |

| Log Kow, ml/gm | Log Koc, ml/g | Vapor Pressure, mm Hg | Henry's Constant, atm cu. m./mol | Aqueous Anaerobic Biodeg. Constant | Aqueous Aerobic Biodeg. Constant | Ground-water Biodeg. Rate aerobic, low | Groundwater, anaerobic high | Removal, Secondary Treatment, % |
|---|---|---|---|---|---|---|---|---|
| 2.84 | 2.52 | 8.8 | 0.00358 | 600 | 150 | 136 | 300 | NA |
| 1.36 | 1.99 | 2530 | 0.003 | 365 | 180 | 56 | 2850 | NA |
| 1.97 | 1.65 | 160 | 0.003 | 28 | 42 | 56 | 1825 | 9600% |
| 1.79 | 1.76 | 182 | 0.042 | 14560 | 140 | 64 | 154 | NA |
| 1.48 | 1.14 | 61 | 0.0011 | 600 | 180 | 100 | 365 | NA |
| 3.52 | 2.44 | 0.6 | 0.0015 | 730 | 180 | 28 | 365 | 99.5 |
| 1.86 | 1.51 | 336 (25) | 0.00938 | | | | | |
| 3.13 | 3.15 | 7 | 0.0066 | 228 | 10 | 6 | 228 | 95-72 |
| 1.38 | 2.13 | 11.6 | 0.00175 | 28 | 7 | 2 | 14 | 22 |
| 1.3 | 0.94 | 349 | 0.00203 | 112 | 28 | 14 | 56 | 94.5 |
| 1.09 | 0.79 | 15 | 0.000149 | 28 | 7 | 2 | 14 | 22 |
| 2.6 | 2.42 | 14 | 0.0153 | 1642 | 365 | 365 | 730 | 86 |
| 2.69 | 1.57 | 28.4 | 0.00594 | 210 | 22 | 7 | 28 | 75 |
| 2.42 | 2.03 | 674 (25) | 0.02 | 1642 | 365 | 321 | 1642 | NA |
| 2.42 | 2.03 | 674 (25) | 0.02 | 1642 | 365 | 321 | 1642 | NA |
| 2.53 | 2.2 | 687 | 0.11 | 1460 | 365 | 365 | 730 | NA |
| 2.49 | 2.03 | 124 | 0.0063 | 1092 | 273 | 140 | 546 | NA |
| 1.36 | 1.99 | 2530 | 1.2 (10) | 730 | 180 | 56 | 2850 | NA |
| 3.12 | 2.22 | 11 | 0.00766 | 365 | 28 | 14 | 365 | NA |
| 1.87 | 1.48 | 1.0 (96) | 7.02E-08 | | | | | |
| 4.2 | 5 | 0.0000002 | 0.00000011 | 389 | 23 | 10 | 389 | 91-70 |
| 1.4 | 2.65 | 0.000345 | 0.00000078 | 224 | 56 | 6 | 112 | 98-20 |
| 2.42 | 2.07 | 0.0621 | 0.00000655 | | | | | |
| 3.29 | 2.97 | 0.087 | 0.00046 | 258 | 20 | 1 | 258 | 98.6-77 |
| 4.46 | 4.36 | 0.0144 (Pa) | 4.0 (Pa) | 799 | 200 | 32 | 402 | 37 |
| 1.48 | 1.43 | 0.2 | 0.00000027 | 28 | 3.5 | 0.5 | 7 | 99.9-90 |
| −0.17 | 0 | 11.4 | 0.00123 | | | | | |
| 0.43 | 0 | 760 | 0.00000658 | | | | | |
| −0.1 | −0.28 | 220 | 0.0000044 | 120 | 28 | 14 | 56 | 99.9-45 |

| CHEMICAL | Formula | Molecular Weight | CAS # | Average Leachate Concentration mg/l (Harris and Gaspar, 1989) | RANGE, Leachate, mg/l (Venkataramani and Ahlert, 1984) | Average Landfill Gas Concentration, ppm | Solubility, mg/l | Density, gm/cc | Boiling Point, deg. C |
|---|---|---|---|---|---|---|---|---|---|
| Benzyl chloride | C7H7Cl | 126.6 | 100-44-7 | | | | 493 | 1.002 | 179.3 |
| Butyric acid | | | 107-92-6 | | 140-5875 (Ole) | | | | |
| Caproic acid | | | 142-62-1 | | 56-600 (Ole) | | | | |
| p-Chloro-m-cresol | CH4OCl | 142.6 | 59-50-7 | | | | | | 235 |
| Dibenzofuran | | | 132-64-9 | | | | | | |
| Diisobutylketone | C9H18O | 142.24 | 108-83-8 | | | | 50 | 0.805 | 168 |
| N,N-Dimethyl-acetamide | C4H9NO | 115.18 | 127-19-5 | | | | INF | 0.937 | 165 |
| Dimethyl phthalate | C10H16O4 | 194.2 | 131-11-3 | | | | 4290 | 1.1905 | 283.8 |
| Dimethyl sulfate | C2H6O4S | 126.13 | 77-78-1 | | | | 28000 | 1.328 | 188.5 |
| Ethanol | | | 64-17-5 | | | | | | |
| Formic acid | CH2O2 | 46.03 | 64-18-6 | | | | INF | 1.22 | 100.7 |
| Furfural | C5H4O2 | 96.1 | 98-01-01 | | | | 83000 | 1.16 | 161.7 |
| Methanol (methyl alcohol) | CH4O | 32.04 | 67-56-1 | | | | | 0.796 | 65 |
| Methyl cellosolve | C3H802 | 76.1 | 109-86-4 | | | | INF | 0.965 | 125 |
| N-Nitroso-methylamine | C2H6N2O | 74.09 | 62-75-9 | | | | INF | 1.006 | 154 |
| N-Nitroso-phenylamine | C12H10N2O | 198.22 | | | | | 35.1 (25) | NA | NA |
| N-Nitroso-n-propylamine | C6H14N2O | 130.19 | | | | | 9900 | 0.916 | 205.9 |
| PCB 1242 | NA | 261 | 1336-36-3 | | | | 0.2 | 1.393 | 325 |
| PCB 1254 | NA | 327 | | | | | 0.05 | 1.505 | 365 |
| Palmitic acid | | | 57-10-3 | | | | | | |
| Picric acid | C6H3N3O7 | 229.11 | 88-89-1 | | | | 14000 | 1.763 | 300 |
| n-Propanol | C3H8O | 60.09 | 71-23-8 | | | | | 0.804 | 97.8 |
| Propionic acid | C3H6 | 74.1 | 79-09-4 | | 182-4375 (Ole) | | | 0.992 | 141 |
| TCDD | C12H4Cl4O2 | 321.98 | 1746-01-6 | | | | 0.0000193 | 1.83 | 412.2 |
| Styrene | C8H8 | 104.15 | 100-42-5 | | | | 300 | 0.906 | 145.2 |
| Valeric acid | C5H10O2 | 102.1 | 109-52-4 | | 112-550 (Ole) | | 37000 | 0.942 | 187 |
| OTHER LANDFILL GAS CHEMICALS: | | | | | | | | | |
| Acrolein | C3H4O | 56.06 | 107-02-8 | | | 0 | 400000 | 0.841 | 52.7 |
| Acrylonitrile | C3H3N | 53.06 | 107-13-1 | | | 0.18 | 80000 | 0.806 | 77.5 |
| Butane | C4H10 | 58 | 106-97-6 | | | 2.08 | 61 | 0.6 | −1 |
| 1-Butanol | C4H10O | 74 | 71-36-3 | | | 2.17 | 77000 | 0.81 | 117.7 |
| 2-Butanol (sec-butyl alcohol) | | | 15892-23-6 | | | | | | |
| Bromochloro-propane | | | 109-70-6 | | | 0.01 | | | |
| Bromodichloro-methane | CHBrCl3 | 163.8 | 75-27-4 | | | 0.45 | 4500 | 1.971 | 90 |

| Log Kow, ml/gm | Log Koc, ml/g | Vapor Pressure, mm Hg | Henry's Constant, atm cu. m./mol | Aqueous Anaerobic Biodeg. Constant | Aqueous Aerobic Biodeg. Constant | Ground-water Biodeg. Rate aerobic, low | Groundwater, anaerobic high | Removal, Secondary Treatment, % |
|---|---|---|---|---|---|---|---|---|
| 2.3 | 2.28 | 0.9 | 0.000304 | 112 | 28 | 0.63 | 12.1 | NA |
| | | | | 112 | 28 | 7 | 28 | NA |
| 2.48 | na | 1.7 | 0.000636 | | | | | |
| −0.77 | 0 | 1.5 | NA | | | | | |
| 2 | 1.63 | 0.01 | 0.00000042 | 28 | 7 | 2 | 14 | 96 |
| −1.24 | 0.61 | 0.5 | 0.00000296 | 112 | 28 | 0.05 | 0.5 | NA |
| | | | | 4.3 | 1.08 | 0.54 | 2.17 | 67 |
| −0.54 | 0 | 35 | 0.000000 | 28 | 7 | 2 | 14 | NA |
| | | | 167 (pH4) | | | | | |
| 0.52 | 0 | 1 | 1.52−3.04 | | | | | |
| $P_{oct} = -0.82$ | | 92 | | 5 | 7 | 1 | 7 | 99-88 |
| | | 6 | | 112 | 28 | 14 | 56 | 65 |
| 0.06 | 1.41 | 2.7 | 0.143 (25) | 730 | 180 | 42 | 365 | 100-74 |
| 3.13 | 2.76 | 0.1 | 2.33E-08 | 730 | 180 | 42 | 365 | 100-74 |
| 1.31 | 1.01 | | 0 | 730 | 180 | 42 | 365 | 100-74 |
| 5.58 | 3.71 | 0.001 | 0.00056 | | | | | |
| 6.47 | 5.61 | 0 | 0.00006 | | | | | |
| 2.03 | NA | 1 | 0.000215 | 12.5 | 180 | 2 | 365 | NA |
| $P_{oct} = 0.34$ | | 14.5 | | | | | | |
| $P_{oct} = 0.25$ | | 2.9 | | | | | | |
| 6.15 | 6.66 | 6.4E-10 | 0 | 2354 | 591 | 836 | 1179 | NA |
| 2.95 | 2.87 | 5 | 0.00261 | 112 | 28 | 28 | 210 | 99-9 |
| $P_{oct} = 0.99$ | | 0.15 | | | | | | |
| −0.1 | −0.28 | 220 | 0.0000044 | 120 | 28 | 14 | 28 | 99.9-45 |
| −0.92 | −1.13 | 100 | 0.00011 | 92 | 23 | 2.5 | 46 | 99.9-75 |
| | | 1823 (25) | | | | | | |
| $P_{oct} = 0.88$ | | 4.4 | | 54 | 7 | 2 | 54 | 99-31 |
| 1.88 | 1.79 | 50 | 0.0024 | | | | | |

| CHEMICAL | Formula | Molecular Weight | CAS # | Average Leachate Concentration mg/l (Harris and Gaspar, 1989) | RANGE, Leachate, mg/l (Venkataramani and Ahlert, 1984) | Average Landfill Gas Concentration, ppm | Solubility, mg/l | Density, gm/cc | Boiling Point, deg. C |
|---|---|---|---|---|---|---|---|---|---|
| Bromoform | CHBr3 | 252.77 | 75-25-2 | | | 0 | 3190 | 2.89 | 149 |
| Bromomethane | CH3Br | 94.95 | 74-83-9 | | | 0 | 900 | 1.73 | 4.6 |
| t-Butyl mercaptan | C4H10S | 90.18 | 75-77-1 | | | 0.03 | 590 | 0.84 | 98 |
| Carbon tetrachloride | CCl4 | 153.82 | 56-23-5 | | | 1.49 | 785 | 1.594 | 76.5 |
| Carbonyl sulfide | COS | 60.07 | 463-58-1 | | | 0.01 | 1000 | 1.24 | −50.2 |
| Chlorobenzene | | | 108-90-7 | | | | | | |
| Chloroethane | C2H5Cl | 64.52 | 75-00-3 | | 0.026 | 1.28 | 5740 | 0.898 | 12.3 |
| Chloromethane | CH3Cl | 51 | 74-87-3 | | 0.009-0.06 | 0.9 | 4000 | 0.991 | −24 |
| Chlorodifluoromethane | F12R12 | 120.92 | 74-75-46 | | | 0.79 | 280 (25) | 1.33 | −29.8 |
| 2-Chloroethylvinylether | C4H8OCl2 | 143.02 | 110-75-8 | | | 0.05 | 10200 | 1.22 | 178 |
| 1,2-Dibromoethane | C2H4Br2 | 187.88 | 106-93-4 | | | 0 | 4.13 (30) | 2.701 | 131.6 |
| Dibromomethane | | | 74-95-3 | | | ? | | | |
| 1,1-Dichloroethylene | C2H2Cl2 | 96.95 | | | | 0.16 | 2460 (g/m3) | 1.218 | 31.9 |
| Dichlorodifluoromethane | | | | | | ? | | | |
| 1,1-Dichloroethylene | | 173 | 75-35-4 | | | | | | |
| 1,2-Dichloroethylene | | 173 | 156-60-5 | | | | | | |
| Dichloroflouromethane | | | 75-43-4 | | | | | | |
| Dichlorotetrafluoroethane | | | 374-07-2 | | | 0.73 | | | |
| 1,2-Dichloropropane | C3H6Cl3 | 112.99 | 78-87-5 | | | 0.07 | 2700 | 1.56 | 96.4 |
| cis-1,3-Dichloropropene | C3H4Cl3 | 110.97 | 542-79-6 | | | 0 | 2700 | 1.089 | 287 |
| trans-1,3-Dichloropropene | C3H4Cl3 | 110.97 | 542-75-6 | | | 0 | 2800 | NA | 112 |
| Dimethylamine | C2H7N | 45.08 | 124-40-3 | | | ? | inf | 0.68 | 7.4 |
| 1,2-Dimethyl benzene (o-xylene) | C2H6? | 106.17 | 1330-20-7 | | | 12.78 | 175 | 0.88 | 144.4 |
| 2,5-Dimethyl furan | | | 625-86-5 | | | 0.89 | | | |
| Dimethyl disulfide | C2H6S2 | 94.19 | 624-92-0 | | | 0.02 | ? | 1.057 | 112 |
| Dimethyl sulfate | C2H6O4S | 126.13 | 77-78-1 | | | ? | 28000 | 1.33 | 188 |

| Log Kow, ml/gm | Log Koc, ml/g | Vapor Pressure, mm Hg | Henry's Constant, atm cu. m./mol | Aqueous Anaerobic Biodeg. Constant | Aqueous Aerobic Biodeg. Constant | Ground-water Biodeg. Rate aerobic, low | Groundwater, anaerobic high | Removal, Secondary Treatment, % |
|---|---|---|---|---|---|---|---|---|
| 2.3 | 2.45 | 4 | 0.0056 | 720 | 180 | 56 | 365 | 68 |
| | | ? | | 112 | 28 | 14 | 38 | NA |
| $P_{oct} = 2.28$ | ? | 35 | 0.00704 | | | | | |
| 2.83 | 2.35 | 90 | 0.024 | 28 | 365 | 7 | 365 | 99 |
| 0 | 0 | ? | | | | | | |
| | | | | 600 | 150 | 136 | 300 | NA |
| 1.43 | 0.51 | 1011 | 0.0111 | 112 | 28 | 14 | 56 | NA |
| NA | NA | 3800 | NA | 112 | 28 | 14 | 56 | NA |
| | | 4250 | | | | | | |
| | | 0.71 | | | | | | |
| | | 10.25 | 0.000629 | 15 | 180 | 19.6 | 120 | NA |
| | | | | 112 | 28 | 14 | 56 | NA |
| | | 500 | | | | | | |
| | | | | 173 | 180 | 56 | 132 | 92-58 |
| | | | | 173 | 180 | 56 | 132 | 92-58 |
| 2.28 | 1.71 | 42 | 0.0023 | 5110 | 3.5 | 334 | 2592 | 0 |
| 1.41 | 1.68 | 25 | 0.0013 | 112 | 28 | 5.5 | 11.3 | NA |
| NA | NA | 34 | NA | 112 | 28 | 5.5 | 11.3 | NA |
| −0.38 | 0 | 1292 | 0.0000177 | 13.2 | 3.3 | 0.17 | 6.58 | 100-93 |
| NA | NA | 5 | NA | 365 | 28 | 14 | 365 | NA |
| $P_{oct} = 1.77$ | | | | | | | | |
| | | | | 112 | 28 | 0.05 | 0.5 | NA |

# References

Ahrens, E.G., G. Farkasdi and I. Ibrahim. 1965. Information Bulletin No. 24. International Research Group on Refuse Disposal.

Aiba, S and M. Shoda. 1969. *J. Ferment. Technol.* 47: 790–794.

Alaska Department of Fish and Game. 2001. Policy on human food and solid waste management and bears in Alaska.

Alexander, M. 1961. *Introduction to Soil Microbiology*. John Wiley & Sons, New York.

Altmann, J. et al. 1993. Body size and fatness of free-living baboons reflect food availability and activity levels. *Am. J. Primatol.* 30(2): 149–161.

al-Yousfi, A. Basel, F.G. Pohland and N.C. Vasuki. 1992. Design of landfill leachate recirculation systems based on flow characteristics, *Proc. Purdue University Industrial Waste Conference* May 11–13, Purdue University, West Lafayette, IN, 191–100.

Anderson, J.M. and I.N. Healey. 1972. Seasonal and interspecific variation in major components of the gut contents of some woodland *Collembola*. *J. Animal Ecol.* 41: 359–368.

Andrews, J.F. 1971. Kinetic models of biological waste treatment. *Biotechnol. Bioeng. Symp.* No. 2. 5–33.

Andreyaev, L.N. and V.V. Biryukov. 1973. Analysis of mathematical models of the effect of pH on fermentation processes and their use for calculating optimal fermentation conditions. *Biotechnol. Bioeng. Symp.* No. 4, Part I: Advances in microbial engineering, John Wiley & Sons, New York, 61–76.

ASME Research Committee on Industrial and Municipal Wastes. (n.d.) *Thermodynamic data for Biomass Materials and Waste Components*. E.S. Domalski, T.L. Jobe and T.A. Milne. (Eds.) American Society of Mechanical Engineers, New York.

Auria, R., A.-C. Aycaguer and J.S. Devinny. 1998. Influence of water on degradation rates for ethanol in biofiltration. *J. Air Waste Manage. Assoc.* 48: 65–70.

Baccini, P.G., Henseler, R. Figi and H. Belevi. 1987. Water and element balances of municipal solid waste landfills. *Waste Manage. Res.* 5. 483–499.

Badyaev, A. 1998. Environmental stress and developmental stability in dentition of the Yellowstone grizzly bears. *Behav. Ecol.* 9: 339–344.

Baetz, B.W. and P.H. Byer. 1989. Moisture control during landfill operation. *Waste Manage. Res.* 7: 259–275.

Bailey, James E. and David F. Ollis. 1986. *Biochemical Engineering Fundamentals*, 2nd ed. McGraw-Hill, New York.

Barlaz, M.A., M.A. Gabr, S. Hussain, A. Rooker and P. Kjeldsen. Closing Gaps in the Regulation of MSW Landfills: Defining the End of the Post-Closure Period and the Future Stability of Leachate Recirculating Landfills.

Batterman, S., I. Padmanabham and P. Milne. 1996. Effective gas-phase diffusion coefficients in soils at varying water content measured using a one-flow sorbent-based technique. *Environ. Sci. Technol.* 30: 770–778.

Bear, J.E. 1972. *Dynamics of Fluids in Porous Media*. American Elsevier, New York.

Bertger, F. and W. Knorre. 1972. *Zhurnal Allgem. Mikrob.* 12(8): 613.

Blake, F.C. 1922. The resistance of packing to fluid flow. *Trans. Am. Inst. Chem. Eng.* 14: 415–421.

Boeckx, P. and O. van Cleemput. 1996. Methane oxidation in a neutral landfill cover soil: influence of moisture content, temperature and nitrogen turnover. *J. Envir. Qual.* 25: 178–183.

Bohn, H.L. and K.H. Bohn. 1999. Moisture in biofilters. *Envir. Prog.* 18(3): 156–161.

Bond, A.B. and J. Diamond. 1992. Population estimates of kea in Arthur's Pass National Park. *Notornis* 39(3): 151–160.

Bonner, J.T. 1948. A study of the temperature and humidity requirements of *Aspergillus niger. Mycologia* 40: 728–738.

Boulet, G., I. Braud and M. Vauclin. 1997. Study of the mechanisms of evaporation under arid conditions using a detailed model of the soil–atmosphere continuum: Application to the EFEDA I experiment. *J. Hydrol.* 193: 114–141.

Brooks, R.H. and A.T. Corey. 1966. Properties of porous media affecting fluid flow. *J. Irrigation Drainage Div., Proc. ASCE.* 92: 62–68.

Brown, A.D. 1978. Compatible solutes and extreme water stress in eucaryotic microorganisms. *Adv. Microbiol. Physiol.* 17: 181–242.

Brunauer, S., P.H. Emmett and E. Teller. 1938. *J. Am. Chem. Soc.* 60: 309.

Brush, P. 1992. Home is where you dig it: Observations on life at the Khe Sanh combat base. *Vietnam Generation* 4(3–4): 96–98.

Burchinal, J.C. 1968. Microbiology and acid production in sanitary landfills. Unpublished report (referenced by Mahloch, 1970).

Burchinal, J.C. and A.A. Wilson. 1966. Sanitary landfill investigation: Final report. U.S. Department of Health, Education and Welfare, Public Health Service, Cincinnati.

Burger, J. 1985. Factors affecting bird strikes at a coastal airport. *Biol. Conserv.* 33: 1–28.

Burges, A. and F. Raw. 1967. *Soil Biology*. Academic Press, New York.

Cahill, A.T. and M.B. Parlange. 1998. On water vapor transport in field soils. *Water Resour. Res.* 34(4): 731–739.

Campbell, G. 1974. A simple method for determining unsaturated conductivity from moisture retention data. *Soil Sci.* 117(6): 311–314.

Cardenas, R.R. and L.K. Wang. 1980. The composting process. In *Handbook of Environmental Engineering: Vol. II: Solid Waste Processing and Recovery,* L.K. Wang and N.C. Periera (Eds.). Humana, Clifton, NJ.

Carman, P.C. 1937. Fluid flow through granular beds. *Trans. Am. Inst. Chem. Eng.* 15: 150–166.

Carpenter, D.O., F.T. Chew, T. Damstra, L.H. Lam, P.J. Landrigan, I. Makalinao, G.L. Peralta and W.A. Suk. 2000. Environmental threats to the health of children: the Asian perspective. *Envir. Health Perspect.* 108(10).

Caurie, M. 1981. Derivation of full range moisture sorption isotherms. In *Water Activity in Foods*, 2nd International Symposium on Properties of Water in Foods, Osaka, Japan, 1978. Academic Press, New York, 63–88.

Chan, Y.S.G., L.M. Chu and M.H. Wong. 1997. Influence of landfill factors on plants and soil fauna — an ecological perspective. *Envir. Pollut.* 97(1–2): 39–44.

Chen, C.S. 1971. *Trans. Am. Soc. Agric. Eng.* 14: 927.

Chhabra, R.P., J. Comiti and I. Machac. 2001. Flow of Newtonian and non-Newtonian fluids in fixed and fluidized beds. *Chem. Eng. Sci.* 56: 1–27.

Christian, J.H.B. 1978. Specific water effects on microbial water relations. In *Water Activity in Foods*. 2nd International Symposium on Properties of Water, Osaka, Japan.

Christiansen, K. 1967a. Bionomics of *Collembola. Ann. Rev. Entomol.* 9: 147–178.

Chung, D.S. and H.B. Pfost. 1967. Adsorption and desorption of water vapor by cereal grains and their products. *Trans. Am. Soc. Agric. Eng.* 10: 549–554.

Clapp, R. and G. Hornberger. 1978. Empirical equations for some soil hydraulic properties. *Wat. Resour. Res.* 14: 601–604.

Coduto, D.P. and R. Huitric. 1990. Monitoring landfill movements using precise instruments. Geotechnics of waste fills to theory and practice. ASTM STP 1070. American Society for Testing and Materials, Philadelphia, 358–370.

Coleman, D.C. 1978. Terrestrial ecosystem. In *Yearbook of Science and Technology.* McGraw-Hill, New York, 359–360.

Coleman, D.C. 1985. Through a PED darkly: An ecological assessment of root–soil–microbial–faunal interactions. In *Plants, Microbes and Animals,* Fitter, A.H. (Ed.), Blackwell Scientific, Oxford, 1–21.

Coleman, D.C. and T.V. St. John. 1987. Underground biology. *Yearbook of Science and Technology*. McGraw-Hill, New York, 26–35.

Collins, M.S. 1969. *Water Relations in Termites. Biology of Termites*, Vol. I, Academic Press, New York, 433–458.

Comiti, J., E. Mauret and M. Renaud. 2000. Mass transfer in fixed beds: proposition of a generalized correlation based on an energetic criterion. *Chem. Eng. Sci.* 55: 5545–5554.

Comiti, J., N.E. Sabiri and A. Montillet. 2000. Experimental characterization of flow regimes in various porous media. III. Limit of Darcy's or creeping flow regime for Newtonian and purely viscous non-Newtonian Fluids. *Chem. Eng. Sci.* 55: 3057–3061.

Comiti, J. and M. Renaud. 1989. A new model for determining mean structure of fixed beds from pressure drop measurements: application to beds with parallelipipedal particles. *Chem. Eng. Sci.* 44(7): 1539–1545.

Cooney, C.F. et al. 1969. *Biotechnol. Bioeng*. 11: 269.

Costello, W.J., P.F. Greenfield and P.L. Lee. 1991. Dynamic modeling of a single-stage high-rate anaerobic reactor. I. Model derivation. *Water Res.* 25: 847–858.

Courtney, P.A. and M.B. Fenton. 1976. The effects of a small rural garbage dump on populations of *Peromyscus leucopus Rafinesque* and other small animals. *J. Appl. Ecol.* 13(2): 413–422.

Cox, H.H.J., F.J. Magielsen, H.J. Dodema and W. Harder. 1996. Influence of the water content and water activity on styrene degradation by *Exophiala jeanselmei* in biofilters. *Appl. Microbiol. Biotechnol.* 45: 851–856.

Crawford, C.S. 1979. *Desert Millipedes: A Rationale for their Distribution*. Springer-Verlag, Heidelberg.

Crossley, D.A., Jr. and D.C. Coleman. 1999. Microarthropods. In *The Handbook of Soil Science*, M. Sumner (Ed.), CRC Press, Boca Raton. C59–C65.

Curry, J.P. 1969. The decomposition of organic matter in soil. II. The fauna of decaying grassland herbiage. *Soil Biol. Biochem.* 1: 259–266.

Curry, J.P. 1979. The arthropod fauna associated with cattle manure applied as slurry to grassland. *Proc. R. Irish Acad.* 76B. 49–71.

Curry, J.P. 1994. *Grassland Invertebrates*. Chapman and Hall, London.

Curry, J.P., D.C.F. Cotton, T. Bolger and V.O. O'Brien, 1979. Effects of landspread animal manures on the fauna of grassland. In *Effluents from Livestock*. Applied Science, London.

Curry, J.P. 1986. Effects of management on soil decomposers and decomposition processes in grassland. In *Microfloral and Faunal Interactions in Natural and Agro-Ecosystems.* Mitchell, M.J. and Nakas, J.P. (Eds.). Martinus Nijhoff, Dordrecht, 349–398.

Das, B.M. 1998. *Principles of Geotechnical Engineering*, 4th ed. PWS Publishing, Boston, MA.

Davey, K.R. 1989. A predictive model for combined temperature and water activity on microbial growth during the growth phase. *J. Appl. Bacteriol.* 67: 483–488.

de Vries, D.A. 1957. Simultaneous transfer of heat and moisture in porous media. *Eos. Trans. Am. Geophys. Union* 39: 909–916.

Delestrade, A. 1994. Factors affecting flock size in the alpine chough *Pyrrhocorax graculus. Ibis* 136(1): 91–96.

Dickinson, C.H. 1974. Litter in soil. In *Biology of Plant Litter Decomposition,* Vol. 2, C.H. Dickinson and G.J.F. Pugh, (Eds.). Academic Press, New York. London.

Dindal, D.L. 1990. Ecology of compost: A public involvement project. SUNY College of Environmental Science and Forestry, Syracuse, NY, 1–12.

Dominguez, P.L. 1992. New research and development strategy for a better integration of pig production in the farming system in Cuba. Swine Research Institute, Havana, Cuba.

Dullien, F.A.L. 1975. Single-phase flow-through porous media and pore structure. *Chem. Eng. J.* 10: 1–34.

Dunger, W. 1968. Die Entwicklung der Bodenfauna auf rekultivieren Kippen un Halden des Braunkohlentagebaues. *Abhandlangen und Berichte de Naturkundemuseums Gorlitz.* 43: 1–256 (Referenced by Weidemann et al., 1980).

Dybbs, F.A. and R.V. Edward. 1984. A new look at porous media fluid mechanics: Darcy to turbulent. In *Fundamentals of Transport Phenomena in Transport Media.* J. Bear and Y. Corapcioglu (Eds.) Martinus Nijhoff, The Hague, Netherlands, 199–256.

Edil, T.B., V.J. Ranguette and W.W. Wuellner. 1990. Settlement of municipal refuse. Geotechnics of waste fills to theory and practice. ASTM STP 1070. American Society for Testing and Materials, Philadelphia, 225–239.

Edlefson, N.E. and A.B.C. Anderson. 1943. Thermodynamics of soil moisture. *Hilgardia,* 15(2): 31–298.

El-Fadel, M, S. Shazbak, E. Saliby and J.Leckie. 1999. Comparative assessment of settlement models for municipal solid waste landfill applications. *Waste Manage. Res.* 1: 347–368.

El-Fadel, M. and R. Khoury. 2000. Modeling settlement in MSW landfills: a critical review. *Crit. Rev. Envir. Sci. Technol.* 30(3): 327–361.

El-Fadel, M., A. Finidikakis and J.V. Leckie. 1989. A numerical model for methane production in managed sanitary landfills. *Waste Manage. Res.* 7(1): 31–42.

Eliassen, R. 1942. Decomposition of landfills. *Am. J. Public Health.* 32: 1029–1037.

Ergun, S. 1952. Fluid flow through packed columns. *Chem. Eng. Progress* 48: 89–94.

Finidikakis, A.N., A. Papelis, C.P. Halvadakis and J.O. Leckie. 1988. Modeling gas production in managed sanitary landfills. *Waste Manage. Res.* 6: 115–123.

Fink, D.H. and R.D. Jackson. An equation for describing water vapor adsorption isotherms of soils. *Soil Sci.* 116(4): 256–261.

Fisher, R.C. 1940. Studies of the biology of the death watch beetle *Xestobium rufovillosum DeG.* III. Fungal decay in timber in relation to the occurrence and rate of development of the insect. *Ann. Appl. Biol.* 27: 545–557.

Folk, R.L. 1968. *Petrology of Sedimentary Rocks.* University of Texas at Austin.

Gabr, M.A., M.S. Hossain and M.A. Barlaz. 2000. Solid Waste Settlement in Landfill with Leachate Recirculation.

Gardner, W.H., R.I. Papendick and G.S. Campbell. 1971. Psychometric measurement of soil water potential: temperature and density effects. *Proc. Soil Sci. Soc. Am.* 35: 8–12.

Garrett, S.D. 1936. Soil conditions and the take-all disease of wheat, *Ann. Appl. Biol.* 23: 667–699.

Garrett, S.D. 1938. Soil conditions and the root-infecting fungi. *Biol. Rev.* 13: 159–185.

Geankopolis, C.J. 1993. *Transport Processes and Unit Operations*, 3rd ed. Prentice-Hall, Englewood Cliffs, NJ.

Gervais, P., P. Molin, W. Grajek and M. Bensoussan. 1988. Influence of the water activity of a solid substrate on the growth rate and sporogenesis of filamentous fungi. *Biotechnol. Bioeng.* 31: 457–463.

Gervais, P. 1990. Water activity: a fundamental parameter of aroma production by microorganisms. *Appl. Microbiol. Biotechnol.* 33: 72–75.

Goleuke, C.G., B.J. Card and P.H. Gauhey. 1954. A critical evaluation of inoculums in composting. *Appl. Microbiol.* 2(1): 45–53.

Gonullu, M.T. 1994. Analytical modeling of organic contaminants in leachate. *Waste Manage. Res.* 12: 141–150.

Gostomski, P.A., J.B. Sisson and R.S. Cherry. 1997. Water content dynamics in biofiltration: The role of humidity and microbial heat generation. *J. Air Waste Manage. Assoc.* 47: 936–944.

Grady, C., P. Leslie, Jr. and H.C. Lim. 1980. Biological wastewater treatment: Theory and applications. *Pollut. Eng. Technol.*, No. 12.

Gray, K.R. and A.J. Biddlestone. 1974. Decomposition of urban wastes. In *Biology of Plant Litter Decomposition*, Vol. 2, G.H. Dickinson and G.F. Pugh. (Eds.). Academic Press, New York, 2: 748.

Gregg, S.J. 1961. *The Surface Chemistry of Solids*, 2nd ed. Reinhold, New York.

Griffin, D.M. 1963. Soil moisture and the ecology of soil fungi. *Biol. Rev.* 38: 141–166.

Griffin, D.M. 1972. *Ecology of Soil Fungi.* Syracuse University Press, Chapman and Hall, London.

Guggenheim, E.A. 1966. *Applications of Statistical Mechanics.* Clarendon Press, Oxford, U.K.

Haack, R.A., D.M. Benjamin and K.D. Haack. 1983. Buprestidae, Cerambycidae and Scolytidae associated with the successive stages of *Agrilus bilineatus (*Coleoptera: Buprestidae) infestation of oaks in Wisconsin. *Great Lakes Entomol.* 16: 47–55.

Halsey, G. 1948. *J. Chem. Phys.* 16: 931.

Halvadakis, C.P. 1983. Methanogenesis in solid waste landfill bioreactors. Ph.D. thesis, Stanford University.

Hansen, K. 1982. Sorption isotherms: A catalogue. Technical Report 162/86, Building Materials Laboratory, Department of Civil Engineering, Technical University of Denmark.

Harding, D.J.L. 1968. Report of the East Malling Research Station for 1967. 169–172.

Harkins, W.D. and G. Jura. 1944. *J. Am. Chem. Soc.* 66: 1366.

Harris, R.F. 1980. Effect of water potential on microbial growth and activity. In *Water Potential Relations in Soil Microbiology*, Soil Science Society of America (SSSA) Special Publication 9, 23–26.

Haug, R.T. 1993. *The Practical Handbook of Compost Engineering*. Lewis Publishers. Chelsea, MI.

Henderson, S.M. 1952. *Agric. Eng*. 33: 29.

Hickin, N.E. 1975. *The Insect Factor in Wood Decay*, 3rd ed., Hutchinson, London.

Hillel, D. 1971. *Soil and Water: Physical Principles and Processes*. Academic Press, New York.

Holden, P.A., J.A. Hunt and M.K. Firestone. 1995. Unsaturated zone gas-phase VOC biodegradation: the importance of water potential. *Purdue Ind. Waste Conf. Proc.* 50: 113–127.

Howard, P.H. 1991. *Handbook of Environmental Fate and Exposure Data for Organic Chemicals*. Lewis Publishers, Chelsea, MI.

Iglesias, H.A. and J. Chirife. 1976. Prediction of the effect of temperature on water sorption isotherms of food materials. *J. Food Technol.* 11: 109–116.

Iglesias, H.A. and J. Chirife. (n.d.) *Handbook of Food Isotherms: Water Sorption Parameters for Food and Food Components*. Academic Press, New York.

Kear, J. 1972. Feeding habits of birds. In *Biology of Nutrition*. R.N.T Fiennes, (Ed.). Pergamon Press, New York. 471–500.

Khalili, N.R., M. Pan and G. Sandi. 2000. Determination of fractal dimensions of solid carbons from gas and liquid phase adsorption isotherms. *Carbon*, 38: 573–588.

Kingsolver, J.G. and W.B. Watt. 1983. Thermoregulatory strategies in Colias butterflies: Thermal stress and the limits to adaptation in temporally varying environments. *Am. Naturalist* 121: 32–55.

Kirsch, E.J. and R.M. Sykes. 1971. Anaerobic digestion in biological waste treatment. *Progress Ind. Microbiol.* 9: 155–237.

Knox, K. and J. Gronow. 1990. A reactor based assessment of co-disposal. *Waste Mgmnt. Res.* 8: 255–276.

Kramer, G.J. 1998. Static liquid hold-up and capillary rise in packed beds. *Chem. Eng. Sci.* 53(16): 2895–2992.

Lamare, S. and M.D. Legoy. 1995. Working at controlled water activity in a continuous process: The gas/solid system as a solution. *Biotechnol. Bioeng.* 45: 387–397.

Landva, A. and J.L. Clark. 1990. Geotechnics of waste fill. In *Geotechnics of Waste Fill: Theory and Practice,* ASTM STP 1070, A. Landva and G.D. Knowles (Eds.). American Society of Testing Materials, Philadelphia, 87.

Langmuir, I. 1918. *J. Am. Chem. Soc.* 40: 1361.

Lawrence, A.W. and P.L. McCarty. 1969. Kinetics of methane fermentation on anaerobic treatment. *J. Water Pollut. Cont. Fed.* 41(2): R1.

Lebrun, P. 1971. Ecology and biocenotics of some settlements of edaphic arthropods. *Memoirs of the Royal Institute of the Natural Sciences of Belgium.*

Lee, G.F. and A. Jones-Lee. 1993. Landfill post-closure care: Can owners guarantee the money will be there? *Solid Waste Power*, 7: 35–39.

Lee, J.J., I.H. Jung, W.B. Lee and J.-O. Kim. 1993. Computer and experimental simulations of the production of methane gas from municipal solid waste. *Water Sci. Technol.* 27(2): 225–234.

Lee, S.R. and H.I. Park. 1999. Discussion of estimation of municipal solid waste landfill settlement by H.I. Ling et al. *ASCE J. Geotech. Geoenvir. Eng.* 125(8): 722–724.

Levenspiel, O. *The Chemical Reactor Omnibook*. Oregon State University Press. 1989.

Li, M., Y. Bando, T. T., K. Yasuda and M. Nakamura. 2001. Analysis of liquid distribution in non-uniformly packed trickle bed with single-phase flow. *Chem. Eng. Sci.* 56: 5969–5976.

Lin, C.Y., T. Noike, H. Furumai and J. Matsumoto. 1989. A kinetic study on the methanogenic process in anaerobic digestion. *Water Sci. Technol.* 21: 175–186.

Ling, H.I., D. Leshchinsky, Y. Mori and T. Kawabata. 1998. Estimation of municipal solid waste landfill settlement. *ASCE J. Geotech. Geoenvir. Eng.* 124(1): 21–28.

Luong, J.H. and B. Volesky. 1983. Heat evolution during the microbial process: estimation, measurement and applications. In *Advances in Biochemical Engineering/ Biotechnology*, Vol. 26, A. Feichter (Ed.), Springer-Verlag, Heidelberg.

Luong, J.H.T. and B. Volesky. 1980. *Can. J. Chem. Eng.* 60: 163.

Mackay, A.D. and E.J. Kladivko. 1985. Earthworms and rate of breakdown of soybean and maize residues in soil. *Soil Biol. Biochem.* 17(6): 851–857.

Mahloch, J. 1970. An investigation of the microbiology of aerobic decomposition of refuse. Ph.D. thesis, University of Kansas Department of Civil Engineering.

Marsh, R.E. and W.E. Howard. 1969. Evaluation of mestranol as a reproductive inhibitor of Norway rats in garbage dumps. *J. Wildlife Manage.*

Marshall, T.J. and J.W. Holmes. 1979. *Soil Physics*. Cambridge University Press, London.

Mata-Alvarez, J. and A. Martinez-Viturtia. 1986. Laboratory simulation of municipal solid waste fermentation with leachate recycle. *J. Chem. Technol. Biotechnol.* 36: 547–566.

Mauret, E. 1995. Mésure des pertes de pression et de la dispersion axiale dans les matelas fibreux: aplication au lavage des pâtes écrues en lits fixés épais (Measurement of losses of pressure and axial dispersion in fibrous mats: application to the washing of the unbleached pastes in thick fixed beds). Thesis. Grenoble, France.

Mauret, E. and M. Renaud. 1997. Transport phenomena in multiparticle systems: 1: limits of applicability of capillary model in high voidage beds: Application to fixed beds of fibers and fluidized beds of spheres. *Chem. Eng. Sci.* 52(11): 1807–1817.

McBean, E.A., F.A. Rovers and G.C. Farquhar. 1995. *Solid Waste Landfill Engineering and Design*. Prentice-Hall, Englewood Cliffs, NJ.

McBride, E.F. 1971. Mathematical treatment of size distribution data. In *Procedures in Sedimentary Petrology*, R.E. Carver (Ed.), Wiley-Interscience, John Wiley & Sons. New York.

McCarty, P.L. 1975. Stoichiometry of biological reactions. *Progr. Water Technol.* 7(1): 157–172.

McDonald, I.F., M. El Sayed and F.A.L. Dullien. 1979. Flow through porous media: The Ergun equation revisited. *Ind. Eng. Chem. Fundamentals* 18: 199–208.

McGowan, K.C., F.G. F.M. Saunders and N.D. Williams. 1987. A microbial model of landfill stabilization.

McMeekin, J.A., R.E. Chandler, P.E. Doe, C.D. Garland, J. Olley, S. Putros and D.A. Ratowsky. 1987. Model for combined effect of temperature and salt concentration/water activity on the growth rate of *staphylococcus xylosus*. *J. Appl. Bacteriol.* 62: 543–550.

McNeil, S. and J.H. Lawton. 1970. *Nature*, 225: 472–474.

McSorley, R. 2000. Sampling and extraction techniques for nematodes. In *Techniques in Nematology Ecology*. Society of Nematologists.

McSorley, R. and J.J. Frederick. 1999. Nematode population fluctuations during decomposition of specific organic amendments. *J. Nematol.* 31: 37–44.

McSorley, R. and J.J. Frederick. 2000. Short-term effects of cattle grazing on nematode communities in Florida pastures. *Nematropica* 30: 211–221.

Medina, M. 1998. Scavenging and integrated biosystems: some past and present examples. In *Integrated Biosystems in Zero Emissions Applications, Proc. Internet Conf. Integrated Biosyst.*, E.-L., Foo and T. Della Senta (Eds.).

Mercer, J.W. and R.K. Waddell. 1992. Contaminant transport in groundwater, *Handbook of Hydrology*, D.R. Maidment (Ed.), McGraw-Hill, New York, 16.1–16.8.

Miller, P.A. 1998. A bioreactor analysis approach to the management of solid waste disposal sites. Ph.D. thesis, Rensselaer Polytechnic Institute.

Milly, P.C.D. 1982. Moisture and heat transport in hysteric inhomogenous porous media: a matric head-based formulation and a numerical model. *Water Resour. Res.* 18(3): 489–498.

Minkevich, I.G. and V.K. Eroshin. 1973. *Folia Microbiol.* 18: 376.

Monod, J. 1949. The growth of bacterial cultures. *Annu. Rev. Microbiol.* 3: 371–394.

Moore, J.C, D.E. Walter and H.W. Hunt. 1988. Arthropod regulation of micro- and mesobiota in below-ground food webs. *Annu. Rev. Entomol.* 33: 419–439.

Morris, D.V. and C.E. Woods. 1990. Settlement and engineering considerations in landfill final cover design. Geotechnics of waste fills to theory and practice. ASTM STP 1070. American Society for Testing and Materials, Philadelphia, 9–21.

Moser, A. 1988. *Bioprocess Technology: Kinetics and Reactors*. Springer-Verlag, New York. 69, 198–200.

Nagy, K.A. 2001. Food requirements of wild animals: Predictive equations for free-living mammals, reptiles and birds. *Nutr. Abstr. Rev. Ser. B* 71: 21R–32R.

Nagy, K.A., I.A. Girard and T.K. Brown. 1999. Energetics of free-ranging mammals, reptiles and birds. *Annu. Rev. Nutr.* 19: 247–277.

Nagy, K.L. 1987. Field metabolic rate and food requirement scaling in mammals and birds. *Ecol. Monogr.* 57(2): 116–128.

Nair, C. 1994. Waste management in emerging industrialized countries. *Green Pages Online J.*

Niven, R.K. 2001. Physical insight into the Ergun and Wen and Yu equations for fluid flow in packed and fluidized beds. *Chem. Eng. Sci.* 52(11): 1807–1817, 56: 1–8.

Noble, J.J. and A.E. Arnold. 1991. Experimental and mathematical modeling of moisture transport in landfills. *Chem. Eng. Commun.* 100: 95–111.

Noble, J.J. and G.M. Nair. 1990. The effect of capillarity on moisture profiles in landfills. *Proc. Purdue Industrial Waste Conference,* Lewis Publishers, Chelsea, MI, 545–554.

Noike, T., G. Endo, J.E. Chang, J. Yaguchi and J. Matsumoto. 1985. Characteristics of carbohydrate degradation and the rate-limiting step in anaerobic decomposition. *Biotechnol. Bioeng.* 27: 1482–1489.

Okamoto, M., O. Fujita, T. Kurosawa, Y. Oku and M. Kamiya. 1992. Natural *Echinococcus multilocularis* infection in a Norway rat, *Rattus norvegicus*, in southern Hokkaido, Japan. *Int. J. Parasitol.* 22(5): 681–684.

Or, D. and J.M. Wraith. 1999. Soil water content and water potential relationships. In *Handbook of Soil Science*, M.E. Sumner (Ed.), CRC Press, Boca Raton.

Oswin, C.R. 1946. *J. Chem. Ind. (London)* 64: 419.

Oweis, I.S., D.A. Smith, R.B. Ellwood and D.S. Greene. 1990. Hydraulic characteristics of municipal refuse. *J. Geotechnol. Eng.*, 116(4): 539–553.

Palmisano, A.C., D.A. Maruscik and B.S. Schwab. 1993. Enumeration of fermentative and hydrolytic microorganisms from three sanitary landfills. *J. Gen. Microbiol.* 139: 387–391.

Papendick, R.I. and G.S. Campbell. 1981. Theory and measurement of water potential. In *Water Potential in Soil Microbiology*. SSSA Special Publication Number 9, Soil Science Society of America, Madison, WI. 1–22.

Park, H., II, S.R. Lee, N.Y. Do. 2002. Evaluation of decomposition effect on long-term settlement prediction for fresh municipal solid waste landfills. *ASCE J. Geotech. Geoenvir. Eng.* 128(2): 107–118.

Pearson, T.H. 1968. The feeding biology of seabird species breeding on the Farne Islands, Northumberland. *J. Animal Ecol.* 37: 521–552.

Petersen, H. and Luxton, M. 1982. A comparative analysis of soil fauna populations and their role in decomposition processes. In *Quantitative Ecology of Microfungi and Animals in Soil and Litter*. Petersen, H. (ed.) *Oikos*, 39: 287–388.

Petersen, L.W., P. Moldrup, Y.H. El-Farhan, O.H. Jacobsen, T. Yamaguchi and D.E. Rolston. 1995, The effect of moisture and soil texture on the adsorption of soil vapors. *J. Envir. Qual.* 24: 752–759.

Petersen, L.W., Y. H. El-Farhan, P. Moldrup, D. E. Rolston and T. Yamaguchi. 1996. Transient diffusion, adsorption and emissions of volatile organic vapors in soils with fluctuating water contents. *J. Envir. Qual.* 25: 1054–1063.

Pfannkuch, H.O. and R. Paulson. 2000. *Grain Size Distribution and Hydraulic Properties.*

Pfeffer, John T. 1974. Temperature effects on anaerobic fermentation of domestic refuse. *Biotechnol. Bioeng*. 16: 771–787.

Pirt, S.J. *Principles of Microbe and Cell Cultivation*. 1975. Blackwell Scientific, Oxford.

Pitt, J.I. and Alisa J. Hocking. 1977. Influence of solute and hydrogen ion concentration on the water relations of some xerophilic fungi. *J. Gen. Microbiol.* 101: 35–40.

Piver, W.T. and Lindstrom, F.T. 1991. Numerical methods for describing chemical transport in the unsaturated zone of the subsurface. *J. Contam. Hydrol.* 8 (N3–4): 243–262.

Powers, L.E. and R. McSorley. 2000. *Ecological Principles of Agriculture.* Delmar Thomson Learning, Albany, NY.

Powrie, W., A.P. Hudson and R.P. Beaven. 2000. Development of sustainable landfill practices and engineering landfill technology. Final report to the Engineering and Physical Sciences Council, Grant Reference GR/L 16149.

Pugh, G.J.F. 1974. Terrestrial fungi. In *Biology of Plant Litter Decomposition*, Vol. 2, C.H. Dickinson and G.J.F. Pugh, (Eds.). Academic Press, New York.: 303–336.

Puncochar, M. and J. Drahos. 2000. Limits of capillary model for pressure drop correlation. *Chem. Eng. Sci.* 55: 3951–3954.

Ragab, R., J. Feyen and D. Hillel. 1982. Effect of the method for determining pore size distribution on prediction of the hydraulic conductivity function and of infiltration. *Soil Sci*. 134(2): 141–145.

Rao, S.K., L.K. Moulton and R.K. Seals. 1977. *Settlement of Refuse Landfills. Geotechnical Practice for Disposal of Solid Waste Materials*. Ann Arbor, MI, 574–599.

Ratkowsky, D.A. and T. Rose. 1995. Modeling the bacterial growth/no-growth interface. *Lett. Appl. Microbiol*. 20: 29–33.

Rees, J.F. 1980. Optimization of methane production and refuse decomposition in landfills by temperature control. *J. Chem. Technol. Biotechnol*. 30: 458–465.

Rees, J.F. and J.M. Grainger. 1982. Rubbish dump or fermenter? Prospects for the control of refuse fermentation to methane in landfills. *Process Biochem*. 17(6): 41.

Reichle, D.E. 1977. The role of invertebrates in nutrient cycling. In *Soil Organisms as Components of Ecosystems.* Lohm, U. and Persson, T. (eds.) *Ecol. Bull. (Stockholm)* 25: 145–156.

Reichle, D.E. 1971. In *Productivity of Forest Ecosystems*, P. Duvigneaud, (Ed.). UNESCO, Paris, 465–477.

Richards, L.A. 1931. capillary conduction of liquids through porous medium. *Physics* 1: 318–333.

Richards, L.A. 1960. Advances in soil physics. *7th International Congress on Soil Science*, Vol. 1, 67–79.

Rodriguez, J.G. and L.D. Rodriguez. 1987. Nutritional ecology of phytophagous mites. *Nutritional Ecology of Insects, Mites, Spiders and Related Invertebrates*, John Wiley & Sons, New York, 177–208.

Röhrs, L.H. and G.E. Blight. 1998. Waste management in developing countries. Report on Workshop, Technical University of Luleå, Sweden.

Rossi, C. and J.R. Nimmo. 1994. Modeling of soil water retention from saturation to oven dryness. *Water Resour. Res*. 30(3): 701–708.

Russell, E.W. 1961. *Soil Conditions and Plant Growth.* Longmans, London.

Sabiri, N.E. 1995. Dynamique des fluides et des transferts. Thesis, Nantes, France.

Saez, A.E. and R.G. Carbonell. 1985. Hydrodynamic parameters for gas–liquid concurrent flow in packed beds. *Am. Inst. Chem. Eng. J.*, 31: 52–62.

Sahimi, M. 1995. *Flow and Transport in Porous Media and Rock: Classical Methods to Modern Approaches*. VCH, Weiheim.

Saravanapavan, T. and G.D. Salvucci. 2000. Analysis of rate-limiting processes in soil evaporation with implications for soil resistance models. *Adv. Water Res.* 23: 493–502.

Schnurer, J., M. Clarholm, S. Bostrom and T. Rosswall. 1986. Effects of moisture on soil microorganisms and nematodes: A field experiment. *Microbiol Ecol.* 12: 217–230.

Schriefer, T. 1981. Regenwurmer (*Lumbricidae*) auf underschiedlich abgedeckten Mulldeponien. Bestandsaufnahme und Besiedlungsmechanismen. *Pedobiologia*, 21. (Referenced by Weidemann et al., 1980).

Schroeder, P.R., T.S. Dozier, P.A. Zappi, B.M. McEnroe, J.W. Sjostrom and R.L. Peyton. 1994. The hydrologic evaluation of landfill performance (HELP) model: Engineering documentation for Version 3. EPA/600/9–94/xxx, U.S. Environmental Protection Agency Risk Reduction Engineering Laboratory, Cincinnati.

Schugerl, K. 1985. *Bioreaction Engineering, Vol. I: Fundamentals, Thermodynamics, Formal Kinetics, Idealized Reactor Types and Operation Modes*. John Wiley & Sons, New York.

Scriber, J.M. 1978. The effects of larval feeding specialization and plant growth form on the consumption and utilization of plant biomass. *Entomol. Exp. Applied.* 24: 694–710.

Seguin, D., A. Montillet and J. Comiti. 1998. Experimental characterization of flow regimes in various porous media. I. limit of laminar flow regime. *Chem. Eng. Sci.* 53(21): 3751–3761.

Seguin, D., A. Montillet, D. Brunjail and J. Comiti. 1996. Liquid–solid mass transfer in packed beds of variously shaped particles at low Reynolds numbers: experiment and model. *Chem. Eng. J.* 63: 1–9.

Senior, E., M. Talaat and M. Balba. 1990. Refuse decomposition. In *Microbiology of Landfill Sites*. CRC Press, Boca Raton.

Sibly, R.M. and R.H. McCleery. 1983. Increase in weight of herring gulls while feeding. *J. Animal Ecol.* 52: 35–50.

Simpson, P.T. and T.F. Zimmie. 2000. Review and analysis of landfill settlement evaluations, hazardous and industrial wastes. *Proc. 32nd Mid-Atlantic Ind. Hazardous Waste Conf.*, J.C. Kilduff, S. Komisar and M. Nyman, (Eds.).

Slansky, F. and J.M. Scriber. 1985. Food consumption and utilization. In *Comprehensive Insect Physiology, Biochemistry and Pharmacology*, Vol. 4, Pergamon, Oxford. 87–163.

Smirnov, M.S. and V.I. Lysenko. 1991. A new method of obtaining the equations for adsorption and desorption isotherms in moisture absorption. *Kolloidnyi Zh.* 53(5): 890–895.

Snow, D., M.H.G. Crichton and N.C. Wright. 1944. Mould deterioration of feedstuffs in relation to humidity of storage. *Ann. Appl. Biol.* 31: 111–116.

Sommers, L.E., C.M. Gilmour, R.E. Wildung and S.M. Beck. 1981. The effect of water potential on decomposition processes in soil. In *Water Potential Relations in Soil Microbiology*, Soil Science Society of America (SSSA) Special Publication 9, Madison, WI. 97–117.

Sowers, G.F. 1973. Settlement of waste disposal fills. *Proc. 8th Int. Conf. Soil Mech. Found. Eng.*, Vol. 2, Part 2, 207–210.

Sparrowe, R.D. 1968. Sexual behavior of grizzly bears. *Am. Midland Naturalist* 80(2): 570–572.

Staaf, H. 1987. Foliage litter turnover and earthworm population in three beech forests of contrasting soil and vegetation types. *Oecologia* 72, 58–64.

Stevenson, R.D. 1985. Body size and limits to the daily range of temperature in terrestrial ectotherms. *Am. Naturalist* 125: 102–117.

Stockdill, S.M.J., 1982. Effects of introduced earthworms on the productivity of New Zealand pastures. *Pedobiologia* 24: 29–35.

Storer, T.I., F.C. Evans and F.G. Palmer. 1944. Some rodent populations in the Sierra Nevada of California. *Ecol. Monogr*. 14(2): 165–192.

Straub, W.A. and D.R. Lynch. 1982. Models of landfill leaching: organic strength. *J Envir. Eng. Div. ASCE* 108(EE2): 251–268.

Tabashnik, B.E. and F. Slansky, Jr. 1987. Nutritional ecology of forb foliage chewing insects. In *Nutritional Ecology of Insects, Mites, Spiders and Related Invertebrates*, John Wiley & Sons, New York, 71–103.

Tahraoui, K. and D. Rho. Biodegradation of BTX vapors in a compost medium biofilter. *Compost Sci. Utilization* 6(2): 13–21.

Taylor, S.A. and G.L. Ashcroft. 1972. *Physical Edaphology: The Physics of Irrigated and Non-Irrigated Soils*. W.H. Freeman, San Francisco.

Teng, H. and T.S. Zhao. 2000. An extension of Darcy's law to non-stokes flow in porous media. *Chem. Eng. Sci.* 55: 2727–2735.

Tevis, L., Jr. 1955. Observations on chipmunks and mantled squirrels in northeastern California. *Am. Midland Naturalist* 53(I) 71–78.

Thibodeaux, L.J. 1996. *Environmental Chemodynamics: Movement of Chemicals in Air, Water and Soil*. 2nd ed. Wiley Interscience, New York, 149.

Thorsen, M., R. Shorten, R. Lucking and V. Lucking. 2000. Norway rats (*Rattus norvegicus*) on Fregate Island, Seychelles: the invasion; subsequent eradication attempts and implications for the island's fauna. *Biol. Conserv*. 96: 133–138.

Todd, D.K. 1980. *Groundwater Hydrology*, 2nd ed., John Wiley & Sons, New York.

Troller, J.A. 1980. Food technology. Activity on microorganisms in food. *J. Appl. Bacteriol.* 76–80.

Tuma, J.J. and M. Abdel–Hady. 1973. *Engineering Soil Mechanics*. Prentice-Hall, Englewood Cliffs, NJ.

Valsaraj, K.J. 2000. *Elements of Environmental Engineering: Thermodynamics and Kinetics*, 2nd ed. Lewis Publishers, Boca Raton.

Van den Berg et al. 1974. Assessment of methanogenic activity in anaerobic digestion: Apparatus and method. *Biotechnol. Bioeng*. 16: 1459–1469.

Van den Berg, C. 1984. Description of water activity of foods for engineering purposes by means of the GAB model of sorption. In *Engineering and Food*, B.M. McKenna (Ed.), Elsevier Science, London.

Van Vliet, P.C.J. 2000. Enchytraeids. Soil fauna, In *The Handbook of Soil Science*, M. Sumner (Ed.). CRC Press, Boca Raton.

Vander Linden, J., S. Filmer, K. Keith, L. Dressler, J. Chen and R. Dyke. 1998. Aversive condition and the Fort Nelson bears. Group Paper, Fort Nelson, BC.

Waldrop, H.A., S.F. Cook, Jr., D.C. Lowrie and G.B. Moment. 1971. Grizzlies: to spare or banish. *Science* 171: 431–433.

Wall, D.K. and C. Zeiss. 1995. Municipal landfill biodegradation and settlement. *ASCE J. Envir. Eng*. 121(3): 214–224.

Waller, D.A. and J.P. La Fage. 1987. Nutritional ecology of termites. In *Nutritional Ecology of Insects, Mites, Spiders and Related Invertebrates*. John Wiley & Sons, New York, 487–532.

Weidemann, G., H. Koehler and T. Schreifer. 1980. Recultivation: A problem of stabilization during ecosystem development. In *Urban Ecology: The Second European Ecological*

*Symposium,* September 8–12, R. Bornkamm, J.A. Lee and M.R.D. Seaward, (Eds.). Berlin.

Weisser, H. 1985. Influence of temperature on sorption equilibria. In *Properties of Water in Foods*, D. Simatos and J.L. Multon (Eds.). Series E: Applied Sciences No. 90, NATO ASI Series, 95–118.

Werner, M.R. and D.D. Dindal. 1987. Nutritional ecology of soil arthropods. In *Nutritional Ecology of Insects, Mites, Spiders and Related Invertebrates*, John Wiley & Sons. New York. 815–836.

Weygoldt, P. 1969. *The Biology of Pseudoscorpions.* Harvard University Press, Cambridge, MA.

Whitford, W.G., D.W. Freckman, N.Z. Elkins, L.W. Parker, R. Parmalee, J. Phillips and S. Tucker. 1981. Diurnal migration and responses to simulated rainfall in desert soil microarthropods and nematodes. *Soil Biol. Biochem.* 13: 417–425.

Williams, N.D., F.G. Pohland, K.C. McGowan and F.M. Saunders. 1987. Simulation of leachate generation from municipal solid waste. Hazardous Waste Engineering Research Laboratory, ORD, USEPA.

Wilson, J.M. and D.M. Griffin. 1979. The effect of water potential on the growth of some soil basidiomycetes. *Soil Biol. Biochem.* 11: 211–212.

Witkamp, M. Seasonal fluctuation of the fungus flora in mull and mor of an oak forest. *Meded. Inst. Toegep. Biol. Onderz. Nat.* 46: 1–51.

Wood, T.G., R.A. Johnson, S. Bacchus, M.O. Shittu and J.M. Anderson. 1982. Abundance and distribution of termites (Isoptera) in a riparian forest in the southern Guinea savanna vegetation zone of Nigeria. *Biotropica*, 14: 25–39.

World Conservation Union, 1996. *African Wildcat: Felis silvestris–Lybica Group.*

Wylie, M.R. and A.R. Gregory. 1955. Fluid flow through unconsolidated porous aggregates. *Ind. Eng. Chem. Proc. Design Dev.* 47: 1379–1388.

Young, A. 1989. Mathematical modeling of landfill degradation. *J. Chem. Technol. Biotechnol.* 46: 189–208.

# Index

## A



## B

## C

## D

## E



## F

## G



## H

## I



## K



## L

## M

## N



## O

## R

## S

## T

## U



## V



## W

## X



## Y